# Impacts of Land Use Pattern in Metropolitan Area

# Impacts of Land Use Pattern in Metropolitan Area

Editors

**Bindong Sun**
**Tinglin Zhang**
**Wan Li**
**Chun Yin**
**Honghuan Gu**

MDPI

Basel • Beijing • Wuhan • Barcelona • Belgrade • Novi Sad • Cluj • Manchester

*Editors*
Bindong Sun
East China Normal University
Shanghai
China

Tinglin Zhang
East China Normal University
Shanghai
China

Wan Li
Zhengzhou University
Zhengzhou
China

Chun Yin
East China Normal University
Shanghai
China

Honghuan Gu
East China Normal University
Shanghai
China

*Editorial Office*
MDPI
St. Alban-Anlage 66
4052 Basel, Switzerland

This is a reprint of articles from the Special Issue published online in the open access journal *Land* (ISSN 2073-445X) (available at: https://www.mdpi.com/journal/land/special_issues/land_use_pattern_metropolitan_area).

For citation purposes, cite each article independently as indicated on the article page online and as indicated below:

| Lastname, A.A.; Lastname, B.B. Article Title. *Journal Name* **Year**, *Volume Number*, Page Range. |
| --- |

**ISBN 978-3-7258-0957-8 (Hbk)**
**ISBN 978-3-7258-0958-5 (PDF)**
**doi.org/10.3390/books978-3-7258-0958-5**

# Contents

land

Article

# Exploring Employment Spatial Structure Based on Mobile Phone Signaling Data: The Case of Shenzhen, China

Yani Lai [1], Zhen Lv [1,*], Chunmei Chen [1] and Quan Liu [2]

[1] Department of Construction Management and Real Estate, Shenzhen University, Shenzhen 518052, China; lai.yani@szu.edu.cn (Y.L.); 2060474009@email.szu.edu.cn (C.C.)
[2] LAY-OUT Planning Consultants Co., Ltd., TAGEN Knowledge & Innovation Center, Shenzhen 518038, China; liuquan@lay-out.com.cn
* Correspondence: 1900474018@email.szu.edu.cn; Tel.: +86-136-4701-2001

**Abstract:** Debate on the shift from a monocentric to polycentric urban structure has been extensive. Polycentricity generally refers to the co-existence of several centers in a city. Existing studies regarding China have mainly focused on the morphological characteristics of urban centers, but few recent studies have focused on functional dimensions of urban centers. Emerging big data sources provide new opportunities to explore the morphological and functional perspectives of urban spatial structure. This study uses mobile phone signaling data and develops a new methodology to measure urban centers' functional centrality. The study area focuses on Shenzhen City, which has rapidly transformed from a village into a metropolitan city in the past few decades. As the first economic special zone in China, Shenzhen has adopted a polycentric urban plan since the beginning of the urbanization process. This study explores the spatial employment structure of the city from the morphological and function dimensions. Based on the findings, this study discusses the role of urban planning in forming an urban spatial structure and provides implications for future planning.

**Keywords:** urban spatial structure; morphological centrality; functional centrality; urban planning; mobile phone signaling data; Shenzhen; China

**Citation:** Lai, Y.; Lv, Z.; Chen, C.; Liu, Q. Exploring Employment Spatial Structure Based on Mobile Phone Signaling Data: The Case of Shenzhen, China. *Land* **2022**, *11*, 983. https://doi.org/10.3390/land11070983

Academic Editors: Bindong Sun, Tinglin Zhang, Wan Li, Chun Yin and Honghuan Gu

Received: 29 April 2022
Accepted: 23 June 2022
Published: 28 June 2022



## 1. Introduction

Rapid urbanization has led to the transformation of urban spatial structures. Many cities in the world have adopted polycentric urban planning as an important spatial strategy toward sustainable development [1], in order to solve urban problems, such as traffic congestion and excessive pressure on resources and the environment [2,3]. A polycentric urban structure is considered a more compact urban form that is conducive to more effective urban space organization [4]. Traditionally, urban spatial structure is measured on the basis of morphological indicators, such as population distribution and urban physical form attributes [5–7]. The theory of 'space of flows', proposed by Castell (1996), provides a new perspective for measurement and emphasizes the importance of urban networks and urban flows. Urban spatial units may be physically separated, but they can be linked by commuting and resource flows [8]. Based on this pool of theoretical literature, some empirical studies have been conducted to measure urban spatial structure from the perspective of functional connections [9,10].

However, related concepts and measurement methods of urban spatial structures are still vague [11,12]. Most of the existing studies on urban spatial structures have focused on western cities, and the research on Chinese urban polycentric spatial structures is still in the preliminary stage. Although many studies have measured the polycentric urban spatial structure from the perspective of morphology, studies on functional connection between centers in a city are limited [3,13]. Investigations on the relationship between morphological and functional centrality are also lacking [14]. This study investigates the urban spatial structure of Shenzhen, a large Chinese city, to fill these gaps. This study analyzes the morphological distribution and functional connections of the employment spatial structure. The relations between the morphological and functional centrality of employment centers are further analyzed. Since it became the first special economic zone in China in the 1980s, Shenzhen has implemented polycentric urban planning for its urbanization. Urban planning and public policies have played important roles in the rapid development of the city. However, the extent to which the polycentric spatial structure has developed is unclear. A better understanding of the urban spatial structure of Shenzhen provides a useful reference for the development of other Chinese cities and basis for the formulation of future urban planning.

The collection of data that represents functional linkages, such as population movements within cities, is a prerequisite for measuring spatial structure from functional perspective. The progress of information and communications technology and emergence of big spatial data, such as bus smart card data, taxi GPS trajectories, and mobile phone signaling data, have provided new possibilities for obtaining information about people's mobility activities [15–19]. Compared with traditional survey data, these new datasets have a much higher accuracy to provide a better tool for measuring urban spatial structure. Especially, mobile phone signaling data has wide coverage that can capture the travellers' travel track information during a whole day and has a unique advantage in the study of urban spatial structure [16,20].

The remainder of this paper is structured as follows. Section Two reviews the related concepts and theories on urban spatial structure and existing methods for identifying and analyzing urban centers. Section Three introduces the research methods and data sources. Section Four analyzes the research results. Section Five discusses the role of urban planning, based on a comparison analysis of the identified urban spatial structure and planned urban spatial structure in the master urban planning of Shenzhen. The last section draws the major conclusions of this paper.

## 2. Literature Review

### 2.1. Concept of the Spatial Structure

A considerable number of debates on urban spatial structure have been conducted. Foley (1964) indicated that the spatial attributes of urban spatial structure included cultural value, functional activities, and physical environment. Weber (1964) illustrated the form and process of the city and pointed out that urban space can be divided into adapted space (e.g., architecture) and channel space (e.g., traffic network). Bounre (1971) used system theory to understand the urban system, which includes three elements, namely urban form (the spatial layout mode of urban elements), urban internal interaction (the internal formation), and urban spatial structure (the organization mechanism). To sum up, the urban spatial structure can be summarized into two aspects: the various distributions of urban elements and functional connection amongst them. According to the literature review, the scale, influence, and spatial distribution of urban centers are the core contents for defining the characteristics of urban spatial structure [11,21,22].

The research on urban centers can be traced back to the central place theory presented by Christaller (1933), which lays the foundation for the definition of urban centers. According to central place theory, the central place refers to the location that provides goods and services for the surrounding residents. However, the definition and measurement of centrality are still vague. Some scholars argue that the essence of the center is a spatial unit with significantly higher attribute value than its surrounding units [23,24]. According to the method for measuring urban center, urban centers can be defined from morphological and functional perspectives [9,11,25]. The indicators for measuring the morphological centrality include regional area, population, density, and the construction of the centrality index. Grounded in the case study in Los Angeles, Giuliano and Small (1991) defined employment centers as having an employment density that is greater than 10 people/acre, as well as a total employment number greater than 10,000 people. Based on the land use map and urban construction data, Yue et al. (2019) identified the high-value areas of development intensity as morphological centers. According to the population and share of employment, Sun and Lv (2020) defined the employment centers as the central clusters, where the employment number met the minimum employment threshold and total amount of employment exceeds 20,000. Based on the literature review, this paper defines the morphological centers as spatial units that have a higher number of employees than the surrounding areas, as well as those that can have an impact on the surrounding areas.

On the one hand, the employment centers are identified to measure the morphological centrality based on the employed population. On the other hand, it measures the centrality of urban spatial units through the dynamic functional connections, such as the amount and density of commuting. The most important urban center is not necessarily the most populous place, but one that is located in the most critical position in the transportation network [10,25,26]. Burger and Meijers (2012) argued that the functional connectivity of urban spatial structure plays a key role in the urban system. They measured the intra-urban centrality based on the shopping and commuting flows. Based on flows, Sarkar et al. (2020) defined three indicators (e.g., trips, density, and accessibility) to measure functional centrality. The larger the index value is, the stronger the centrality is. Wei et al. (2020) measured the importance of each spatial unit by relative centrality, based on check-in and -out taxi GPS data. The functional centrality of the center is defined as the strong functional connections between the spatial unit and surrounding area, and it is in a key position in the urban network.

At the beginning of an urbanization process, the urban core area is the main space carrier that reflects a monocentric urban structure [27,28]. With the development of the city, the city expands outward because of land scarcity, traffic congestion, and pollution in the urban core area [29,30]. To adapt to the development of cities, western countries began to formulate policies of polycentric spatial structure. For instance, the urban structure of many western countries, such as Chicago [22], Finland [10], and England [2], have the characteristics of polycentric urban spatial structures. Previous studies on polycentric spatial structures focused on the geographical distribution characteristics of cities with different scales from a morphological perspective [5,28,31]. However, recent studies have emphasized that a better understanding of spatial structures would also include attention paid to the functional connections amongst different nodes in the urban system [2,32]. Research in this area usually includes the measurement of flows between the centers of a city. Each (sub) center is considered to be connected with other (sub) centers through multidirectional flows [10,25].

Drawing on the experience of western countries, China has introduced the polycentric development of urban planning [33] and issued different polycentric urban planning at various urban scales, such as the regional [34] and city scales [5]. Many foreign studies have measured centrality from the morphological and functional perspectives and explored the relationship between them. Most foreign cities tend to be more functionally polycentric than morphologically polycentric [10,25]. However, existing studies regarding China have mainly focused on the morphological. Studies on the measurement of functional centrality by functional connection within cities are few. Shenzhen, which is representative of Chinese cities, has implemented polycentric urban planning since its establishment. Therefore, we use the case study of Shenzhen to reflect on morphological and functional perspectives.

### 2.2. Network City Approach

Some scholars apply the theory of the 'space of flows' proposed by Castells (1996) to analyze the urban spatial structure from the perspective of the network city [35,36]. According to Castells (1996), the spatial units may be separated based on their physical forms, but they can be connected with each other through different kinds of flows. Based on the theory of the space of flows and the concept of the network city, social network analysis has been applied to measuring functional centrality [37]. Green (2007) proposed the principle of functional polycentricity and defined the ratio of the total number of actual connections and potential connections in the network as network density. Wang et al. (2020)

used GPS taxi trajectory data to construct a commuter network and applied social network analysis to measure the functional centrality with four indicators, namely, network density, intermediate centrality, point centrality potential and center periphery index. The higher these indexes are, the stronger their functional centrality will be. These studies reflect that network density and centrality are two important indicators for social network analysis.

Other methods can also be used to measure the polycentricity of the urban spatial structure. The evaluation of the relative balance between centers according to the 'importance' of centers are crucial by the number of centers, the ratio method and the rank-size method. The number of centers can reflect the distribution of centers and the development degree of polycentricity most intuitively through the number of centers. The more the number of centers is, the higher the development degree of polycentricity will be [5,38]. The ratio method directly reflects the relationship between main centers and subcenters by comparing their attribute values. The closer the ratio is to 1, the more balanced the size distribution between main centers and subcenters is, and the higher the degree of polycentricity will be [10]. In recent studies, the rank-size method is the most common method for measuring polycentricity. This method reflects the relative status of each center in the urban system. The flatter the slope of the fitting line is, the higher the degree of polycentricity is [25,39].

## 3. Research Data and Methods

### *3.1. Study Area and Data Sources*

Shenzhen is one of the most prosperous cities in China, and it is located south of Guangdong Province. This research takes Shenzhen as the study area for the following reasons. First, Shenzhen provides a good example for studying urban development in China because it is the first special economic zone that has developed from a small fishing village into one of the most prosperous cities in the past few decades. Second, since the beginning of its urbanization process, Shenzhen has implemented polycentric urban planning as an important spatial development strategy. However, systematic studies on the spatial structure of the city are limited. Third, the Shenzhen government proposed a polycentric urban spatial structure with two main urban centers, five subcenters, and eight community-level centers in the 'master plan of Shenzhen City (2010–2020)' (hereinafter referred to as the '2010 master plan'). The examination of the effect of this planning guidance on the formation of urban spatial structures is helpful for formulating better strategies to improve urban planning in the future. As of October 2019, Shenzhen has 10 administrative districts, with a total area of 1997.47 square kilometers, as well as a built-up area of 927.96 square kilometers and population of 13.4388 million. The administrative region of Shenzhen, excluding all areas within the ecological control line, is defined as the study area.

This study collects the mobile phone signaling data (MPSD) of Shenzhen from China Unicom (the largest telecommunication company in China) from 1 June to 30 June 2019. We used the data of June 2019 in our study. June is a normal working month. No significant change was found in the mobility behavior or distribution characteristics of citizens in Shenzhen during this period. Thus, the data of June 2019 are appropriate to be used for the identification of urban spatial structure in 2019 [40]. The MPSD is generated when the mobile phone user is in the event of a call, SMS, or mobile location. The data include the time and location information of a person. This information can be used to deduce the travel path of users. As long as the user turns on the mobile phone, the travel information can be captured. Our actual sampling consisted of 4,467,500 people. Considering the market share of China Unicom, the age structure difference of mobile phone users, and other factors, this paper expands the sample and excludes the situation of non-human number cards, as well as one person with multiple cards, and obtains the expanded resident population. After sample enlargement, we obtained 19.825 million samples. The number of users in different districts in the MPSD has high similarity with that of the survey data in the Shenzhen statistical yearbook. This finding suggests that MPSD is appropriate for analysis.

The residence and employment locations are identified according to the time characteristics of employment and residence. The locations of residence correspond to places where mobile phone users stay in the same place for more than 4 h from 9:00 p.m. to 8:00 a.m. the next day and were observed for more than 15 days in a month. The employment locations correspond to places where mobile phone users stay in the same place for more than 4 h from 9:00 a.m. to 5:00 p.m. and were observed for more than 15 days in a month. Then, the population in employment and residence areas are defined as the employment and residential populations, respectively. To ensure a more accurate analysis, this paper allocates the mobile phone users into 250 m $\times$ 250 m grids.

The OD data are constructed by taking the place of residence as the original point and place of employment as the destination; thus, the residence–employment functional flows between spatial units are obtained. For example, mobile phone users live in place A and work in place B; when they go to work and go home, the functional connections between A and B are observed. After counting these functional connections, we can obtain the inflows and outflows of each spatial unit. Each spatial unit is used as a node, and flows are used as connections to build a direct urban network. The research area is divided into 18,226 grids, as research units with 2,313,529 flows. This study excludes the grids without connections with other grids and flows, whose starting and ending points are in the same grid. A total of 16,761 grids and 2,152,384 flows are, finally, identified for analysis.

### 3.2. Identification of Employment Centers

The spatial auto-correlation analysis method is used to identify the spatial clustering, which can reflect the spatial attributes of different areas [10,39]. This method divides spatial agglomeration into four categories, namely high-high, high-low, low-low, and low-high. The high-high agglomeration area is identified as a hot spot area, a high-value aggregation area surrounded by other high values areas. In this study, employment centers are identified on the basis of employed density. First, we use the ArcGIS 10.3 software to carry out local spatial auto-correlation analysis on the employment density of spatial units, select the significance of 0.01 for hot spot analysis, eliminate the insignificant areas, and choose the high-high gathering areas as candidate centers. Then, a cut-off value is applied to eliminate the small and practically insignificant spatial clusters. To make the cut-off value sensitive to local variation in each area, the cut-off value is defined in relative terms, where areas having an employment less than 0.5% of Shenzhen's total employment are excluded [1,5]. Considering that a center should have significant impact on the surrounding areas, identified areas with more than 10 grids [11,41] are regarded as the final employment centers. After that, according to the employment population of the identified centers, combined with the natural discontinuity method, we divided the centers into three levels.

### 3.3. Measuring Functional Centrality

Social network analysis has been recently adopted to measure functional centrality [26,37]. Most existing studies have only considered a single factor in shaping the centricity of key nodes, such as network density, the number of nodes, and commuter traffic. Few studies have paid attention to the directions of functional connections because of data limitations. To overcome these limitations, this study not only considers the commuting traffic of node connections but also uses the number of nodes connected and directions of commuting traffic to measure functional centrality. We combine network density [26] and degree centrality [41,42] as important measures to quantify the functional connections amongst nodes and measuring functional centrality.

Theoretically, the functional connections of a node in the urban network may occur in the following situations: (1) a node is connected with a plurality of nodes, but the connection is weak (Figure 1a); (2) the connection between nodes is very strong, but the number of connected nodes is few or the inflows and outflows are very different (Figure 1b); (3) nodes are not only connected with multiple nodes but also have strong connection strength (Figure 1c).

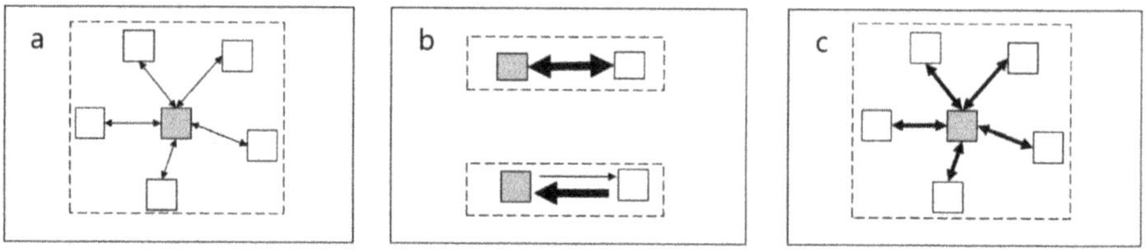


**Figure 1.** Different relations amongst nodes in an urban network. (**a**) A node is connected with a plurality of nodes, but the connection is not strong; (**b**) The connection between nodes is very strong, but there are few connected nodes or the inflows and outflows are very different; (**c**) Nodes are not only connected with multiple nodes, but also have strong connection strength.

This study uses the relative degree centrality of social network analysis to measure the frequency between nodes (Equation (1)), where $N_i^{in}$ is the number of all nodes connected with node i as the ending point (indegree), and $N_i^{out}$ is the number of all nodes connected with node i as the starting point (outdegree). Then, we divide the relative degree centrality of each node by the maximum value to normalize (Equation (2)).

$$\text{Relative Degree } (AD_i) = \left(N_i^{in} + N_i^{out}\right)/(2n-2), \tag{1}$$

$$AD_i' = AD_i/AD_{imax}. \tag{2}$$

We divide the inflow/outflow population of node i by the number of nodes that flows into/out of node i to indicate the connection strength between node i and other nodes; then, we take the mean value (Equation (3)). Then, we divide the commuting density of each node by the maximum value to normalize (Equation (4)).

$$\text{Commuting Density } (CDi) = [(POP_i^{in}/N_i^{in}) + (POP_i^{out}/N_i^{out})]/2 \tag{3}$$

$$CD_i' = CD_i/CD_{imax}. \tag{4}$$

We assigned 50% weight to the relative degree centrality and commuting density of each node to construct the FCI to measuring functional centrality (Equation (5)). The larger the FCI value of the center is, the stronger the functional centrality is, which implies that the center is in an important position in the urban network. After that, according to the FCI value of the identified center, combined with the natural discontinuity method, we divided the center into three levels.

$$FCi = \frac{1}{2}AD'_i + \frac{1}{2}CD_i' \tag{5}$$

## 4. Employment Spatial Structure of Shenzhen

### *4.1. Morphological Characteristics of Employment Centers*

From a morphological perspective, we have identified eight employment centers (Figure 2). Based on the employment population, the centers are graded by a natural discontinuity method. Figure 2 shows the morphological levels of the identified centers, among which, 1–8 is the area order of centers: 1 is Futian–Luohu center, 2 is Kejiyuan center, 3 is Chegongmiao center, 4 is Songhe center, 5 is Fuhai center, 6 is Aviation City center, 7 is Longhua center, and 8 is Dengliang center. The results show that the eight employment centers include one first-level morphological center (with an employment population of 2,001,404), one second-level morphological center (with an employment population of 1,072,720), and six third-level morphological centers (with an employment population of less than 270,317). The employment centers are mainly distributed in the central and western parts of Shenzhen. No center has been identified in the east, indicating that the development of the employment centers in the east of Shenzhen are behind those of the central and western regions. Four centers, namely Futian–Luohu, Kejiyuan, Chegongmiao, and Dengliang, are distributed inside the special economic zone (SEZ); the four remaining centers (i.e., Songhe, Longhua, Fuhai, and Aviation City) are located outside SEZ. The first- and second-level morphological centers are distributed in Futian–Luohu and Nanshan within SEZ. The third-level center around the two centers suggests that the first- and second-level centers play a leading role in the development of the surrounding areas. The area of the employment centers outside the SEZ is smaller and more scattered than that inside the SEZ.

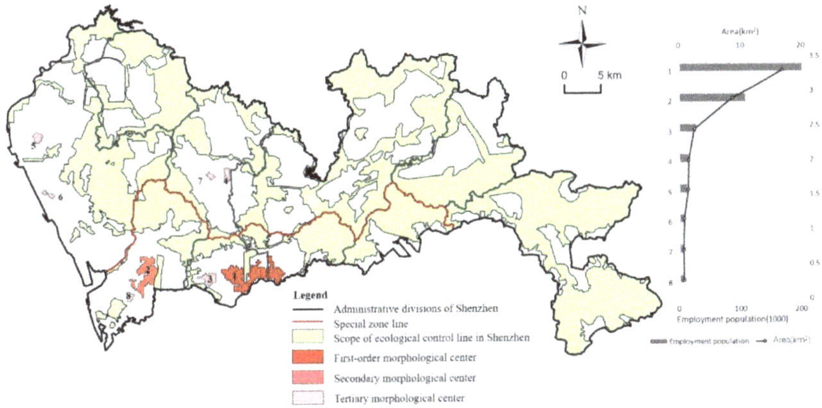


**Figure 2.** Identified employment centers.

Overall, in the SEZ centers, the total number of employment is 3,018,169, with an average employment density of 94,588 people/km$^2$ and total area of 28.648 square kilometers. In the non-SEZ centers, the total number of employees is 277,353, with an average employment density of 65,568 people/km$^2$ and total area of 4.251 square kilometers. The number of employees, average employment density, and centers' areas inside the SEZ are much larger than those outside the SEZ. The Futian–Luohu center has the largest area of 17.008 square kilometers, accounting for 51.70% of the total area of all the centers, as well as the largest number of employees, 1,820,679, accounting for 55.25% of the total employment of the centers. The Kejiyuan center takes the second place, with an area of 8.640 square kilometers, accounting for 26.26% of the total area of all the centers, with the number of employed people at 931,073, accounting for 28.25% of the total number of employed people of all the centers. Regarding the third-level centers, Chegongmiao has a higher number of employees (222,837) and a higher area (2.375 km$^2$) than other centers.

### 4.2. Functional Centrality of Employment Centers

Based on the FCI value, the identified employment centers are graded by the natural discontinuity method (summary statistics for the variables shown in Table 1). Figure 3 shows the functional centrality levels of the identified centers. The results show that the eight employment centers include two first-level functional centers (FCI values of 0.996 and 0.561), three second-level functional centers (FCI values of 0.177, 0.135, and 0.135), and three third-level functional centers (FCI values of 0.069, 0.066, and 0.077). From the perspective of spatial distribution, the first-level functional centers are distributed in Futian District and Nanshan District, whereas the second- and third-level functional centers are scattered, showing that, as they are affected by the location advantages of Nanshan District and Futian District, Futian–Luohu center, and Kejiyuan center, they have a strong functional connection within the whole region and the greatest influence on other areas. Furthermore, the functional centrality of the centers distributed inside and outside the special zone is quite different, which means that the functional connection of each employment center is unbalanced. In other words, the functional centrality in the special zone is stronger than outside the special zone, and the employment centers distributed inside the special zone have a stronger functional influence. From the level of each center, we can find differences in the morphological and functional levels of some centers, such as the Kejiyuan, Chegongmiao, Songhe, and Aviation City centers. Specifically, the morphological level of Kejiyuan center is the second, whereas the functional level is the first. The morphological level of the Chegongmiao, Songhe, and Aviation City centers are third, whereas the functional level is second, showing a mismatch between the employment population aggregation and functional connections of these centers. The distributed employment population can generate stronger commuter flows, which leads to the functional level of these centers, being one level higher than the morphological level.

**Table 1.** Summary statistics for the variables to construct FCI.

| Centers | $N_{in}$ | $N_{out}$ | $C_i$ | $C_i'$ | $POP_{in}$ | $D_{in}$ | $POP_{out}$ | $D_{out}$ | $D_i$ | $D_i'$ | FCI |
|---|---|---|---|---|---|---|---|---|---|---|---|
| Futian–Luohu | 1571 | 1462 | 0.090 | 0.994 | 2,001,404 | 1,274.0 | 1,287,879 | 881.1 | 644,380 | 0.999 | 0.996 |
| Kejiyuan | 1293 | 1114 | 0.072 | 0.789 | 1,072,720 | 830.0 | 430,239 | 386.2 | 215,313 | 0.334 | 0.561 |
| Chegongmiao | 387 | 384 | 0.023 | 0.252 | 270,317 | 699.3 | 130,086 | 339.1 | 65,213 | 0.101 | 0.177 |
| Songhe | 301 | 327 | 0.019 | 0.206 | 134,312 | 446.9 | 82,431 | 252.4 | 41,342 | 0.064 | 0.135 |
| Fuhai | 278 | 338 | 0.018 | 0.202 | 117,751 | 423.1 | 86,600 | 256.2 | 43,428 | 0.067 | 0.135 |
| Aviation City | 148 | 161 | 0.009 | 0.101 | 73,415 | 495.4 | 46,555 | 288.9 | 23,422 | 0.036 | 0.069 |
| Dengliang | 142 | 147 | 0.009 | 0.095 | 68,006 | 479.3 | 47,538 | 323.3 | 23,931 | 0.037 | 0.066 |
| Longhua | 143 | 154 | 0.009 | 0.097 | 66,189 | 462.1 | 71,756 | 465.1 | 36,111 | 0.056 | 0.077 |

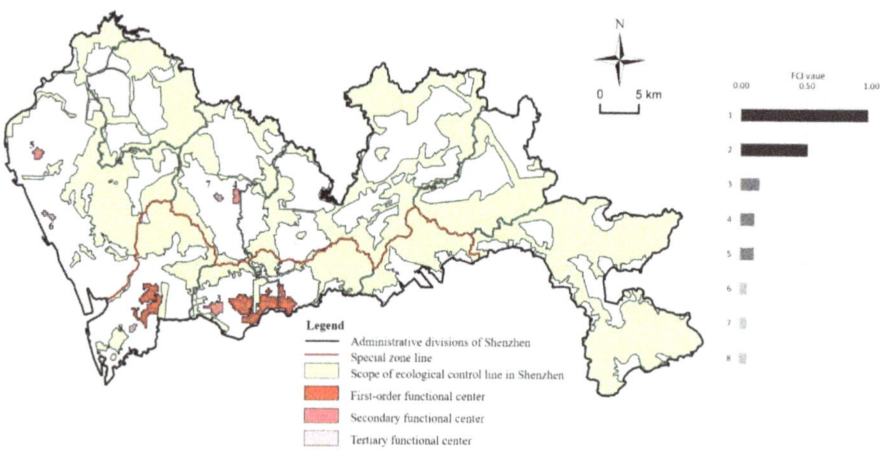


**Figure 3.** Functional centrality of the identified centers.

Based on the OD data, we further investigated the functional centrality of the first-level functional centers, including Futian–Luohu center and Kejiyuan center. Residence-employment connections whose starting and ending points are all distributed in Futian–Luohu center and Kejiyuan center are excluded. Figures 4 and 5 show the influence area of Futian–Luohu center and Kejiyuan center, respectively. According to the results, the Futian–Luohu employment center has many functional connections to the whole city. Many people lived in the midwest of Shenzhen and some people in the east worked in Futian–Luohu employment center. The affected area is mainly located in four districts: Nanshan District, Bao'an District, Longhua District and Longgang District. Nanshan District is the closest to Futian–Luohu center in space, and has the advantage of a short commuting distance. Many residential areas, such as urban villages, are found near the special zone line in Bao'an District and Longhua District, showing that Futian–Luohu center provides many jobs for the population lived in these areas. The influence area is mainly distributed along the metro line in the westernmost part of Bao'an District and Longgang District.

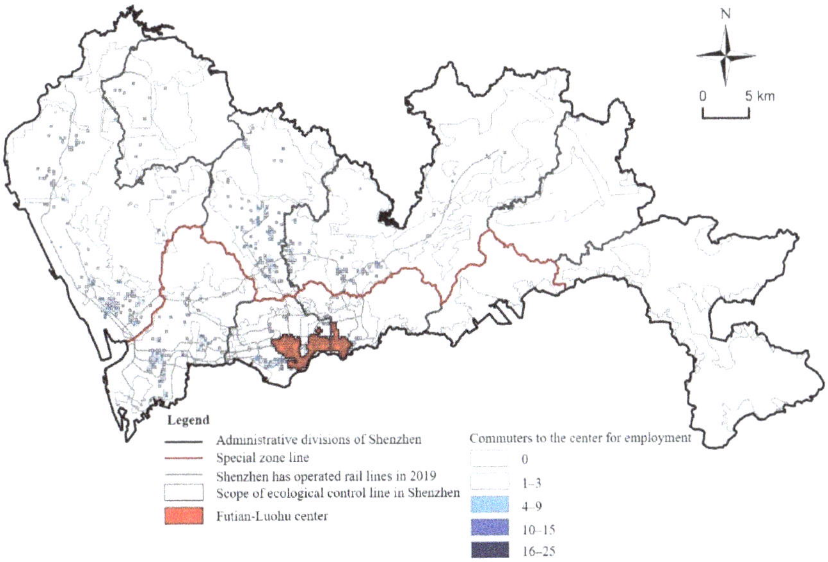


**Figure 4.** The influence area of Futian–Luohu center.

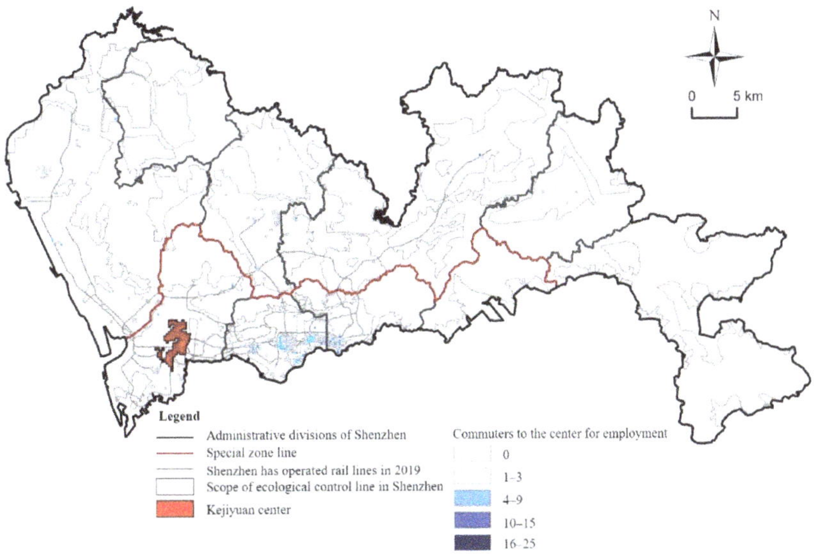


**Figure 5.** The influence area of Kejiyuan center.

According to Figure 5, although the functional centralities of the Kejiyuan and Futian–Luohu centers are at the first level, the influence area of Kejiyuan center is much smaller than that of Futian–Luohu center. The influence area of Kejiyuan center is mainly distributed at the south of Kejiyuan center, the boundary area of the Futian and Luohu Districts, as well as scattered in Longhua District and west of Bao'an District. According to the results, Kejiyuan center has a strong connection to Futian–Luohu area, indicating a spatial interaction between the Futian–Luohu and Kejiyuan centers. Many science parks and high-tech zones are located in Kejiyuan center and have attracted many people residing in Futian and Luohu to work in this employment center.

### 4.3. Comparison of Morphological Centrality and Functional Centrality

We use R1 to measure the morphological centrality of each employment center, which is the ratio of employment population of each center to the sum of the employment population of all employment centers. We use R2 to measure the functional centrality of each employment center, which is the ratio of the FCI value of each center to the sum of the FCI value of all employment centers. Then, by comparing the ratio of R1/R2 and 1, we can evaluate the strength of morphological and functional centrality of each employment center. R1/R2 > 1 means that the morphological centrality of the center is stronger than its functional centrality. R1/R2 < 1 means that the functional centrality of the center is stronger.

Figure 6 shows the results of comparing the morphological and functional centrality of each employment center. The result shows that the ratio of R1/R2 of the Futian–Luohu and Kejiyuan centers is greater than 1, meaning the morphological centrality is stronger than the functional centrality. The remaining centers are morphological tertiary employment centers, showing that the functional centrality is stronger than the morphological centrality. Furthermore, influenced by the employment population gathering, the larger the center is, the more employed people can be gathered, which leads to the morphology centrality being stronger than the functional centrality.

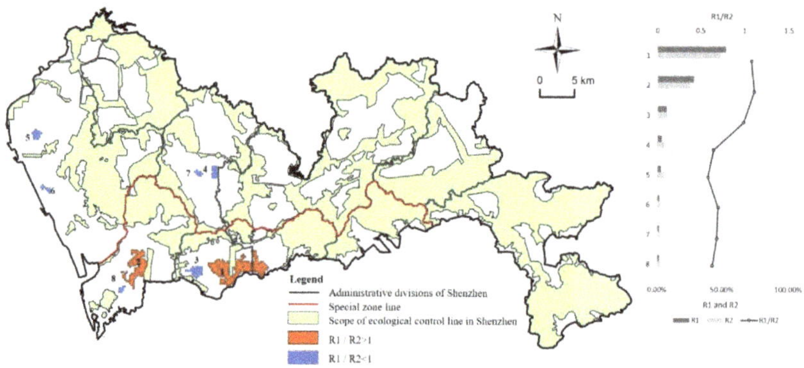


**Figure 6.** Comparison of morphological centrality and functional centrality.

This study curve fit the employment population and FCI value of different employment centers to measure the distribution equilibrium degree of each employment center's morphological and functional centrality. The method is applied as follows. The logarithm is taken as the horizontal axis for the ranking of each center. Then, the logarithm of the employment population and FCI values of each center are taken as the vertical axis for linear fitting. Next, the logarithm is taken. According to Figure 7, the fitted curve is shaped like an "L", and the absolute value of the slope is greater than 1, which shows the unbalanced morphological and functional centrality of the centers, as well as a monocentric employment spatial structure. In other words, Futian–Luohu center still concentrates most on the employment population and is in the dominant position of the employment–residence network in the city. We further compared the slope of the two fitting lines to understand the distribution of the morphological and functional centrality of the identified employment centers. The results show that the distribution of morphological centrality in employment centers is even more concentrated than that of functional centrality. According to the results, the employment spatial structure of Shenzhen is still monocentric in the morphological and functional perspectives.

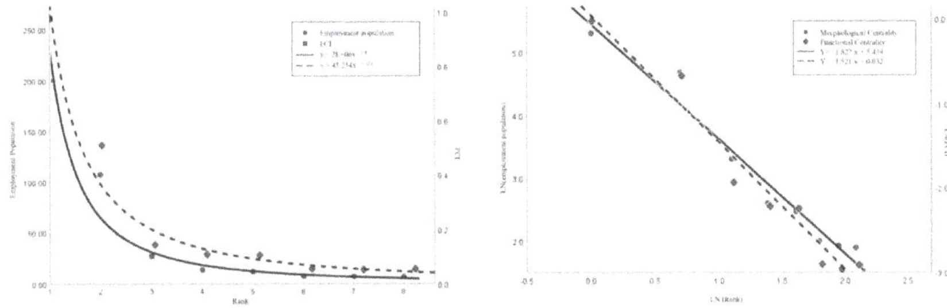

**Figure 7.** Relationship between functional and morphological centrality.

## 5. The Role of Urban Planning in the Formation of Spatial Structure

China's urban development is deeply affected by the intervention of urban planning and relevant policies [43]. Since its establishment of a special economic zone, the Shenzhen government has adopted polycentric development as an important spatial strategy. The Shenzhen 2010 master plan proposed building a '2 + 5 + 8' three-level polycentric urban spatial structure system that includes two main centers, five subcenters, and eight cluster centers (Figure 8). Our study shows that the master plan has played an important role in the formation of the spatial structures of Shenzhen. Figure 9 shows the center, subcenters, and cluster centers that were formed, or not formed, during the planning period. The urban development of Shenzhen is generally in line with the planning. The identified eight employment centers are all located within the planned urban centers. This finding reflects Sorensen's (2001) argument that the powerful structure plan contributed to the trend of urban spatial structure development. Strong government intervention measures and planning policies can promote the growth of employment centers in metropolitan areas, in order to cope with urban sprawl [44]. For example, Futian–Luohu and Kejiyuan were planned to become a R&D center that developed high-tech industries and became the hub of industrial clusters in Shenzhen. As a consequence, a large number of enterprises have been attracted to be located in Futian–Luohu and Kejiyuan, thereby promoting employment and local economic development.

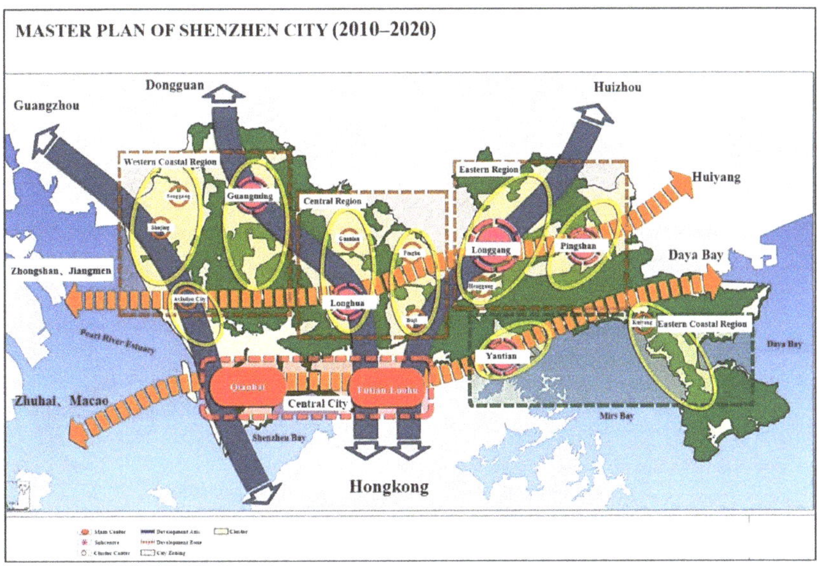


**Figure 8.** The Shenzhen master plan (2010–2020).

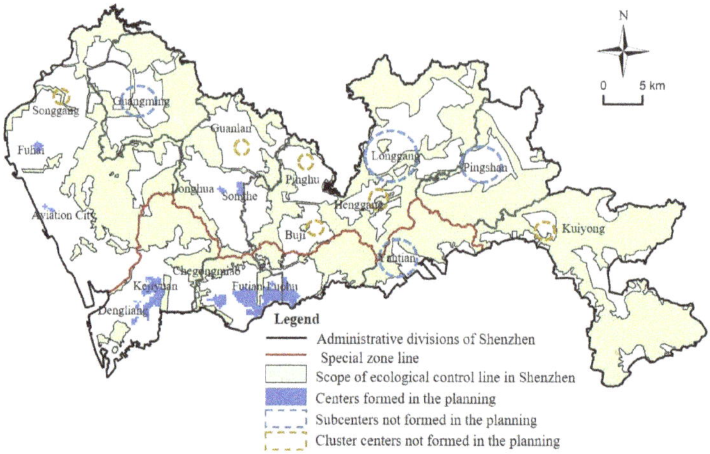


**Figure 9.** Classification of centers compared to the planning.

However, the planned polycentric spatial structure has not yet formed. The previous results have revealed that the spatial structure is still monocentric from the morphological and functional perspectives. The planned centers have not yet been formed in many areas, such as Yantian, Guangming, Pingshan, and Kuichong. These areas have a single industrial structure and relatively backward economic development. For example, Yantian's development is restricted by the functional orientation of the city. It is expected to develop a port logistics industry with few types of industries. In addition, a large part of these areas (e.g., Yantian and Kuichong) are located in the ecological control line, wherein development and construction are not allowed. As such, economic and spatial development are constrained, thereby hindering these areas from becoming centers. The public investment in service infrastructure and large-scale urban projects has a direct impact on the positioning of settlements and activities and is one of the driving factors for the formation and change in spatial structure [45]. Therefore, in order to improve the implementation performance of urban planning, strengthening infrastructure construction and land development in these areas in the future is necessary to help form the planned urban centers.

## 6. Conclusions

This study uses mobile phone signaling data to explore Shenzhen's employment spatial structure from the morphological and functional dimensions. Eight employment centers have been identified, all of which are located in the city's central and western parts. The analysis shows the differences between each center's morphological and functional centrality. It shows that the two biggest centers' morphological centrality(Futian–Luohu center and Kejiyuan center) are stronger than their functional centrality. Both of them are located in the SEZ area. On the contrary, the other centers' morphological centrality are weaker than their functional centrality. Most of these centers are located in the non-SEZ area. The findings suggest that, although Shenzhen has implemented polycentric urban planning since its foundation, its employment spatial structure is still monocentric in the morphological and functional terms. Further investigation shows that the distribution of morphological centrality in employment centers is even more concentrated than that of functional centrality.

Based on the results, this work discusses the role of urban planning in the forming of spatial structures in Shenzhen and provides implications for future urban planning. We find that the master plan has played an important role in the formation of the polycentric spatial structures of Shenzhen. Globally, the shift of the urban structure from monocentric to polycentric has been widely recognized in the literature. Several cities in Europe, America, and Japan have formed polycentric urban spatial structures [7,46,47]. Recently, studies on the relationships between morphological and functional urban spatial structures have been performed. These studies reflect that some western countries (e.g., Finland and Netherlands) are more polycentric, in terms of the functional perspective, compared to the morphological perspective [10,25].

Like western cities, many Chinese cities have adopted polycentric urban planning as an important spatial strategy for sustainable development. However, the extent to which these Chinese cities have been polycentric remains inadequately explored. This study explores the polycentricity of Shenzhen from the morphological and functional perspectives. The finding shows that, as influenced by the significant advantages of the Futian–Luohu center, most employment population and employment–residence connections are still concentrated in the Futian–Luohu center. Although eight employment centers have been identified, the employment spatial structure of Shenzhen remains monocentric, from the morphological and functional perspectives. Shenzhen's polycentric urban planning has not yet guided the city to form polycentric urban spatial structures.

Furthermore, this study contributes to the methodological approach. The method of measuring centrality has been improved. Previous studies often use a single factor, such as traffic density or the number of nodes, to measure functional centrality. Based on mobile phone signaling data, the present study combines social network analysis with GIS for analysis. The number of connected nodes, traffic density, and direction of population flow have been considered, in order to construct the FCI to measure functional centrality, thus providing a new perspective for analyzing the functional contact characteristics of the centers. Moreover, this study shows that the mobile phone signaling data, which can reflect the track of people in the city, is valuable for identifying the urban centers and reflecting on the spatial structure. These findings contribute to the recent studies using new and big datasets for urban analysis. Although the empirical analysis has focused on Shenzhen city, the proposed approach can be used to identify other cities, as long as the relevant data are available. The FCI index constructed in this study considers the number of connected nodes, traffic density, and direction of population flow to measure the functional centrality of the identified urban centers. Thus, the application of this research method requires datasets covering the information about the studied population and origin-destination of their commutes.

This study has some limitations that point to the directions for the future research. First, this study only explores the urban spatial structure of Shenzhen in 2019. However, investigating the evolution of the spatial structure during a certain period can help us understand how the city has been developed step-by-step. By examining the evolution of the spatial structure for many years, we can reveal whether the planning policy affects the overall trend of urban development and provide a policy-based explanation for the spatial evolution trajectory. Second, this research only explores the polycentric spatial structure of Shenzhen City, but the performance and efficiency of the formed spatial structure are still unknown. More research is required to understand the performance of polycentric spatial structures for Chinese cities and whether it is conducive to sustainable development.

**Author Contributions:** Conceptualization, Y.L. and Z.L.; methodology, Y.L. and Z.L.; software, Z.L. and C.C.; validation, Y.L. and Q.L.; formal analysis, Z.L. and C.C.; investigation, Y.L. and Q.L.; resources, Y.L.; data curation, Z.L. and C.C.; writing—original draft preparation, Z.L. and C.C.; writing—review and editing, Y.L. and Q.L.; visualization, Y.L. and Z.L.; supervision, Y.L. and Q.L.; project administration, Y.L.; funding acquisition, Y.L. All authors have read and agreed to the published version of the manuscript.

**Funding:** This research was funded by National Natural Science Foundation of China (72174122), National Natural Science Foundation of Guangdong Province (2022A1515011816), and Shenzhen Science and Technology Plan (20200813170728001).

**Acknowledgments:** We thank the National Natural Science Foundation of China, National Natural Science Foundation of Guangdong Province, and Shenzhen Science and Technology Plan for funding this study. We would also like to thank the three anonymous reviewers for their constructive comments.

**Conflicts of Interest:** The authors declare no conflict of interest.

## References

1. Liu, Z.; Liu, S. Polycentric development and the role of urban polycentric planning in China's mega cities: An examination of Beijing's metropolitan area. *Sustainability* **2018**, *10*, 1588. [CrossRef]
2. Burger, M.J.; de Goei, B.; van der Laan, L.; Huisman, F.J.M. Heterogeneous development of metropolitan spatial structure Evidence from commuting patterns in English and Welsh city-regions, 1981–2001. *Cities* **2011**, *28*, 160–170. [CrossRef]
3. Liu, X.; Derudder, B.; Wang, M. Polycentric urban development in China: A multi-scale analysis. *Environ. Plan. B Urban Anal. City Sci.* **2017**, *45*, 953–972. [CrossRef]
4. Sorensen, A. Subcentres and Satellite Cities: Tokyo's 20th Century Experience of Planned Polycentrism. *Int. Plan. Stud.* **2001**, *6*, 9–32. [CrossRef]
5. Huang, D.; Liu, Z.; Zhao, X.; Zhao, P. Emerging polycentric megacity in China: An examination of employment subcenters and their influence on population distribution in Beijing. *Cities* **2017**, *69*, 36–45. [CrossRef]
6. Zhang, T.; Sun, B.; Li, W.; Dan, B.; Wang, C. Polycentricity or dispersal? The spatial transformation of metropolitan Shanghai. *Cities* **2019**, *95*, 102352–102360. [CrossRef]
7. Heider, B.; Siedentop, S. Employment suburbanization in the 21st century: A comparison of German and US city regions. *Cities* **2020**, *104*, 102802–102815. [CrossRef]
8. Hall, P. Looking Backward, Looking Forward: The City Region of the Mid-21st Century. *Reg. Stud.* **2009**, *43*, 803–817. [CrossRef]
9. Sarkar, S.; Wu, H.; Levinson, D.M. Measuring polycentricity via network flows, spatial interaction and percolation. *Urban Stud.* **2020**, *57*, 2402–2422. [CrossRef]
10. Vasanen, A. Functional Polycentricity: Examining Metropolitan Spatial Structure through the Connectivity of Urban Sub-centres. *Urban Stud.* **2012**, *49*, 3627–3644. [CrossRef]
11. Liu, X.; Wang, M. How polycentric is urban China and why A case study of 318 cities. *Landsc. Urban Plan.* **2016**, *151*, 10–20. [CrossRef]
12. Sun, T. A longitudinal study of changes in intra-metropolitan employment concentration in Beijing: Decentralisation, reconcentration and polycentrification. *Urban Stud.* **2019**, *5*, 748–765. [CrossRef]
13. Yue, W.; Liu, Y.; Fan, P. Polycentric Urban Development: The Case of Hangzhou. *Environ. Plan. An Econ. Space* **2010**, *42*, 563–577. [CrossRef]
14. Chen, W.; Golubchikov, O.; Liu, Z. Measuring polycentric structures of megaregions in China: Linking morphological and functional dimensions. *Environ. Plan. B Urban Anal. City Sci.* **2020**, *48*, 2272–2288. [CrossRef]
15. Hu, Y.; Han, Y. Identification of Urban Functional Areas Based on POI Data: A Case Study of the Guangzhou Economic and Technological Development Zone. *Sustainability* **2019**, *11*, 1358. [CrossRef]
16. Liu, Y.; Fang, F.; Jing, Y. How urban land use influences commuting flows in Wuhan, Central China: A mobile phone signaling data perspective. *Sustain. Cities Soc.* **2020**, *53*, 101914–101924. [CrossRef]
17. Xiao, Y.; Wang, Y.; Miao, S.; Niu, X. Assessing polycentric urban development in Shanghai, China, with detailed passive mobile phone data. *Environ. Plan. B Urban Anal. City Sci.* **2021**, *48*, 2656–2674. [CrossRef]
18. Song, Y.; Huang, B.; Cai, J.; Chen, B. Dynamic assessments of population exposure to urban greenspace using multi-source big data. *Sci. Total Environ.* **2018**, *634*, 1315–1325. [CrossRef]
19. Song, Y.; Huang, B.; He, Q.; Chen, B.; Wei, J.; Mahmood, R. Dynamic assessment of PM2.5 exposure and health risk using remote sensing and geo-spatial big data. *Environ. Pollut.* **2019**, *253*, 288–296. [CrossRef]
20. Ratti, C.; Frenchman, D.; Pulselli, R.M.; Williams, S. Mobile Landscapes: Using location data from cell phones for urban analysis. *Environ. Plan. B* **2005**, *33*, 727–748. [CrossRef]
21. Kloosterman, R.C.; Musterd, S. The Polycentric Urban Region: Towards a Research Agenda. *Urban Stud.* **2001**, *38*, 623–633. [CrossRef]
22. McDonald, J.F.; Prather, P.J. Suburban Employment Centres: The Case of Chicago. *Urban Stud.* **1994**, *31*, 201–218. [CrossRef]
23. McMillen, D.P. Nonparametric Employment Subcenter Identification. *J. Urban Econ.* **2001**, *50*, 448–473. [CrossRef]
24. Leslie, T.F. Identification and Differentiation of Urban Centers in Phoenix Through a Multi-Criteria Kernel-Density Approach. *Int. Reg. Sci. Rev.* **2010**, *33*, 205–235. [CrossRef]
25. Burger, M.; Meijers, E. Form Follows Function? Linking Morphological and Functional Polycentricity. *Urban Stud.* **2012**, *49*, 1127–1149. [CrossRef]
26. Green, N. Functional Polycentricity: A Formal Definition in Terms of Social Network Analysis. *Urban Stud.* **2007**, *44*, 2077–2103. [CrossRef]
27. Huang, D.; Liu, Z.; Zhao, X. Monocentric or Polycentric? The Urban Spatial Structure of Employment in Beijing. *Sustainability* **2015**, *7*, 11632–11656. [CrossRef]

28. Taubenboeck, H.; Standfuss, I.; Wurm, M.; Krehl, A.; Siedentop, S. Measuring morphological polycentricity—A comparative analysis of urban mass concentrations using remote sensing data. *Comput. Environ. Urban Syst.* **2017**, *64*, 42–56. [CrossRef]
29. Hamidi, S.; Ewing, R. A longitudinal study of changes in urban sprawl between 2000 and 2010 in the United States. *Landsc. Urban Plan.* **2014**, *128*, 72–82. [CrossRef]
30. Qiang, Y.; Xu, J.; Zhang, G. The shapes of US cities: Revisiting the classic population density functions using crowdsourced geospatial data. *Urban Stud.* **2019**, *57*, 2147–2162. [CrossRef]
31. Kloosterman, R.C.; Lambregts, B. Clustering of economic activities in polycentric urban regions: The case of the Randstad. *Urban Stud.* **2001**, *38*, 717–732. [CrossRef]
32. Liu, W.; Wu, W.; Thakuriah, P.; Wang, J. The geography of human activity and land use: A big data approach. *Cities* **2020**, *97*, 102523–102534. [CrossRef]
33. Li, W.; Sun, B.; Zhang, T.; Zhang, Z. Panacea, placebo or pathogen? An evaluation of the integrated performance of polycentric urban structures in the Chinese prefectural city-regions. *Cities* **2022**, *125*, 103624–103632. [CrossRef]
34. Sun, B.; Li, W.; Zhang, Z.; Zhang, T. Is polycentricity a promising tool to reduce regional economic disparities? Evidence from China's prefectural regions. *Landsc. Urban Plan.* **2019**, *192*, 103367–103377. [CrossRef]
35. Batten, D. Network Cities: Creative Urban Agglomerations for the 21st Century. *Urban Stud.* **1995**, *32*, 313–327. [CrossRef]
36. Reades, J.; Calabrese, F.; Ratti, C. Eigenplaces: Analysing cities using the space–time structure of the mobile phone network. *Environ. Plan. B Plan. Des.* **2009**, *36*, 824–836. [CrossRef]
37. Zhang, P.; Zhao, Y.; Zhu, X.; Cai, Z.; Xu, J.; Shi, S. Spatial structure of urban agglomeration under the impact of high-speed railway construction: Based on the social network analysis. *Sustain. Cities Soc.* **2020**, *62*, 102404–102414. [CrossRef]
38. Zhong, C.; Schläpfer, M.; Arisona, S.M.l.; Batty, M.; Ratti, C.; Schmitt, G. Revealing centrality in the spatial structure of cities from human activity patterns. *Urban Stud.* **2017**, *54*, 437–455. [CrossRef]
39. Yue, W.; Wang, T.; Liu, Y.; Zhang, Q.; Ye, X. Mismatch of morphological and functional polycentricity in Chinese cities An evidence from land development and functional linkage. *Land Use Policy* **2019**, *88*, 104175–104176. [CrossRef]
40. Yang, X.; Fang, Z.; Yin, L.; Li, J.; Zhou, Y.; Lu, S. Understanding the Spatial Structure of Urban Commuting Using Mobile Phone Location Data: A Case Study of Shenzhen, China. *Sustainability* **2018**, *10*, 1435. [CrossRef]
41. Wei, L.; Luo, Y.; Wang, M.; Cai, Y.; Su, S.; Li, B.; Ji, H. Multiscale identification of urban functional polycentricity for planning implications: An integrated approach using geo-big transport data and complex network modeling. *Habitat Int.* **2020**, *97*, 102134–102149. [CrossRef]
42. Wang, T.; Yue, W.; Ye, X.; Liu, Y.; Lu, D. Re-evaluating polycentric urban structure: A functional linkage perspective. *Cities* **2020**, *101*, 102672–102682. [CrossRef]
43. Lin, G.C.S. State Policy and Spatial Restructuring in Post-reform China, 1978–1995. *Int. J. Urban Regional Res.* **1999**, *23*, 670–696. [CrossRef]
44. Shearmur, R.; Alvergne, C. Regional Planning Policy and the Location of Employment in the Ile-De-France. *Urban Aff. Rev.* **2003**, *39*, 3–31. [CrossRef]
45. Hoyler, M.; Kloosterman, R.C.; Sokol, M. Polycentric Puzzles—Emerging Mega-City Regions Seen through the Lens of Advanced Producer Services. *Reg. Stud.* **2008**, *42*, 1055–1064. [CrossRef]
46. Kane, K.; Hipp, J.R.; Kim, J.H. Los Angeles employment concentration in the 21st century. *Urban Stud.* **2018**, *55*, 844–869. [CrossRef]
47. Liu, K.; Murayama, Y.; Ichinose, T. Using a new approach for revealing the spatiotemporal patterns of functional urban polycentricity: A case study in the Tokyo metropolitan area. *Sustain. Cities Soc.* **2020**, *59*, 102176–102191. [CrossRef]

*land*

MDPI

*Article*

# Effects of Polycentricity on Economic Performance and Its Dependence on City Size: The Case of China

**Bindong Sun [1,2,3,4,5], Tinglin Zhang [1,2,3,4,5,*], Wan Li [6] and Yan Song [7]**

1 Research Center for China Administrative Division, Shanghai 200241, China
2 Institute of Eco-Chongming, Shanghai 202162, China
3 The Future City Lab, East China Normal University, Shanghai 200062, China
4 The Center for Modern Chinese City Studies, East China Normal University, Shanghai 200062, China
5 School of Urban and Regional Science, East China Normal University, Shanghai 200241, China
6 Business School, Zhengzhou University, Zhengzhou 450052, China
7 Department of City and Regional Planning, University of North Carolina at Chapel Hill, Chapel Hill, NC 27599-3140, USA
* Correspondence: tlzhang@re.ecnu.edu.cn

**Abstract:** Polycentric planning strategies have often failed to achieve the expected effects. The ensuing uncertainty associated with the desirability of polycentric strategies is also reflected in the early literature which offers no clear conclusion about whether the polycentricity affects economic performance and how. This paper aims at offering a clear conclusion about it, especially its dependence on city size. Against this backdrop, we conceptualize polycentricity as a process of reclustering after decentralization to reevaluate its impact on performance. To this end, we use the city proper level Chinese Economic Census (2004, 2008, and 2013) and apply a fixed-effects panel model, the results of which show that the dependence of the urban economy on spatial structure is contingent on city size. More specifically, both decentralization and clustering (and therefore the polycentric structure) facilitate economic performance only when cities reach a certain size. We use our findings as the basis for outlining an emergent research agenda for urban polycentricity.

**Keywords:** polycentricity; city size; economic performance; optimal city size; China

**Citation:** Sun, B.; Zhang, T.; Li, W.; Song, Y. Effects of Polycentricity on Economic Performance and Its Dependence on City Size: The Case of China. *Land* **2022**, *11*, 1546. https://doi.org/10.3390/land11091546

Academic Editor: Roger White

Received: 25 July 2022
Accepted: 6 September 2022
Published: 12 September 2022



## 1. Introduction

Suburbanization or decentralization, which is fueled by the expansion of city size, has become one of the most important characteristics of urban spatial transformations worldwide. Urban planners have been seeking spatial adjustment strategies to meet the challenges caused by the agglomeration diseconomies associated with such expansion and often advocate for polycentric spatial structures. Polycentricity, they argue, is supposed to reduces the negative agglomeration effects that occur once employment is no longer centralizes in the main city center, and it facilitates cities in regaining positive momentum when employment reclusters in subcenters. However, polycentric planning practices have long been met with skepticism because they usually fail to achieve the expected effects.

Therefore, the prevailing but usually ineffective practice of polycentric planning strategy has brought about an urgent demand for research on the relationship between polycentricity and economic performance. Unfortunately, a consensus on this issue has not yet been reached (see the meta-analysis by Li et al., 2022) [1]. The reasons for this lack of consensus may stem from the following three unsolved problems. First, is the effect of spatial structure on economic performance contingent on city size, especially at the metropolitan level? Although empirical results on the total effects of polycentricity on the economy have been enriched in many countries, such as the USA [2], Korea [3], or China [4,5], sufficient attention has not been paid to the moderating effects of city size. Second, what is the underlying mechanism for the economic performance of polycentricity?

The two theoretical paths, namely decentralization [6,7] and "borrowed size" [8–10] have not been supported by empirical evidence. Third, how to deal with the effects of the reverse causality of the economic performance on polycentricity? The results achieved by cross-sectional or OLS models [2,4,11,12] are likely to be biased.

Our paper aims to expand our understanding of the effect of urban spatial structure on economic performance. In particular, we delve into its linkages with city size and whether size exerts a moderating effect on the relationship between economic performance and spatial structure at the Chinese city proper level. To provide empirical evidence for two theoretical paths of polycentricity (decentralization and "borrowed size"), this analysis unfolds polycentricity in the process of reclustering after decentralization, a process that has generally neglected in the previous empirical literature. Furthermore, we improve on commonly used cross-sectional models and historical instrumental variables by adopting a two-way fixed-effects panel model and more effective instrumental variables for a two-stage least squares (TSLS) model.

## 2. Literature Review and Hypotheses

### 2.1. Concept of Urban Spatial Structure

The concept of urban spatial structure, despite its centrality in the urban economics and regional science literature, remains ambiguous. In this paper, we focus mainly on the morphological spatial structure rather than the functional structure, which involves the networks between centers [13,14]. Moreover, we measure spatial structure with data on the distribution of employment rather than with other human activity data such as population [11], land use [15], or nighttime light [16,17]. The main reason for this choice is that our research aims at addressing the spatial frictions associated with the labor market.

However, despite the multiplicity of morphological definitions, there is considerable overlap between the key dimensions and the initial conceptualization that Anas et al. (1998) synthesized in their work [18]. They proposed that urban spatial structure can be defined along two dimensions, namely, the degree of centralization versus decentralization and the degree of clustering versus dispersion. Anas et al. (1998) thus suggested two dimensions to describe urban spatial structures, specifically, centralization and clustering [18]. Lee (2007) elaborated on and clarified Anas et al.'s (1998) conceptualization and named these dimensions centralization and concentration [19]. Centralization is the extent to which employment is concentrated near a central business district (CBD). Concentration, similar to the dimension "the degree of clustering versus dispersion," measures the extent to which employment is either clustered in a few nodes or dispersed. Thus, polycentricity is the combined result of decentralization and clustering [18,19]. Accordingly, we illustrate this process of polycentricity in Figure 1.

### 2.2. Relation between Economic Performance and Urban Spatial Structure

The distribution of urban employment determines the spatial organization of employment centers and how they are connected with each other. This structure, in turn, may have serious implications for urban prosperity depending on whether it benefits productivity. Polycentricity may contribute to economic performance via the following two potential mechanisms: (1) decentralization reduces the agglomeration diseconomy of the main centers [6,7]; and (2) clustering in subcenters recovers the positive externalities through "borrowed size" [8–10]. However, decentralization has the potential to harm the economy when it fails to compensate for the reduction in the agglomeration economies of the main center. Therefore, we group the existing empirical studies that examine the influence of spatial structure on economic performance into three classes: those that study the spatial dimensions of polycentricity, those that study centralization, and those that study clustering.

(1) The literature on the spatial dimensions of polycentricity tends to focus on the direct relationship between polycentricity (as measured by the balance in the distribution of city/center sizes) and urban economic performance. However, differences in oper-

ationalization have led to inconsistent results. For example, some studies identified a positive relationship between polycentricity and urban economic performance among US metropolitan areas [2] or Korea [3], while Lee and Gordon (2007) found no significant relationship [12]. The results for Chinese cities are heterogeneous across different study scales. Zhang, Sun, and Li (2017) found support for a positive association between polycentricity and economic performance within a sample of city proper areas [4]. However, Sun and Li (2016) and Li, Sun, and Zhang (2018) rebutted their argument and found contrary evidence at the municipal administrative area level [16,20]. Furthermore, Wang, Derudder, and Liu (2019) showed that intra-urban monocentricity and inter-urban polycentricity are linked with higher levels of labor productivity [5].

(2) The relationship between centralization and urban economic performance exhibits similar inconsistencies. Cervero (2001) and Veneri and Burgalassi (2011) assessed the effects of centralization and reported a positive impact of centralization on labor productivity [21,22]. However, Glaeser and Khan (2004) examined the influence of the percentage of employees within 3 miles of the CBD on urban economic performance and found that a 10% increase in decentralization led to, on average, 2.7% growth in GDP per capita [23].

(3) The literature on the relationship between clustering and urban economic performance is the only branch of the literature to provide seemingly consistent results. Fallah, Partridge and Olfert (2011) found that urban sprawl is negatively related to average labor productivity [24]. Qin and Liu (2015) obtained consistent results by using nighttime light data from a Chinese prefecture-level city to calculate the same index as Fallah, Partridge, and Olfert (2011) [24,25]. Furthermore, Liu, Chen, and Liu (2020) defined urban compactness using Landsat data and found that it was negatively correlated with urban GDP [15].

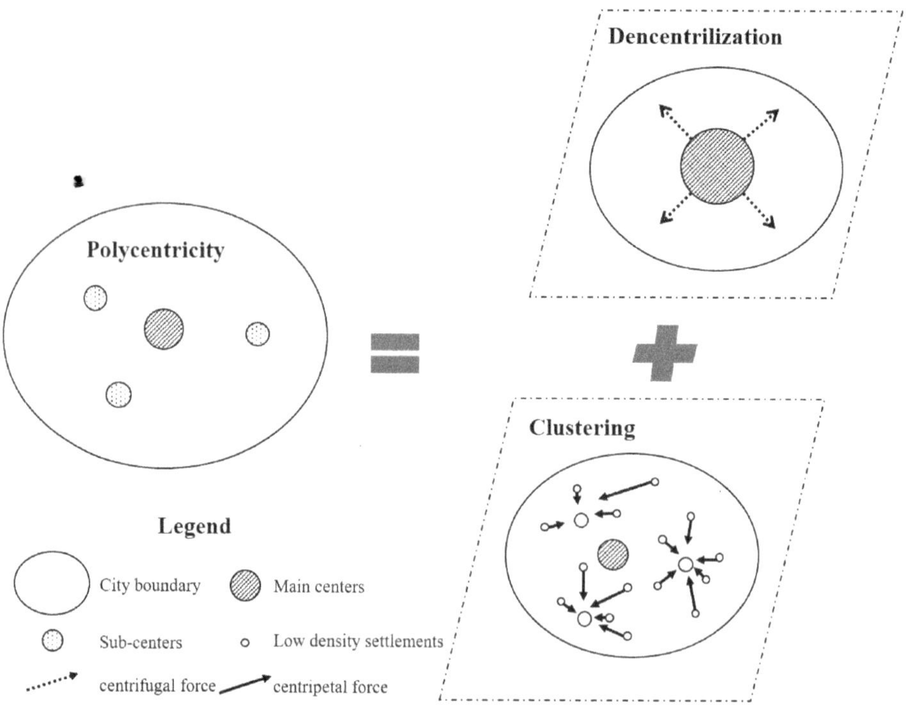


**Figure 1.** Two dimensions of polycentricity (decentralization and clustering).

### 2.3. Moderating Effect of City Size

Cities at different stages of development may have different optimal spatial structures. Our main theoretical prediction is that as cities pass a certain size threshold and continue to grow, the positive effects of polycentricity increase. This is because polycentricity tips the balance of net economic gains away from centralization and toward clustering. In other words, in large cities, as the cost of centralization increases with size and with the high concentration of activities in one location, the net benefits of clustering outside the main center overtake those of centralization.

Friedman (1966) proposed a theory of stages of spatial organization consistent with this hypothesis [26]. He expounded on a sequential process for the interaction between spatial structure and development. According to him, as economic development and city-size growth occur, the spatial structure transitions from a low-level spatial equilibrium to a monocentric structure and then shifts to a high-level spatial equilibrium with a polycentric structure. His theoretical conjecture has been supported by both simulations [27] and empirical studies [28].

A polycentric structure improves economic performance more in larger cities according to the theoretical predictions outlined above. However, empirical analyses that have directly examined the mediating effect of city size have not always supported these theoretical conjectures. In the city proper or metropolitan scale on which our study focuses, neither Zhang, Sun, and Li's (2017) work nor Lee and Gordon's (2007) study identified a significant interaction effect of city size on the relationship between economic performance and polycentricity [4,12]. However, Meijers and Burger (2010) found that the contribution of polycentricity to productivity is greater in smaller metropolitan areas than in larger metropolitan areas [2]. At the larger scale, Li and Liu (2018) as well as Sun and Li (2016) found that the relation between economic performance and polycentricity varied with different population sizes or densities at the prefecture city level [11,20], while Wang, Sun, and Zhang (2020) found evidence that polycentricity also boosts positive economic performance when regions have a larger population at the level of city cluster [29].

Otherwise, we also find some attempts to identify the moderating effect of city size between other dimensions of urban spatial structure and economic performance. Lee and Gordon (2007) revealed that more dispersion leads to higher growth rates as metropolitan areas grow [12], while Meijers and Burger (2010) and Li and Liu (2018) did not find any evidence of such an effect in the relationship between dispersion and economic performance by using population or density as interaction factors [2,11].

### 2.4. Existing Gaps

First, unlike theoretical advancements, the heterogeneity in the economic efficiency of polycentricity at the city proper or metropolitan level has not been strongly emphasized in the empirical literature. In particular, evidence concerning the moderating effect of city size on the relationship between urban spatial structure and economic performance is mixed and needs more robust empirical testing [2,4,10,16,19,20].

Second, until recently, very few studies on economic performance and spatial structure considered the full process of polycentric development along its two dimensions—the local reclustering of employment or the population after decentralizing away from the main city center. Studies have tended to borrow the framework of Meijers and Burger (2010) and to consider polycentricity to be a dimension of the distribution of economic activity [2]. Such approaches are crude when we want to know which channel plays a significant role in determining the economic effects of polycentricity. It is critical to distinguish whether polycentricity leads to economic gains because it diminishes the externalities from agglomeration through decentralization, because it diminishes agglomeration externalities through the gains from the positive externalities associated with reclustering, or because of both mechanisms. A complete exposition of polycentric evolution involves decentralization away from the CBD and then reclustering in several high-density areas, which has not been carefully tested.

Third, the estimation methods that most researchers have relied on to identify the causal association between spatial structure and economic performance have shortcomings. Limited by data availability, the bulk of early research results comes from cross-sectional models [2,4,10,12] and is likely biased because of omitted time-independent variables. Furthermore, many studies have examined the spatial distribution of the population rather than of employment [2,10,16]. The spatial distribution of the population has a less direct logical connection with economic output (especially as measured by GDP) than that of employment, given that spatial frictions mainly arise in the labor market.

We provide an integrated treatment of these gaps, and our starting point is to provide rigorous evidence on the relationship between the spatial structure of employment and economic performance, particularly at the level of the city proper in the Chinese context. In addition, we endeavor to connect the empirical work more strongly with geographers' and economists' theory of the evolutionary stages of spatial structure by emphasizing the moderating effect of city size. We extend the analysis to tests of different aspects of the hypothesized polycentricity process by including the two dimensions of decentralization and reclustering. We are aware of the endogeneity issues related to these tests and therefore propose a two-way fixed-effects panel model and instrumental variables for the TSLS models.

### 2.5. Our Hypotheses

We present a set of theoretical hypotheses to guide the empirical work and to establish clear expectations. Polycentricity is the result of a shift in the balance between the centrifugal forces that cause decentralization and the centripetal forces that lead to reclustering. We adopt the seminal framework established by Anas et al. (1998) and Lee (2007) to capture this dynamic [18,19]. To reiterate, the two dimensions of the metropolitan-level spatial structure are centralization versus decentralization and clustering versus dispersion (Figure 2). We can classify urban spatial structures into the following four types based on these two dimensions: monocentric (centralized and dispersed); sprawling (decentralized and dispersed); polycentric (decentralized and clustered); and monocentric with obvious subcenters (centralized and clustered). Using these four spatial structure types instead of simple dichotomies makes the differentiation of the processes at work both easier and clearer.

The relation between economic performance and the urban spatial structure is a dynamic balance between the positive and negative externalities of agglomeration. Polycentricity can be understood as an effective way to reduce agglomeration costs because it involves urban decentralization. In addition, economic productivity can be enhanced due to the reclustering of the population or of employment in subcenters through the mechanisms of sharing, matching, and learning [30]. On the other hand, polycentric structures also have the potential to harm labor productivity, given that decentralization may damage the agglomeration economies of the main center and lead to increasing transaction costs between centers.

The trade-off between agglomeration economies and agglomeration diseconomies is highly dependent on city size. We posit that which urban spatial structure is most efficient varies with city size. As cities grow, the negative externalities from agglomeration in the main center exceed the positive externalities. Decentralization can help reduce these agglomeration diseconomies, and reclustering can lead to gains through the scale effect among subcenters. Therefore, as long as cities are small enough that the benefits of agglomeration economies arising from the high level of concentration in the main city center exceed the corresponding costs, polycentricity likely undermines economic gains. In other words, it may not be possible for a city to reach a sufficiently large scale that the losses incurred from decentralization are outweighed by the positive agglomeration effects generated by subcenters (i.e., the metropolitan economy may be too small to be divided, and therefore, the sum of the parts is not greater than the centralized whole). Based on this theoretical argument, we predict that only when decentralization and reclustering are combined in sufficiently large cities can polycentricity foster greater economic development.

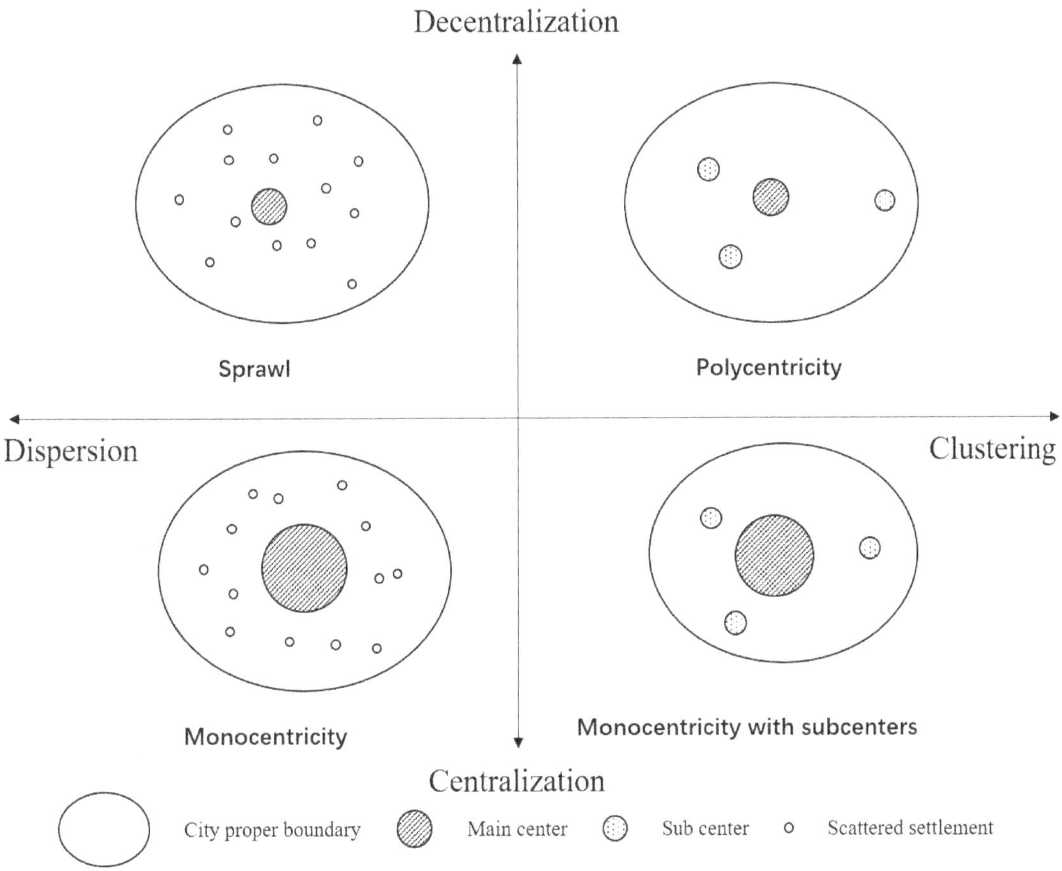


**Figure 2.** Spatial structure dimensions.

## 3. Methods and Data

Our methodological framework is organized as follows (Figure 3). First, we collect the multi-source data and determine the research scale in Section 3.1. Second, we clarify the concept of urban spatial structure and expound its measurement for our research in Section 3.2. Finally, we carry out the performance analysis by proceeding with the basic model, robustness test, and discussions (the models are listed in Section 3.3, and the results are shown in Section 4).

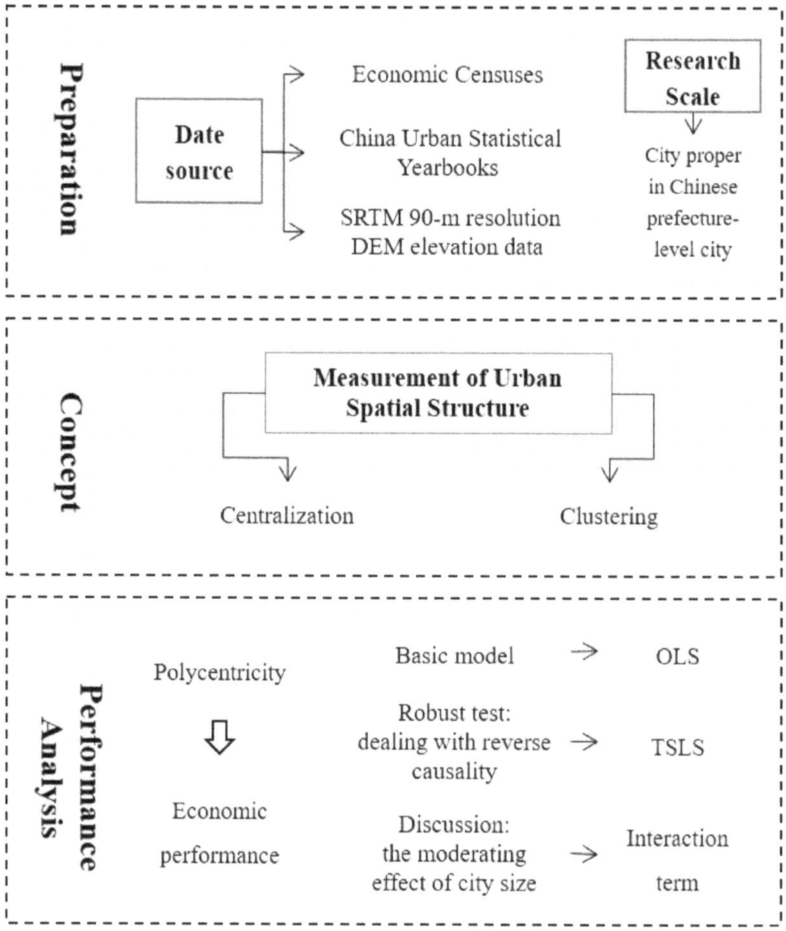


**Figure 3.** Methodological framework.

*3.1. Data Sources and Research Scale*

The accessibility of the labor market plays an important role in the association between urban spatial structure and economic performance because commuting creates friction between economic activities and space. Therefore, the distribution of labor is a key variable for economic productivity. We draw upon the Economic Censuses of 2004, 2008 and 2013 to calculate the urban spatial structure. The Economic Census is unique in China due to its provision of microlevel enterprise data, it contains detailed information on legal manufacturing and service entities in China, such as their postal code and the number of employees at each firm. Using this information, we can obtain the number of employees in each postal code zone for further calculation.

In addition to the availability of employment data with finer geographical scales, China is a quite suitable case for this study. China has many cities of all sizes, which could provide rich evidence for testing the effect of the heterogeneity of city size on the economic performance of polycentricity. Further, a reliable academic examination is urgently needed for the future development of ubiquitous polycentric planning practice in China.

As to the research scale, metropolitan areas are the preferred spatial unit for the analysis of economic performance because they are economically integrated areas. As a metropolitan area is not precisely defined in China, the concept of an administrative city (prefecture-level city) was more commonly used in the early literature. However, prefecture-level cities not only include highly urbanized subunits (cities proper, named Shixiaqu) but also contain a certain area of peripheral, semirural, and rural areas (county area, named Xian). Thus, we choose the city proper in the prefecture-level city as our second-best choice in terms of research scale as it is comparable to the metropolitan scale. The administrative divisions of 2008 (the middle year of our study time span) are taken as the standard and thus our full sample totals 287.

The choice of the analytical unit scale to be used for the spatial structure calculations is also of great importance. A unit that is too large averages important differences in the employment distribution within the area. To obtain sufficient precision with the available data, we choose postal code zones as our unit of analysis and sum the number of jobs in each unit to calculate our urban spatial structure indices.

### 3.2. *Measurement of Urban Spatial Structure*

We operationalize centralization and clustering by using the indices proposed in existing studies (summarized in Table 1). Centralized cities can feature dispersion or clustering outside the CBD, and clusters may be located near the CBD or away from the CBD. Compared with using polycentricity directly [2,4,5], we split the polycentricity into the dimension of decentralization and clustering. This approach gives up gaining the direct results of the extent to which polycentricity improves economic performance. However, to further investigate the mechanism and contribute to the early literature, we believe that it is worth using two dimensions to describe polycentricity.

Figure 4 illustrates the calculation process and symbol interpretation of the centralization and clustering indices.

All the indices are time-variant, as both the location of the CBD and the density of employment in each postal code vary across years. Similar to Wheaton (2004) [31], we use a virtual urban area boundary that contains 98 percent of all employment and excludes mostly low-density areas in outlying locations.

The centralization–decentralization dimension contains location information: it measures the extent to which employment is centralized near the CBD. For the purpose of creating our indices, the CBD is defined as the postal code with the highest employment density. The modified Wheaton index (MWI) and modified weighted average distance from the CBD (MADC) both measure how quickly the cumulative share of employment increases from the CBD to the urban edge [31,32]. The larger these two indices are, the more centralized the city is. A value of 1 indicates that all employment is concentrated in the center. When calculating the MWI, all postal code zones should be sorted by their distance to the CBD from nearest to farthest.

**Table 1.** Indices for the two dimensions of polycentricity.

| **Centralization indices** | |
|---|---|
| Modified Wheaton index [31] | $MWI = \frac{\sum_{i=1}^{n} E_{i-1} DCBD_i - \sum_{i=1}^{n} E_i DCBD_{i-1}}{DCBD^*}$ |
| Modified weighted average distance from the CBD [32] | $MADC = 1 - \sum_{i=1}^{n} \frac{e_i}{E} * \frac{DCBD_i}{DCBD^*}$ |
| **Clustering indices** | |
| Delta index [32,33] | $DELTA = \frac{1}{2} \sum_{i=1}^{n} \left| \frac{e_i}{E} - \frac{a_i}{A} \right|$ |
| Gini coefficient [34,35] | $GINI = \sum_{i=1}^{n} E_i A_{i-1} - \sum_{i=1}^{n} E_{i-1} A_i$ |

Notes: The zip code zone with the highest density of employment in each city is defined as the CBD. Symbols: ei: number of employed persons in zone i; E: total metropolitan employment; ei/E: share of metropolitan employment in zone i; Ei: cumulative share of employment in zone i; ai: land area of zone i; A: total metropolitan land area; ai/A: share of metro land area in zone i; Ai: cumulative share of land area in zone i; DCBDi: distance between zone i and the CBD; DCBD*: distance between the outermost zone and the CBD (city proper radius); n: number of zones (zip code zones).

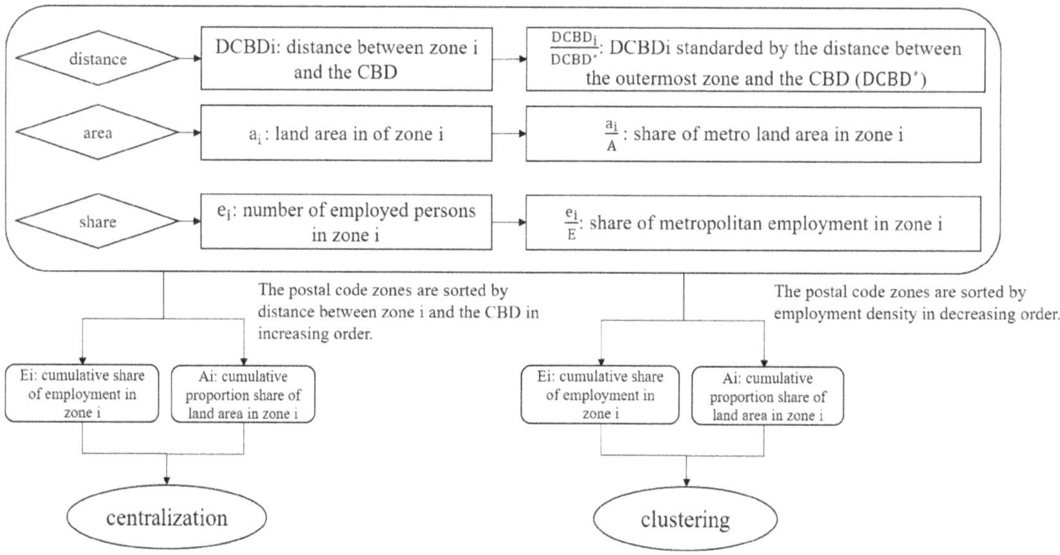


**Figure 4.** Calculation progress of centralization and clustering.

The clustering–dispersion dimension contains the information on the degree of agglomeration in the urban spatial structure. It captures the extent to which metropolitan employment is disproportionately located in areas with different densities and is measured by the Gini coefficient and Delta index in our paper [32–35]. The postal code zones are sorted by employment density in decreasing order.

### 3.3. Models

We rely on the Cobb–Douglas production function, which uses physical capital (measured as the physical capital stock per worker, $\frac{K}{L}$) and human capital (measured as the number of middle school students per $10^4$ persons, $\frac{H}{L}$) as the most important production factors for economic growth. In addition to these production factors, economic productivity also influences economic performance by determining the efficiency with which these production factors are used. Therefore, the degree of government intervention (G), which is measured as the ratio of government consumption to GDP, is added to our model. Agglomeration has also traditionally been considered a key determinant of economic productivity. Population (POP) is a conventional aspect of agglomeration. We also add the quadratic term for population to identify the "optimal city size" with decreasing returns. Our main concern is the urban spatial structure variables (STU, including centralization and clustering), which we consider to be the structure of agglomeration. The econometric model is specified as follows:

$$\ln\left(\frac{GDP}{L}\right)_{it} = \pi + \alpha \ln\left(\frac{K}{L}\right)_{it} + \beta \ln\left(\frac{H}{L}\right)_{it} + \gamma G_{it} + \delta \ln(POP)_{it} + \varepsilon[\ln(POP)_{it}]^2 + \epsilon \ln(STU)_{it} + \theta_i + \vartheta_t + \mu_{it} \quad (1)$$

where $\theta_i$ and $\vartheta_t$ are time and individual fixed effects, respectively. In Equation (1), STU includes two dimensions, namely, centralization and clustering. The coefficient on centralization is expected to show whether decentralization reduces the negative externalities of agglomeration and is thus helps improve labor productivity. The clustering term tests whether clustering is associated with better economic performance. Descriptive statistics for both the dependent and the independent variables are shown in Table 2.

**Table 2.** Descriptive statistics of the variables.

| Variable Name | Description | Mean | S.D. | Min | Max |
|---|---|---|---|---|---|
| $\ln\left(\frac{GDP}{L}\right)$ | GDP per worker in yuan (ln) | 11.62 | 0.527 | 9.527 | 13.95 |
| $\ln\left(\frac{K}{L}\right)$ | Physical capital stock per worker in $10^4$ yuan (ln) | 2.516 | 0.579 | 0.469 | 4.787 |
| $\ln\left(\frac{H}{L}\right)$ | Number of middle school student per $10^4$ persons | 6.868 | 0.395 | 4.904 | 8.017 |
| G | Ratio of government consumption to GDP | 100% | 0.113 | 0.0720 | 0.0200 |
| ln POP | Population in $10^4$ persons | 4.558 | 0.774 | 2.654 | 7.499 |
| lnMWI | Centralization index 1 | 0.484 | 0.169 | −0.69 | 0.688 |
| lnMADC | Centralization index 2 | 0.579 | 0.084 | 0.111 | 0.688 |
| lnDELTA | Clustering index 1 | 0.461 | 0.109 | 0.0250 | 0.646 |
| lnGINI | Clustering index 2 | 0.517 | 0.111 | 0.0250 | 0.666 |

Consistent with the theoretical assumptions proposed by geographers and economists, the urban spatial structure becomes polycentric as the urban population grows. Therefore, we consider the moderating effect of city size on the causal link between the urban spatial structure and labor productivity by introducing interaction terms into Equation (1). The interaction terms test whether and how the partial effect of the spatial structure on economic performance depends on the size of the urban population, as shown in Equation (2). For example, $\rho > 0$ implies that an increase in population yields a higher increase in labor productivity for more centralized/clustered cities and vice versa.

$$\begin{aligned}\ln\left(\tfrac{GDP}{L}\right)_{it} &= \pi + \alpha\ln\left(\tfrac{K}{L}\right)_{it} + \beta\ln\left(\tfrac{H}{L}\right)_{it} + \gamma G_{it} + \delta\ln(POP)_{it} + \varepsilon[\ln(POP)_{it}]^2 + \epsilon\ln(STU)_{it} + \rho\ln(STU)_{it} \times \ln POP + \theta_i + \vartheta_t + \mu_{it} \\ &= \pi + \alpha\ln\left(\tfrac{K}{L}\right)_{it} + \beta\ln\left(\tfrac{H}{L}\right)_{it} + \gamma G_{it} + \delta\ln(POP)_{it} + \varepsilon[\ln(POP)_{it}]^2 + \epsilon\ln(STU)_{it} + [\epsilon + \rho\ln POP] \times \ln(STU)_{it} + \theta_i + \vartheta_t + \mu_{it}\end{aligned} \quad (2)$$

## 4. Empirical Results

### 4.1. Basic Models

Table 3 shows the results of our baseline models, which use OLS regressions with individual and time fixed effects. The results for Models 1 and 3 indicate that the spatial dimensions of centralization/clustering do not appear to be directly associated with higher economic productivity. In Models 2 and 4, we introduce interaction terms (lnPOP × lnMWI and lnPOP × lnDELTA) to test the moderating effect of city size on the relation between economic performance and the urban spatial structure. In Model 4, the positive and significant coefficient on the interaction between lnPOP × lnDELTA confirms one of our hypotheses: in large cities, having more clusters boosts economic development. In Model 5, we introduce both the centralization and clustering indices and the interaction terms between spatial structure and city size. The interaction effect between population and centralization is significant and negative. This indicates that decentralization can indeed diminish negative agglomeration effects and improve urban productivity. The positive and significant influence of the interaction term between lnPOP and lnDELTA implies that population size increases the effect of clustering on urban productivity. More specifically, decentralization and clustering (i.e., a polycentric spatial structure) appear to be more helpful for urban economic performance only when the city reaches a certain population size. Another source of concern is that the effect of city population on economic performance is an inverted U shape, which confirms the existence of an optimal city size. All other significant control variables have the expected signs.

**Table 3.** OLS regressions with time and city fixed effects.

| | (1) | (2) | (3) | (4) | (5) |
|---|---|---|---|---|---|
| **Dependent Variable: ln(GDP/L)** | **FE** | **FE** | **FE** | **FE** | **FE** |
| ln(K/L) | 0.5816 *** | 0.5828 *** | 0.5817 *** | 0.5807 *** | 0.5860 *** |
| | (0.035) | (0.035) | (0.035) | (0.035) | (0.034) |
| ln(H/L) | 0.0830 ** | 0.0834 ** | 0.0841 ** | 0.0869 *** | 0.0873 *** |
| | (0.033) | (0.034) | (0.034) | (0.033) | (0.033) |
| G | −0.7232 * | −0.7189 * | −0.7224 * | −0.7120 * | −0.6919 * |
| | (0.375) | (0.374) | (0.375) | (0.377) | (0.374) |
| lnPOP | 1.1806 *** | 1.2396 *** | 1.1816 *** | 0.9285 ** | 1.0014 *** |
| | (0.317) | (0.372) | (0.316) | (0.362) | (0.355) |
| lnPOP × lnPOP | −0.0993 *** | −0.1027 *** | −0.0993 *** | −0.0883 *** | −0.0961 *** |
| | (0.030) | (0.032) | (0.029) | (0.031) | (0.032) |
| lnMWI | −0.0283 | 0.2361 | | | 1.3465 * |
| | (0.079) | (0.619) | | | (0.693) |
| lnPOP × lnMWI | | −0.0619 | | | −0.3322 ** |
| | | (0.139) | | | (0.166) |
| lnDELTA | | | −0.0248 | −1.4951 * | −2.7848 *** |
| | | | (0.137) | (0.858) | (1.062) |
| lnPOP × lnDELTA | | | | 0.3338 * | 0.6525 *** |
| | | | | (0.189) | (0.246) |
| Time FE | Y | Y | Y | Y | Y |
| City FE | Y | Y | Y | Y | Y |
| Constant | 6.5128 *** | 6.3167 *** | 6.4983 *** | 7.3782 *** | 7.1902 *** |
| | (0.933) | (1.127) | (0.918) | (1.085) | (1.058) |
| Observations | 734 | 734 | 734 | 734 | 734 |
| R-squared | 0.873 | 0.873 | 0.873 | 0.874 | 0.875 |
| Number of cities | 273 | 273 | 273 | 273 | 273 |
| Hausman test Prob > chi$^2$ | 0.000 | 0.000 | 0.000 | 0.000 | 0.000 |

Note: Robust standard errors are in parentheses. *** $p < 0.01$, ** $p < 0.05$, * $p < 0.1$. A total of 103 observations (city × year) are missing due to the lack of zip code information in the economic census data, and another 24 observations are missing due to the lack of relevant control variables in the China Urban Statistical Yearbooks (287 × 3 − 103 − 24 = 734). The Hausman test strongly indicates the rejection of the null hypothesis; therefore, a fixed effects model should be adopted instead of a random effects model.

### 4.2. Robustness Tests

The endogeneity of polycentricity in relation to better economic performance or to the omission of key variables are salient concerns and have the potential to significantly bias the coefficients. First, urban spatial structure and labor productivity are potentially related in two directions. The positive correlation between them may stem from the fact that cities with higher productivity are more likely to be decentralized and clustered. Second, although we included as many relevant control variables as possible and used a two-way fixed-effects model to control for unobservable time and city effects, some relevant variables may still be missing from the regressions.

Ordinary least squares (OLS) can suffer from the potential bias caused by reverse causality and omitted variables. TSLS estimation is a common method for reducing this potential bias. Therefore, we conduct a TSLS estimation by using an instrumental variable that is correlated with the potentially endogenous urban spatial structure but not with labor productivity.

Inspired by previous related research [36], topographic data could be a valid source of instruments for urban spatial structure. Thus, we use the SRTM 90-m resolution DEM elevation data gathered by the National Aeronautics and Space Administration (NASA) and the National Imagery and Mapping Agency (NIMA) to obtain the average slope of each postal code area in each Chinese city proper.

However, rather than directly adopting the average slope of terrain roughness, we designed a group of more relevant instruments. Generally, firms prioritize building in areas where the slope is less steep. In contrast, areas with steep inclines have high construction and usage costs; thus, they have a lower potential for becoming centers of employment concentration. Therefore, a less steep postal code area, i.e., a postal code area with a lower average slope, could probably attract more employment. We use (90°—the average slope of each postal code area) as a measure of potential employment to replace the actual employment in the corresponding postal code area. Then, based on the formulas for the urban spatial structure indices, we replace actual employment with each location's potential employment (90°—average slope) and then use these new indices as instrumental variables (IVs) for the urban spatial structure. A higher value for these IVs could be positively related to a higher level of centralization or concentration. The IVs are also time-variant, as the locations of the CBDs change over time.

Regarding the validity of the instruments, the slope of the surface is a natural feature and thus highly exogenous in relation to economic activities. Furthermore, we use topographic data from 2000, which are very unlikely to have been influenced by economic development after 2004; in addition, because of the height measurements taken by the SRTM 90-m resolution DEM elevation instruments are precise to approximately 16 meters (approximately the height of a five-story building), we can conclude that our slope measurements are unlikely to be affected by the built environment.

In Table 4, the Cragg–Donald F statistic shows that our instruments are relevant in most of our models. However, limited information maximum likelihood (LIML) estimation, which is less sensitive to weak IVs, is used in Table 5 to reduce the negative impact of weak IVs.

**Table 4.** The first stage of the IV regressions.

| | lnMWI | lnMADC | lnDELTA | lnGINI |
|---|---|---|---|---|
| First-stage coefficients on the IVs | 0.3014 *** | 2.7629 *** | 0.3525 *** | 1.1047 *** |
| | (0.0319) | (0.7746) | (0.0386) | (0.4376) |
| Shea partial $R^2$ | 0.1645 | 0.0273 | 0.1559 | 0.0139 |
| Anderson canon. corr. LM statistics | 75.848 *** | 12.591 *** | 71.690 *** | 6.396 ** |
| Cragg–Donald Wald F statistics | 89.209 | 12.720 | 83.419 | 6.374 |

Note: Robust standard errors are in parentheses. *** $p < 0.01$, ** $p < 0.05$.

Table 5 confirms the results that we obtained from the OLS model. As the spatial structure variables are not shown to be endogenous by Durbin–Wu–Hausman tests, we conclude that the OLS estimations are more efficient. However, we present the TSLS results here as a robustness test.

**Table 5.** TSLS regression results.

| | (1) | (2) | (3) | (4) | (5) |
|---|---|---|---|---|---|
| **Dependent Variable: ln(GDP/L)** | **TSLS** | **TSLS** | **TSLS** | **TSLS** | **TSLS** |
| lnPOP | 1.1767 *** | 1.9634 *** | 1.1790 *** | −1.1453 | −1.4068 |
| | (0.345) | (0.503) | (0.354) | (1.731) | (1.791) |
| lnPOP × lnPOP | −0.0991 *** | −0.1448 *** | −0.0993 *** | 0.0017 | 0.0146 |
| | (0.033) | (0.040) | (0.034) | (0.090) | (0.101) |
| lnMWI | −0.1272 | 3.4557 ** | | | −0.6968 |
| | (0.215) | (1.592) | | | (2.874) |
| lnPOP × lnMWI | | −0.8217 ** | | | 0.1124 |
| | | (0.380) | | | (0.642) |

**Table 5.** *Cont.*

| | (1) | (2) | (3) | (4) | (5) |
|---|---|---|---|---|---|
| **Dependent Variable: ln(GDP/L)** | **TSLS** | **TSLS** | **TSLS** | **TSLS** | **TSLS** |
| lnDELTA | | | −0.6979 | −14.2572 | −14.9520 ** |
| | | | (0.862) | (9.214) | (7.302) |
| lnPOP × lnDELTA | | | | 3.0653 | 3.2583 * |
| | | | | (2.144) | (1.663) |
| Others | Y | Y | Y | Y | Y |
| Time FE | Y | Y | Y | Y | Y |
| City FE | Y | Y | Y | Y | Y |
| Observations | 714 | 714 | 714 | 714 | 714 |
| R-squared | 0.872 | 0.860 | 0.864 | 0.795 | 0.777 |
| Number of cities | 253 | 253 | 253 | 253 | 253 |
| Hausman Prob > chi2 | 1.0000 | 0.7408 | 0.9990 | 0.9407 | 0.9708 |

Note: Robust standard errors are in parentheses. *** $p < 0.01$, ** $p < 0.05$, * $p < 0.1$. Twenty observations were removed because of they were the only observation in their group.

In addition, we also find that the results are mostly robust to alternative urban spatial structure indices (Table 6).

**Table 6.** Robustness to urban spatial structure indices.

| | (1) | (2) | (3) |
|---|---|---|---|
| **Dependent Variable: ln(GDP/L)** | **FE** | **FE** | **FE** |
| lnPOP | 1.1375 ** | 0.9224 ** | 1.1242 *** |
| | (0.446) | (0.361) | (0.422) |
| lnPOP × lnPOP | −0.0978 *** | −0.0909 *** | −0.0955 *** |
| | (0.032) | (0.031) | (0.032) |
| lnMADC | −0.3233 | | 2.1723 |
| | (1.377) | | (1.813) |
| lnPOP × lnMADC | 0.0495 | | −0.5319 |
| | (0.306) | | (0.426) |
| lnGINI | | −1.6431 * | −2.7654 ** |
| | | (0.864) | (1.273) |
| lnPOP × lnGINI | | 0.3561 * | 0.6315 ** |
| | | (0.193) | (0.301) |
| Others | Y | Y | Y |
| Time FE | Y | Y | Y |
| City FE | Y | Y | Y |
| Constant | 6.7249 *** | 7.4470 *** | 6.6762 *** |
| | (1.486) | (1.092) | (1.377) |
| Observations | 734 | 734 | 734 |
| R-squared | 0.873 | 0.874 | 0.875 |
| Number of cities | 273 | 273 | 273 |

Note: Robust standard errors are in parentheses. *** $p < 0.01$, ** $p < 0.05$, * $p < 0.1$.

### 4.3. Discussion

The estimates suggest that the interactions between urban spatial structure and labor productivity are heterogeneous with respect to population size. Most models support the idea that in large cities, a decentralized and clustered structure performs better, which confirms that the relation between economic performance and polycentricity depends on the urban population size. Recalling our second hypothesis, there are two potential mechanisms for the larger economic influence of polycentric structures: decentralization diminishes the negative externalities of agglomeration, and reclustering in subcenters re-establishes the positive externalities through "borrowed size". Our results confirm both mechanisms. With the growth of the city population, whether employment is more clustered or dispersed matters just as much to urban economic performance as whether clustering occurs near the CBD.

To put our findings into context and enrich the academic and practical guidelines on the evolution of urban spatial structure, we provide the following discussion.

#### 4.3.1. Discussion 1: Comparing Our Results with Those of Previous Studies

To the best of our knowledge, this paper is the first to verify the theoretical predictions at the city proper scale that polycentric structure improves economic performance more in larger cities. Referring to the previous works on a comparable scale, some argue that the effects of polycentricity on economic performance do not depend on city size [4,12], while others raise conclusions opposite ours [2].

First, one of the reasons for the different results could be the methods used. Due to data availability, we adopt panel data models instead of the cross-sectional models, as were used in previous works. The cross-sectional models are not as reliable as panel models because they are more likely to be biased.

Second, the different context of the study case could also affect the results. Our results contrast with those obtained by Meijers and Burger (2010) [2], who argued and empirically showed that polycentricity resulted in better economic performance in small metropolitan areas than in large metropolitan areas. They explained that the functional connections between urban subcenters in small metropolitan areas were denser than those in larger areas. This inconsistency may originate from differences in the developmental stages of the samples. Metropolitan areas in the United States are already mature; therefore, the influence of agglomeration economies and diseconomies is more balanced. Since the 1970s, American metropolitan areas have evolved toward polycentric spatial structures that are functional rather than morphological. In contrast, China is in the midst of fast-paced urbanization, and the share of the urban population has only recently surpassed that of the rural population. At this stage, the urban morphological spatial structure is evolving rapidly, and the surplus between agglomeration economies and diseconomies plays an important role in labor productivity.

Nevertheless, our results are in line with some findings on larger scales [10,20,29].

#### 4.3.2. Discussion 2: City Size Threshold for a Positive Influence from Polycentricity

We attempt to find the city size threshold at which the economic influence of polycentricity changes from negative to positive. We specify the interactions between the urban spatial structure (centralization and clustering) and urban population size to capture potential heterogeneity (Model 5 in Table 3). We then calculate the marginal effect of lnSTU's contribution to economic performance as $\beta_{lnSTU} + \beta_{lnSTU \times lnPOP} \times$ lnPOP. As $\beta_{lnSTU}$ and $\beta_{lnSTU \times lnPOP}$ are opposite in sign in all our models, the sign of lnSTU ($\beta_{lnSTU} + \beta_{lnSTU \times lnPOP} \times$ lnPOP) changes from negative to positive or from positive to negative as the population grows (see Figure 5). Simply put, when we set the estimated coefficient for lnSTU (i.e., $\beta_{lnSTU} + \beta_{lnSTU \times lnpop} \times$ lnpop) equal to zero, we obtain the critical point for this change. As Table 7 and Figure 5 show, centralization and dispersion (monocentricity) better facilitate economic performance only in small cities with fewer than approximately 600,000 residents. However, decentralization and clustering (polycentricity) are better structures for cities with more than 700,000 residents. Furthermore, the city size threshold for centralization is lower than that for clustering, which implies that for better economic performance, decentralization should occur before clustering in the urban structure evolutionary process.

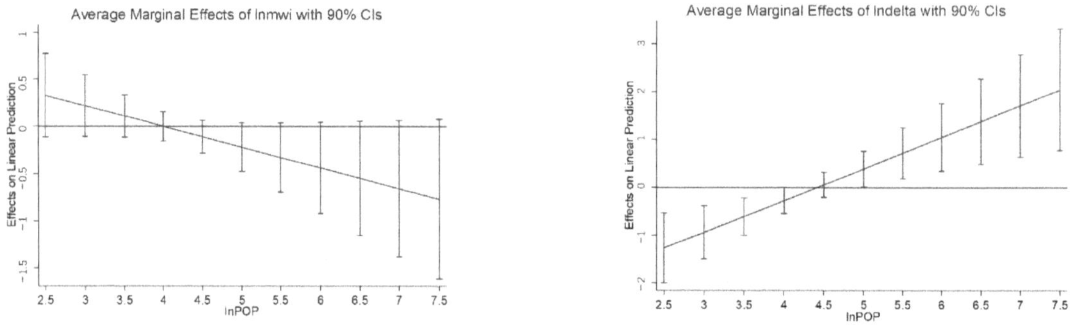


**Figure 5.** Marginal effects of the spatial structure (based on Model 5 in Table 3).

**Table 7.** Population threshold for the economic effects of the spatial structure.

| | **Table 3 Model 5** |
|---|---|
| β(lnMWI) | 1.3465 |
| β(lnpop × lnMWI) | −0.3322 |
| the threshold of city size for MWI | 575,800 |
| β(lnDELTA) | −2.7848 |
| β(lnpop × lnDELTA) | 0.6525 |
| the threshold of city size for DELTA | 713,700 |

According to a study on "ghost towns" using nighttime light data, unsuccessful new towns appear quite frequently around small-sized cities, such as Jiayuguan, Zhangye, Jiuquan, and Fangchengang [37]. We find that the populations of these cities are usually below 600,000 residents, which coincides with our findings. Thus, polycentricity strategies are planned too far ahead for small cities. Instead, monocentricity (centralization and clustering) could be better choice for these small cities.

#### 4.3.3. Discussion 3: Optimal City Size Constrained by Different Spatial Structures

By adding the quadratic form of lnPOP and the interaction terms between lnPOP and the urban spatial structure variables, we can calculate the peak population point (P*, henceforth) that represents the optimal city size, as constrained by different spatial structures. Maximizing GDP per worker and holding the other control variables constant gives a peak size of

$$P^* = \frac{\beta_{\mathrm{lnPOP}} + \beta_{\mathrm{lnSTU \times lnpop}} \times \mathrm{lnSTU}}{2 \times \beta_{\mathrm{lnPOP}^2}}$$

As the mediating effect of city size on centralization is negative and that on clustering is positive, the peak size is larger for decentralized and clustered cities. Calculated on the basis of the estimates from Model 5 in Table 3, the results presented in Table 8 and Figure 6 support our hypothesis and indicate the peak points where GDP per worker is maximized for each quartile of the spatial structure indices in 2013. The peak population size increases as cities become more decentralized and clustered. To simplify the comparison, we define two hypothetical cities. The first is polycentric with a MWI value in Q1 and a DELTA value in Q3 (decentralized and clustered). The second city is monocentric and has its MWI value in Q3 and its DELTA value in Q1 (centralized and dispersed). The optimal population size in the polycentric city, under the chosen specifications, is twice as large as that in the monocentric city (584.13/253.86 ≈ 2.3).

**Table 8.** Peak city population size (in 10,000) relative to GDP per worker (2013).

| Peak Population Size | | Dispersed | | Clustered |
|---|---|---|---|---|
| | | DELTA in Q1 | DELTA in Q2 | DELTA in Q3 |
| Decentralized | MWI in Q1 | 340.72 | 424.08 | 584.13 |
| | MWI in Q2 | 286.60 | 356.72 | 417.04 |
| Centralized | MWI in Q3 | 253.86 | 324.48 | 379.36 |

Note: Q1, Q2, and Q3 represent the lower quartile, median, and upper quartile, respectively.

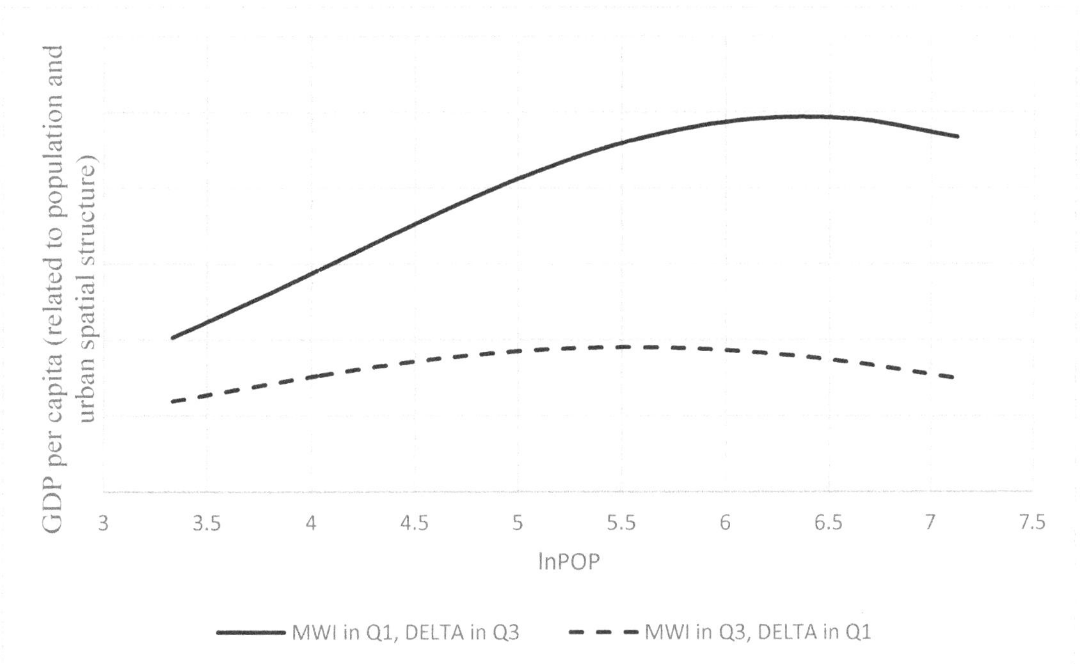


**Figure 6.** The inverted U shape in the relationship between size and economic performance for cities with different spatial structures.

As a validation of our results, we collect the estimated Chinese optimal city size raised by early works (Table 9). These numbers are very close to our findings.

**Table 9.** Optimal city size in the early literature.

| Authors (Year) | Optimal City Size (in 10,000 Persons) |
|---|---|
| Wang and Xia (1999) [38] | 100–400 |
| Chen and Jiang (2002) [39] | 100–400 |
| Ma and Song (2003) [40] | 100–200 |
| Au and Henderson (2006) [41] | 54.4–144 |
| Liu (2007) [42] | 270 |
| Zhang and Xie (2017) [43] | 200–500 |

#### 4.3.4. Discussion 4: The Economic Significance of Urban Spatial Structure

Based on the results from Model 5 in Table 3, we aim to calculate the economic significance of urban spatial structure transformation, namely, how much profit is accrued when cities become more decentralized (a decrease in MWI) and more clustered (an increase in DELTA). Thus, we choose the following five Chinese cities with different population sizes as our study cases: Jiayuguan (200,000), Suizhou (500,000), Weinan (1 million), Wuhan (5 million), and Tianjin (8 million).

Table 10 confirms that the economic performance that results from the transformation of the urban spatial structure varies based on the urban population size. In small cities such as Jiayuguan (200,000), each 1% reduction in centralization results in a 1.1 thousand yuan decrease in GDP per capita, and a 1% increase in clustering also results in a 2.6 thousand yuan decrease in GDP per capita. In Suizhou (500,000), the loss values are 0.04 and 0.2 thousand yuan per capita, respectively, which are both smaller than those in Jiayuguan. However, in larger cities, the decentralization and clustering processes create an increase in economic benefits, and larger cities earn more. The effects of both the 1% decrease in centralization and the 1% increase in clustering in Tianjin (8 million) are double those in Wuhan (5 million). These values are all of economic significance and thus cannot be ignored. These economically significant outcomes are also confirmed when we consider a change of one standard deviation.

**Table 10.** Economic benefits of spatial structure in different sized cities.

| City | Jiayuguan | Suizhou | Weinan | Wuhan | Tianjin |
|---|---|---|---|---|---|
| Population (million persons) | 0.2 | 0.5 | 1 | 5.1 | 8.2 |
| GDP per capita (thousand yuan/person) | 311 | 93.8 | 157.2 | 195.2 | 293.1 |
| Change in GDP per capita with a 1% decrease in centralization (thousand yuan/person) | −1.1 | −0.04 | 0.3 | 1.4 | 2.6 |
| Change in GDP per capita with a 1% increase in clustering (thousand yuan/person) | −2.6 | −0.2 | 0.3 | 2.5 | 4.7 |
| Change in GDP per capita with a decrease of one standard deviation in centralization (thousand yuan/person) | −18.5 | −0.7 | 0 | 0.2 | 43.7 |
| Change in GDP per capita with a decrease of one standard deviation in clustering (thousand yuan/person) | −28.1 | −2.4 | 3.7 | 27.4 | 50.9 |

## 5. Conclusions and Policy Implications

This paper provides a deeper understanding of and more robust evidence for the link between spatial structure and economic performance in the Chinese city proper, specifically in terms of the two dimensions of polycentricity, namely, centralization and clustering. Based on the China Economic Census Database that covers 2004, 2008, and 2013, we use a two-way fixed-effects panel model to examine the aforementioned causal link. After controlling for the main characteristics of the cities that may influence both urban spatial structure and labor productivity, we find that polycentricity contributes more to economic performance in larger cities. In contrast, building strong CBDs is a more effective way to promote urban economic development in small cities during the initial stages of development. In addition, we find that the optimal population size increases when cities transform from a monocentric structure to a polycentric structure.

This finding is particularly relevant in China, where a number of cities have focused on shifting employment away from the main centers and developing subcenters. Our findings suggest that polycentricity strategies are effective policy instruments for addressing the limitations imposed by urban growth, such as congestion and pollution. At the same time, decentralization and clustering constrain the population growth process and substantially influence the optimal population size. Polycentricity is reasonable and even desirable for large cities but not for small cities. Urban planners should be cognizant of the costs of developing multicenter plans in small cities.

However, some things remain to be carried out for future research. First, as well as economic performance, environmental and social performance are also worth paying attention to. Second, individual behavior analysis on how employees choose between possible locations and thus how the individual choice effects the economic performance are promising with the popularization of big data.

**Author Contributions:** Conceptualization, B.S. and T.Z.; methodology, T.Z. and W.L.; data curation, T.Z.; writing—original draft preparation, B.S., T.Z., W.L. and Y.S.; writing—review and editing, B.S. and T.Z.; supervision, B.S. and Y.S.; funding acquisition, B.S. and T.Z. All authors have read and agreed to the published version of the manuscript.

**Funding:** This research was funded by National Natural Science Foundation of China (42001183, 42071210, 41901184); the Fundamental Research Funds for the Central Universities (2022ECNU-XWK-XK001, 2019ECNU-HLYT023); Major Program of National Social Science Foundation of China (17ZDA068); Shanghai Pujiang Program (2020PJC030).

**Data Availability Statement:** Not applicable.

**Conflicts of Interest:** The authors declare no conflict of interest.

## References

1. Li, W.; Sun, B.; Zhang, T.; Zhang, Z. Polycentric spatial structure and its economic performance: Evidence from meta-analysis. *Reg. Sci.* **2022**, 1–15. [CrossRef]
2. Meijers, E.J.; Burger, M.J. Spatial structure and productivity in US metropolitan areas. *Environ. Plan. A* **2010**, *42*, 1383–1402. [CrossRef]
3. Kwon, K.; Seo, M. Does the polycentric urban region contribute to economic performance? The case of korea. *Sustainability* **2018**, *10*, 4157. [CrossRef]
4. Zhang, T.; Sun, B.; Li, W. The economic performance of urban structure: From the perspective of Polycentricity and Monocentricity. *Cities* **2017**, *68*, 18–24. [CrossRef]
5. Wang, M.; Derudder, B.; Liu, X. Polycentric urban development and economic productivity in china: A multiscalar analysis. *Environ. Plan. A* **2019**, *51*, 1622–1643. [CrossRef]
6. Fujita, M.; Thisse, J.F.; Zenou, Y. On the endogeneous formation of secondary employment centers in a city. *J. Urban Econ.* **1997**, *41*, 337–357. [CrossRef]
7. Parr, J. The polycentric urban region: A closer inspection. *Reg. Stud.* **2004**, *38*, 231–240. [CrossRef]
8. Camagni, R.P.; Salone, C. Network urban structures in northern Italy: Elements for a theoretical framework. *Urban Stud.* **1993**, *30*, 1053–1064. [CrossRef]
9. Capello, R. The city network paradigm: Measuring urban network externalities. *Urban Stud.* **2000**, *37*, 1925–1945. [CrossRef]
10. Meijers, E.J.; Burger, M.J. Stretching the concept of 'borrowed size'. *Urban Stud.* **2017**, *54*, 269–291. [CrossRef]
11. Li, Y.; Liu, X. How did urban polycentricity and dispersion affect economic productivity? A case study of 306 Chinese cities. *Landsc. Urban Plan.* **2018**, *173*, 51–59. [CrossRef]
12. Lee, B.; Gordon, P. Urban spatial structure and economic growth in US metropolitan areas. In Proceedings of the 46th Annual Meetings of the Western Regional Science Association, Newport Beach, CA, USA, 21–24 February 2007.
13. Meijers, E.J.; Burger, M.J.; Hoogerbrugge, M.M. Borrowing size in networks of cities: City size, network connectivity and metropolitan functions in Europe. *Pap. Reg. Sci.* **2016**, *95*, 181–198. [CrossRef]
14. Burger, M.J.; Meijers, E.J. Agglomerations and the rise of urban network externalities. *Pap. Reg. Sci.* **2016**, *95*, 5–15. [CrossRef]
15. Liu, Y.; Chen, X.; Liu, D. How does urban spatial structure affect economic growth? Evidence from landsat data in China. *J. Econ. Issues* **2020**, *54*, 798–812. [CrossRef]
16. Li, W.; Sun, B.; Zhang, T. Spatial structure and labour productivity: Evidence from prefectures in China. *Urban Stud.* **2019**, *56*, 1516–1532. [CrossRef]
17. Xiao, W.; Liu, W.; Li, C. Can the urban spatial structure accelerate urban employment growth? Evidence from China. *Growth Change* **2021**. [CrossRef]
18. Anas, A.; Arnott, R.; Small, K.A. Urban spatial structure. *J. Econ. Lit.* **1998**, *36*, 1426–1464.
19. Lee, B. "Edge" or "edgeless" cities? Urban spatial structure in US metropolitan areas, 1980 to 2000. *J. Reg. Sci.* **2007**, *47*, 479–515. [CrossRef]
20. Sun, B.; Li, W. City Size Distribution and Economic Performance: Evidence from City-Regions in China. *Sci. Geogr. Sin.* **2016**, *36*, 328–334.
21. Cervero, R. Efficient urbanisation: Economic performance and the shape of the metropolis. *Urban Stud.* **2001**, *38*, 1651–1671. [CrossRef]
22. Veneri, P.; Burgalassi, D. *Spatial Structure and Productivity in Italian NUTS-3 Regions*; Università Politecnica delle Marche, Dipartimento di Scienze Economiche e Sociali: Ancona, Italy, 2011.
23. Glaeser, E.L.; Kahn, M.E. Sprawl and urban growth. In *Handbook of Regional and Urban Economics*; Elsevier: Amsterdam, The Netherlands, 2004; Volume 4, pp. 2481–2527.
24. Fallah, B.N.; Partridge, M.D.; Olfert, M.R. Urban sprawl and productivity: Evidence from US metropolitan areas. *Pap. Reg. Sci.* **2011**, *90*, 451–472. [CrossRef]
25. Qin, M.; Liu, X. Does urban sprawl lead to urban productivity losses in China? Empirical study based on nighttime light data. *J. Financ. Econ.* **2015**, *41*, 28–40.

26. Friedmann, J. *Regional Development Policy: A Case Study of Venezuela*; MIT Press: Cambridge, MA, USA, 1966.
27. Fujita, M.; Ogawa, H. Multiple equilibria and structural transition of non-monocentric urban configurations. *Reg. Sci. Urban Econ.* **1982**, *12*, 161–196. [CrossRef]
28. McMillen, D.P.; Smith, S.C. The number of subcenters in large urban areas. *J. Urban Econ.* **2003**, *53*, 321–338. [CrossRef]
29. Wang, Y.; Sun, B.; Zhang, T. Do polycentric urban regions promote functional spillovers and economic performance? Evidence from China. *Reg. Stud.* **2020**, *56*, 63–74. [CrossRef]
30. Duranton, G.; Puga, D. Micro-foundations of urban agglomeration economies. In *Handbook of Regional and Urban Economics*; Elsevier: Amsterdam, The Netherlands, 2004; Volume 4, pp. 2063–2117.
31. Wheaton, W.C. Commuting, congestion, and employment dispersal in cities with mixed land use. *J. Urban Econ.* **2004**, *55*, 417–438. [CrossRef]
32. Galster, G.; Hanson, R.; Ratcliffe, M.R.; Wolman, H.; Coleman, S.; Freihage, J. Wrestling sprawl to the ground: Defining and measuring an elusive concept. *Hous. Policy Debate* **2001**, *12*, 681–717. [CrossRef]
33. Massey, D.S.; Denton, N.A. The dimensions of residential segregation. *Soc. Forces* **1988**, *67*, 281–315. [CrossRef]
34. Gordon, P.; Richardson, H.W.; Wong, H.L. The distribution of population and employment in a polycentric city: The case of Los Angeles. *Environ. Plan. A* **1986**, *18*, 161–173. [CrossRef]
35. Small, K.A.; Song, S. Population and employment densities: Structure and change. *J. Urban Econ.* **1994**, *36*, 292–313. [CrossRef]
36. Liu, X.; Li, S.; Qin, M. Urban spatial structure and regional economic efficiency: The choice of urbanization path mode in China. *Manag. World* **2017**, *1*, 51–64. (In Chinese)
37. Dong, L.; Pan, J.; Feng, Y.; Wang, W. Spatial difference pattern of house vacancy in china from nighttime light view. *Econ. Geogr.* **2017**, *37*, 62–69, 176. (In Chinese)
38. Wang, X.; Xia, X. Optimize city scale and promote economic growth. *Econ. Res.* **1999**, *9*, 22–29. (In Chinese)
39. Chen, W.; Jiang, H. City Size Economics and Policy. *Financ. Econ.* **2000**, *4*, 67–70. (In Chinese)
40. Ma, S.; Song, L. Analysis and Comparative Study on Development of City Scales of China. *Stat. Res.* **2003**, *7*, 30–34. (In Chinese)
41. Au, C.C.; Henderson, J.V. Are Chinese Cities Too Small? *Rev. Econ. Stud.* **2006**, *73*, 549–576. [CrossRef]
42. Liu, B. *The Empirical Research on the Optimum City Size in China*; Northeast Normal University: Changchun, China, 2007. (In Chinese)
43. Zhang, J.; Xie, Y.; Qian, F. Optimal City Size in China:An Extended Empirical Study from the Perspective of Energy Consumption. *China City Plan. Rev.* **2017**, *26*, 22–28.

*land*

MDPI

*Article*

# Plus-InVEST Study of the Chengdu-Chongqing Urban Agglomeration's Land-Use Change and Carbon Storage

**Chaoyue Wang [1], Tingzhen Li [1], Xianhua Guo [1,*], Lilin Xia [1], Chendong Lu [2] and Chunbo Wang [1]**

[1] Research Institute of Three Gorges, Chongqing Three Gorges University, Chongqing 404100, China
[2] Institute of Strategic Planning, Chinese Academy of Environmental Planning, Beijing 100043, China
* Correspondence: guoxianhua@sanxiau.edu.cn

**Abstract:** Based on China's "carbon neutrality" strategy, this study explores the relationship between land-use/cover change and temporal and spatial changes of ecosystem carbon storage in urban agglomerations. Using the Plus-InVEST model, the projected spatial patterns of land use in the Chengdu-Chongqing urban agglomeration in 2030 under natural development and ecological protection scenarios were simulated and predicted, and the characteristics of carbon storage, together with its spatio-temporal dynamics, were evaluated under two scenarios. Results show that: (1) From 2000 to 2020, forests, water areas, construction areas, and unused land continued to increase, while the area of cropland and grassland decreased continuously. During the last 20 years, carbon storage in urban agglomeration showed an increasing trend, with an overall increase of $24.490 \times 10^6$ t. (2) Compared with the natural development scenario, forest land, grassland, and water area in 2030 under the ecological protection scenario exhibits a substantial change; the area of construction land is limited; and an ecological spatial effect is reflected. (3) Compared to 2020, carbon storage under natural development and ecological protection scenarios decreased by $50.001 \times 10^6$ t and $49.753 \times 10^6$ t in 2030, respectively. The stability of carbon storage under the ecological conservation scenario was significantly higher than that under the natural development scenario. Therefore, under the ecological protection scenario, as a result of the coordinated land use of Chengdu-Chongqing, the functions of various regions can be coordinated and carbon storage losses can be mitigated.

**Keywords:** carbon storage; PLUS model; InVEST model; land use; urban agglomeration

**Citation:** Wang, C.; Li, T.; Guo, X.; Xia, L.; Lu, C.; Wang, C. Plus-InVEST Study of the Chengdu-Chongqing Urban Agglomeration's Land-Use Change and Carbon Storage. *Land* **2022**, *11*, 1617. https://doi.org/10.3390/land11101617

Academic Editors: Bindong Sun, Tinglin Zhang, Wan Li, Chun Yin and Honghuan Gu

Received: 12 August 2022
Accepted: 17 September 2022
Published: 21 September 2022



## 1. Introduction

During the last few decades, the global carbon cycle has received considerable attention due to the storage of carbon in terrestrial eco-systems [1,2]. The primary driver behind changes in carbon storage in ecological processes is variability in land-use type [3]. Carbon sequestration capacity varies considerably out all over land-use types. Ecological processes store carbon in plants and soils, which are influenced by changes in land use [4,5]. Currently, industrial development are causing a massive development of urban land areas [6]. The rise in developed land and resulting loss of natural vegetation have had a significant impact on regional carbon storage, posing a serious threat to sustainability and the supply of regional ecological processes [7]. Timely and effective assessments of regional carbon storage affected by urban agglomeration construction and development are crucial to maintain carbon storage services while enhancing other ecosystem services [8,9]. Thus, the sustainable development of urban agglomerations can be improved by providing information to enable the coordination of land use [9–11].

Land-use/cover change (LUCC) impacts carbon storage using field investigations and modeling [12]. This is a complex process that has both spatial and temporal aspects [13]. Several models, including Conversion of Land Use and its Effects at Small Region Extent (CLUE-S) [14] and the Land Use Scenario Dynamics (LUSD) model [15], are suitable for assessing urban areas. Additionally, the Cellular Automata-Markov (CA-Markov) model has gained popularity for simulating LUCC in a variety of situations and produces accurate results [16,17]. In one study, Liang et al. (2021) combined a CA-Markov model with an InVEST model to assess the impact of land-use change on global key ecological carbon stocks [18,19]. In addition, the Simulation of Future Land Use (FLUS) model has been applied in scenario analyses to a certain extent due to its different operation mode relative to the CA model [20]. A combination of FLUS and the InVEST model was used by Deng et al. [21], Liu et al. [22], and Gao [23] to examine the relationship between future land use and carbon storage in the future. However, utilizing the land expansion analysis technique, a patch-generated land-use change simulation tool refers to a network data may more accurately assess the reasons behind diverse land-use changes (LEAS). The model PLUS includes a multiseed growth mechanism (CARS) that better simulates patch-level changes across multiple land uses, enabling the appraisal of multiple land-use types [24]. Depending on the use of the LEAS and CARS modules [25], under several anticipated future scenarios, the PLUS model can also provide an accurate assessment of how urban expansion affects carbon storage in land ecosystems.

In the upper reaches of the Yangtze River, the Chengdu-Chongqing urban agglomeration is located in an ecological barrier area. In response to the rapid loss of cultivated land resources due to the expansion of urban and rural construction land as well as occupying ecological land, this urban agglomeration faces severe challenges when it comes to production, living, and ecological spaces [26,27]. It is thus very important to explore evolution, simulation, and scenario prediction in this region. This paper examines the potential impacts of future urban agglomeration development on regional carbon storage. Our study examines land change from a territorial spatial evolution perspective taking into account the impacts of natural, social, economic, and transportation factors. We quantitatively simulated regional land-use change in urban agglomerations between 2020 and 2030 as well as determined whether different spatial regulation scenarios might have a significant impact on regional carbon storage in Chengdu-Chongqing. The planning space should be used for a variety of spatial regulation purposes. The objective of the study is to explore urban agglomerations effectively and alleviate known contradictions between urban development and environmental conservation by attempting to explore urban agglomeration development and alleviate the known contradictions between urban development and environmental protection.

## 2. Materials and Methods

### 2.1. Study Area

The Chengdu-Chongqing urban agglomeration, which has its centers in Chengdu and Chongqing, is a crucial platform for the growth of the western area and a vital ally for the Yangtze River Economic Belt and an important area for China to promote new-type urbanization. The agglomeration includes 15 cities in Sichuan province and 29 districts (counties) in Chongqing. As shown in Figure 1, the permanent population in this area was 97,709,900 in 2021, accounting for 6.8% of the national population, and its economic aggregate in the same year accounted for 6.5% of the national total. Chengdu and Chongqing influence the surrounding areas by virtue of their relative economic strength. The inflow of Chongqing population into Chengdu accounted for 4.32%, while the inflow of Chengdu population into Chongqing accounted for 7.68%, demonstrating a "dual flow" development of both cash flow and traffic flow.

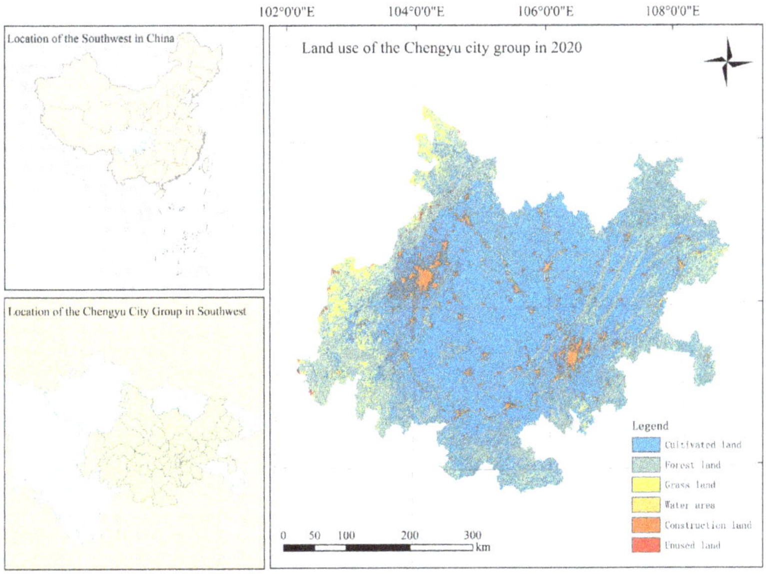


**Figure 1.** Map of the study area.

## 2.2. Data Acquisition and Processing

For the simulation of land use and carbon storage in this article, the following data used here are briefly described, including the land use and carbon storage simulation data used in this paper. Given the accessibility of remotely sensed data, Landsat pictures with a pixel size of 30 m for the years 2000, 2010, and 2020 (containing Thematic Mapper, Enhanced Thematic Mapper, and Operational Land Imager) were obtained from the Geographic Information Cloud site (http://www.gscloud.cn, accessed on 6 June 2021) to categorize land use and cover. Land-use types were divided into 6 categories and 25 subcategories [28]. Resource and Environment Science and Data Center (http://www.resdc.cn/, accessed on 12 June 2021) provided the digital elevation model (DEM), slope, gross domestic product (GDP), and population data. DEM and slope data were processed at a spatial resolution of 30 m, while GDP and population data were processed at a spatial resolution of 1 km. Point of Interesting(POI), river, and night light data were also obtained from RESDC. Our road network data was derived from OpenStreetMap (https://www.openstreetmap.org/, accessed on 2 July 2021). With a spatial resolution of 100 m, annual mean temperatures and annual mean precipitation data were collected from World Clim (https://www.worldclim.org/, accessed on 16 August 2021) [29].

In ArcGIS 10.8 (ArcGIS 10 series created by Esri (Redlands, CA, USA)), a unified spatial resolution of 100 m $\times$ 100 m was set for all of the above-mentioned spatial data, adopting the Albers geographic coordinate system. Driving factors can be divided into the following four categories (Figure 2): terrain, climate, location, and social and economic. Aspect, slope, and elevation are topographic factors. Temperature and precipitation are examples of climatic factors. Location factors include distance to rivers, roads at all levels, and schools. Distances were calculated using the ArcGIS Euclidean distance tool. Socioeconomic factors include GDP per capita, population density, and nighttime lighting conditions [30,31].

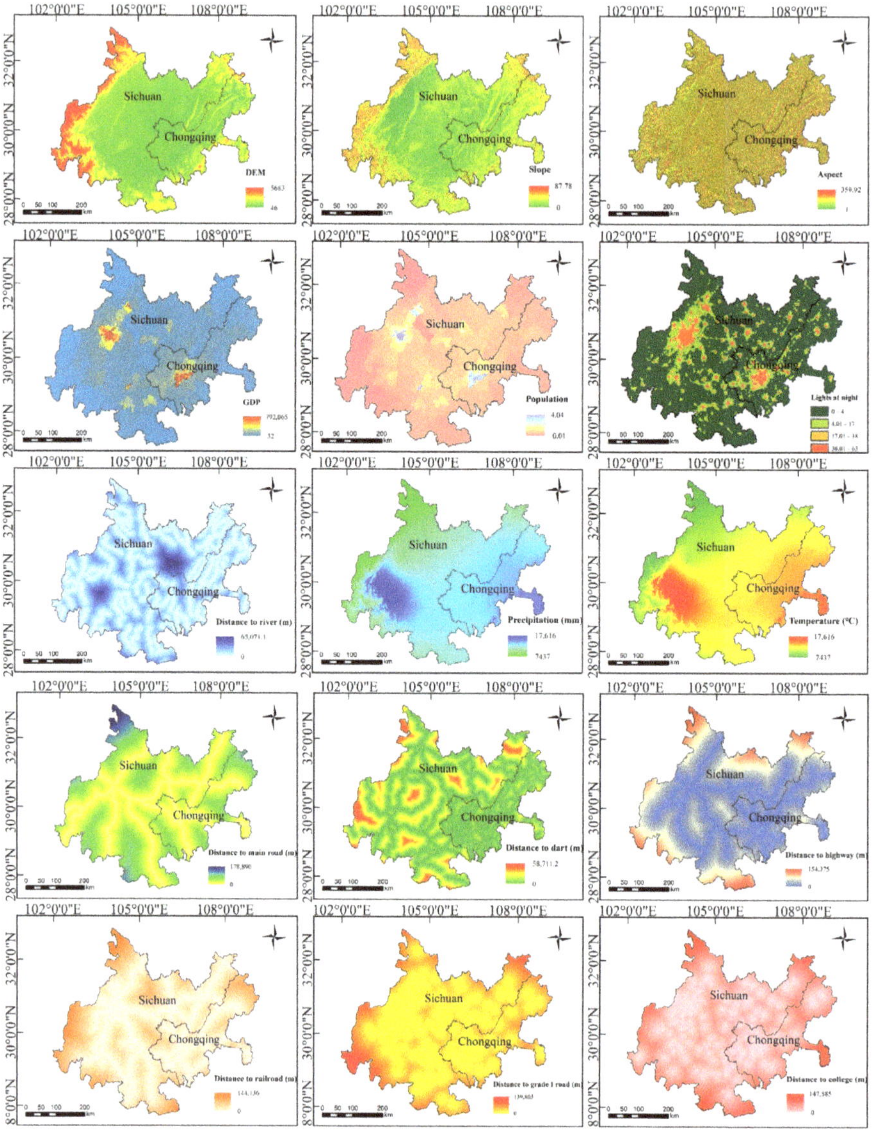


**Figure 2.** Main drivers behind land usage in the urban agglomeration of Chengdu and Chongqing.

## 2.3. Research Methods

### 2.3.1. PLUS Model

A model of land-use change called PLUS is based on patches of grid data. The modeling can replicate diverse changes in land use and properly characterize them at the component level. Two modules make up the PLUS model: Based on several random patch seeds, CARS is a CA model and LEAS a land extension analysis technique [24]. The LEAS module may harvest and sampling land expansion between two periods of land-use change, utilizing the random forest algorithm to mine and determine the likelihood that different land uses will emerge as well as the percentage of driving variables that each land use will contribute. The CARS module simulates autonomous plaque production under the constraint of development probability by combining the mechanisms of random seed formation and cutoff decline. Based on the PLUS model, the LEAS module was used to analyze land expansion from 2000 to 2020. Subsequently, the demand for land use by 2030 was estimated using the Markov chain algorithm. Under two distinct 2030 development scenarios, the CARS module was used to simulate and predict land-use changes.

The development risk surfaces $P_{i,k}^{d=1}$ for land use and overall likelihood $OP_{i,k}^{d=1,t}$ may be calculated using the Monte Carlo method when $\Omega_{i,k}^{t}$ is 0 [32], as follows:

$$OP_{i,k}^{d=1,t} = \begin{cases} P_{i,k}^{d=1} \times (r \times \mu_k) \times D_k^t \text{ if } \Omega_{i,k}^t = 0 \text{ and } r < P_{i,k}^{d=1} \\ P_{i,k}^{d=1} \times \Omega_{i,k}^t \times D_k^t \text{ all others} \end{cases} \quad (1)$$

where r varies between 0 and 1; the threshold for creating new land-use patchwork for land-use type k is represented by $\mu_k$, which the user chooses. $\Omega_{i,k}^t$ is the percentage of land-use k that makes up the area around cell i; and $D_k^t$ denotes the gap between present and future land-use demands at iteration t. $\tau$ is used to evaluate the nominated land use c, which is selected by the roulette wheel, if land-use c is more common than land-use k:

$$\text{If } \sum_{k=1}^{N} \left|G_c^{t-1}\right| \sum_{k=1}^{N} \left|G_c^{t}\right| < \text{ Step Then, } l = l+1 \quad (2)$$

$$\begin{cases} \text{Change } P_{i,c}^{d=1} > \tau \text{ and } TM_{k,c} = 1 \\ \text{No change } P_{i,c}^{d=1} \le \tau \text{ or } TM_{k,c} = 0 \end{cases} \tau = \delta^l \times r1 \quad (3)$$

where Step refers to the step size needed by the PLUS model to roughly represent the land-use requirement; as $\delta$ is the decay factor for $\tau$, which ranges from 0 to 1, the decay factor is set by the expert; l is the total number of decay steps, and r1 is a normal distributions stochastic variable with a mean of 1 and a range of 0 to 2. The transition matrix, $TM_{k,c}$, determines whether land-use type k may change to type c [24,33].

The interaction between various land-use types and various land-use divisions within the neighborhood is another neighborhood component [24], which can be said in the following manner:

$$\Omega_{p,k}^t = \frac{\sum_{N\times N} \text{con}\left(C_p^{t-1} = k\right)}{N \times N - 1} \times w_k \quad (4)$$

where $\Omega_{p,k}^t$ is the local effect factor for the cell p at time t and is the entire amount of cells that land type k occupied in the Moore neighborhood window of $N \times N$ in the previous iteration $t-1$; and $w_k$ represents the neighborhood factor parameter of each land-use type. The neighborhood factor parameter ranges from 0 to 1, with a value proportional to land expansion capacity [34]. The land-use factor parameters in this paper are primarily based on current situations and future development trends of land use in the study area (Table 1).

**Table 1.** Neighborhood factor parameters.

| Land Use Type | Cultivated Land | Forest | Grassland | Water | Construction Land | Unused Land |
|---|---|---|---|---|---|---|
| Natural development neighborhood factor | 0.07 | 0.11 | 0.01 | 0.29 | 1 | 0.09 |
| Ecological protection neighborhood factor | 0.07 | 0.31 | 0.10 | 0.34 | 0.95 | 0.09 |

### 2.3.2. Validation of Model Accuracy

The applicability and reliability of the model for forecasting changes in land use and cover were assessed using quantifiable correctness and the kappa coefficient. The overall agreement between simulation findings and observation data is tested using the kappa value. Kappa values greater than 0.75 signify good simulation accuracy. Taking 2010 as the base period data, the paper uses the above methods to simulate 2020 land-use patterns, then cross-checks the simulation graph of 2020 and the current situation graph of 2020. Calculating the kappa coefficients is as follows:

$$\text{Kappa} = \frac{OA_O - OA_E}{(1 - OA_E)}, OA_O = \left(\sum_{k=1}^{n} OA_{kk}\right)/N \tag{5}$$

where $OA_O$ is the classification's overall accuracy and denotes the likelihood that each random sample's simulation outcome would match the data on land use. $OA_E$ is the likelihood that the simulation's findings match the data on current land use; the number n represents the number of types of land use; N is the overall sample count; the quantity of samples that were accurately identified for the k type of land use is called $OA_{kk}$. The range of values for the kappa coefficient is −1 to 1, with a higher number indicating a more appropriate prediction.

#### 2.3.3. Setting the Scene

Natural development scenario (NDS): In light of the land-use development trend between 2000 and 2020, With the Markov chain, it was possible to determine the demand for land usage in 2030 underneath the historical development trend (Table 2) [35,36]. According to historical changes, cultivated land has become grassland or construction land, so we set it to 1. Since it is unlikely to turn to other ground classes, it is set to 0. A similar situation exists for woodlands and arable lands. Despite its particularity, construction land cannot be converted to other land classes, so it is set at 0. In the past, unused land has more often migrated to OTHER land classes than to the rest of the land class.

**Table 2.** Natural development scenario cost matrix.

| Land Use Type | Cultivated Land | Forest | Grassland | Water | Construction Land | Unused Land |
|---|---|---|---|---|---|---|
| Cultivated land | 1 | 1 | 0 | 0 | 1 | 0 |
| Forest | 1 | 1 | 0 | 0 | 1 | 0 |
| Grassland | 1 | 1 | 1 | 0 | 1 | 0 |
| Water | 1 | 1 | 0 | 1 | 1 | 0 |
| Construction land | 1 | 0 | 0 | 0 | 1 | 0 |
| Unused land | 1 | 1 | 1 | 1 | 1 | 1 |

Ecological protection scenario (EPS): The EPS's goal is to improve the safeguards for ecological regions including grasslands and forests. According to CP, the development of Chengdu-Chongqing Urban Agglomeration Development Plan, the conversion of wetland to built-up area, pasture, forest, and farming were all strictly regulated (2016–2020) (Table 3). In comparison with natural development, conversion of woodlands and grasslands to the rest of the land class represents the biggest difference. Aside from construction land, other types of land are more easily converted to woodlands and grasslands, and conversion between them is also easier. The probability of woodland and grassland being set to 1 increases as a result.

**Table 3.** Budget matrix for ecological conservation scenarios.

| Type of Land Usage | Cultivated Land | Forest | Grassland | Water | Construction Land | Unused Land |
|---|---|---|---|---|---|---|
| Cultivated land | 1 | 1 | 0 | 0 | 1 | 0 |
| Forest | 1 | 1 | 1 | 1 | 1 | 1 |
| Grassland | 1 | 1 | 1 | 1 | 1 | 1 |
| Water | 1 | 1 | 0 | 1 | 1 | 0 |
| Construction land | 1 | 0 | 0 | 0 | 1 | 0 |
| Unused land | 1 | 1 | 1 | 1 | 1 | 1 |

### 2.3.4. InVEST Model

Using the InVEST model, a regional carbon storage evaluation was conducted. Further, it was investigated if spatial management might successfully stop the loss of region carbon storage [34]. In order to calculate the total carbon storage in a region, the following calculations were made:

$$C_T = \sum_{i=1}^{n} C_{i\_T} = \sum_{i}^{n} A_i \times (C_{i\_above} + C_{i\_below} + C_{i\_dead} + C_{i\_soil}) \quad (6)$$

where $C_T$ stands for region net carbon storage, $C_{i\_T}$ for i land-use type's carbon storage, $A_i$ for i land-use type's area, and $C_{i\_above}$, $C_{i\_below}$, and $C_{i\_dead}$ but instead $C_{i\_soil}$ for i land-use type's above-ground, below-ground, dead organic matter, and soil carbon densities, respectively.

The carbon density of various land-use types is the fundamental element of the InVEST model, and it is based on previous research findings that have been modified in accordance with the characteristics of the Chengyu Cities Group (Table 4).

**Table 4.** Carbon density based on land-use/cover type included in the InVEST model (t/ha).

| Land-Use Type | Aboveground Carbon Density | Underground Carbon Density | Density of Soil Carbon | Carbon Density of Dead Organic Materials | Sources |
|---|---|---|---|---|---|
| Cultivated land | 38.70 | 80.70 | 92.90 | 1.00 | [37–39] |
| Forest | 55.56 | 144.87 | 206.45 | 3.50 | [39–41] |
| Grassland | 29.30 | 52.90 | 135.00 | 1.00 | [37–40] |
| Water | 21.40 | 73.10 | 113.00 | 1.00 | [41,42] |
| Construction land | 3.30 | 87.30 | 115.30 | 0 | [42,43] |
| Unused land | 22.60 | 136.90 | 171.80 | 0 | [38,42] |

## 3. Results

### *3.1. LUCC Dynamics during 2000–2020*

From 2000 to 2020, land use in the Chengdu-Chongqing urban agglomeration was dominated by cultivated land, accounting for more than 57% of the total land area of the urban agglomeration. Woodland occupied more than 29% of the total land area; however, the areas of grassland, water, construction land, and unused land were relatively small, accounting for only 10% of the total area of urban agglomeration land (Table 5). During the past 20 years, land use has changed in different ways, among which the largest change reflects the area of construction land. Increases in unused land, water, building, and forests, are in the increments of 1473, 495, 4393, and 96 km$^2$, respectively. The largest percentage increase occurred for construction land, with an increase of 58.82%. The areas of arable land and grassland decreased by 3609 and 2848 km$^2$, respectively. From 2000 to 2010, under the influence of regional development orientation, continuous urbanization led to the rapid expansion of urban and rural construction land, whereas cultivated land gradually decreased. In addition, the policy of "returning farmland to forest" piloted in the Chengdu-Chongqing region restored forestland area, which was another important reason for the decreasing area of cultivated land. From 2010 to 2020, with further urbanization, urban agglomerations become increasingly large, resulting in a further decrease in the area of cultivated land. Additionally, the expansion of the "returning farmland to forest" project encouraged the ongoing expansion of the forest, while the arable area continued to decrease.

**Table 5.** Area and percentage of the study area's various land-use classifications through 2000 to 2020.

| Land Use Type | 2000 | | 2010 | | 2020 | | Area Change ($km^2$) |
|---|---|---|---|---|---|---|---|
| | Area ($km^2$) | Percentage (%) | Area ($km^2$) | Percentage (%) | Area ($km^2$) | Percentage (%) | |
| Cultivated land | 122,591 | 58.84 | 121,014 | 58.08 | 118,982 | 57.10 | −3609 |
| Forest | 60,696 | 29.13 | 61,812 | 29.66 | 62,169 | 29.84 | 1473 |
| Grassland | 18,944 | 9.09 | 17,021 | 8.17 | 16,096 | 7.73 | −2848 |
| Water | 2839 | 1.36 | 3080 | 1.48 | 3334 | 1.60 | 495 |
| Construction land | 3076 | 1.48 | 5120 | 2.46 | 7469 | 3.58 | 4393 |
| Unused land | 211 | 0.10 | 310 | 0.15 | 307 | 0.15 | 96 |

### *3.2. Analysis of Prediction Results of Various Land Use Situations*

In an urban agglomeration, agricultural area, forest areas, grass, water area, and unoccupied land are all expected to shrink by 361, 599, 438, 488, and 20 $km^2$ by 2030, respectively, under the natural outcome measurement (NDS in Figure 3). Conversely, construction land area has a projected increase of 25.52%. According to the direction of land-use area transfer (Figure 3), arable land, grassland, and water area will mainly be converted to construction land, whereas forest land and unused land will be evenly transferred to other land types. Although the change range of construction land is the largest, it seldom changes to other land types, and its increase mainly results from the transfer of large areas of cultivated land.

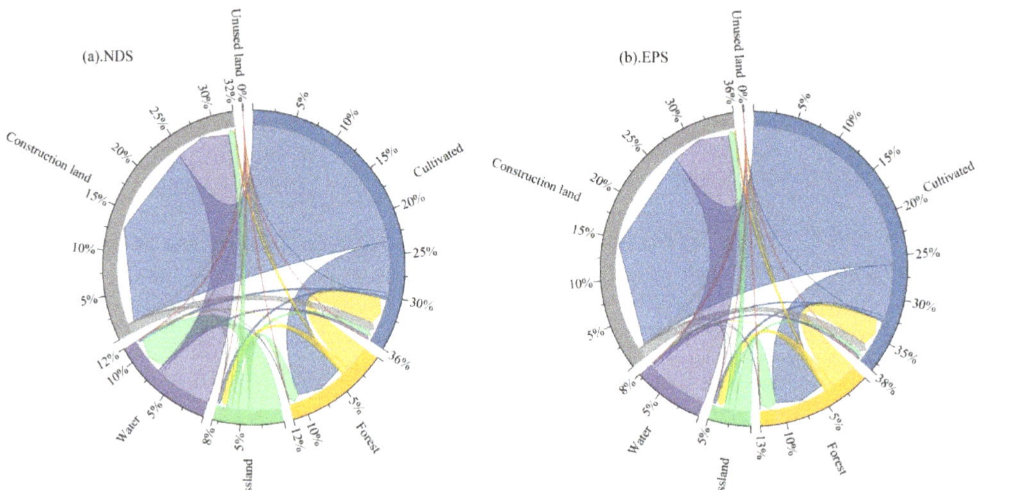


**Figure 3.** Chord diagram of land-use transfer.

Compared with 2020, under the ecological protection scenario, woodland, grassland, and construction land will continue to increase by 101, 345, and 1906, respectively (EPS in Figure 3), While the amount of water, agricultural lands, and undeveloped land will all fall by 2100, 232, and 20, respectively. According to Figure 3, most cultivated land is converted to forest and construction land, most forest land is changed to farmland and grasslands, most grassland is transformed to forestry land and water, and any unused land is evenly distributed to various land types. Relative to the natural development scenario, the change trend of cultivated land, forest land, grassland, and water area under the ecological protection scenario undergoes great changes (Table 6). This primarily results from the use of land for construction to meet the ecological security patterns of urban agglomeration in the near future. The proportion of forest land, grassland, and water area in the total area of urban agglomeration increased significantly from −0.96%, −2.72%, and −14.64% to 0.16%, 2.14%, and −6.96%, respectively.

**Table 6.** Area of each region in 2030 under the concept of environmental preservation and natural development, and its ratio to 2020.

| Land Use Type | 2020 | | 2030 | | | | Change from 2020 to 2030 | | | |
|---|---|---|---|---|---|---|---|---|---|---|
| | | | NDS | | EPS | | NDS | | EPS | |
| | Area ($km^2$) | Proportion (%) | Area ($km^2$) | Proportion (%) | Area ($km^2$) | Proportion (%) | Area ($km^2$) | Rate (%) | Area ($km^2$) | Rate (%) |
| Cultivated land | 118,982 | 57.10 | 118,621 | 56.93 | 116,882 | 56.10 | −361 | −0.30 | −2100 | −1.77 |
| Forest | 62,169 | 29.84 | 61,570 | 29.55 | 62,270 | 29.89 | −599 | −0.96 | 101 | 0.16 |
| Grassland | 16,096 | 7.73 | 15,658 | 7.52 | 16,441 | 7.89 | −438 | −2.72 | 345 | 2.14 |
| Water | 3334 | 1.60 | 2846 | 1.37 | 3102 | 1.49 | −488 | −14.64 | −232 | −6.96 |
| Construction land | 7469 | 3.58 | 9375 | 4.50 | 9375 | 4.50 | 1906 | 25.52 | 1906 | 25.52 |
| Unused land | 307 | 0.15 | 287 | 0.13 | 287 | 0.13 | −20 | −6.52 | −20 | −6.52 |

### 3.3. Accuracy Verification and Driving Factor Contribution Analysis

Land-use data from 2010 and 2020 were used as examples to simulate changes in land use based on Markov's predictions for every land-use level in 2020. The results were then compared to the actual values for 2020 to assess the PLUS model's simulation accuracy (Figure 4). The method was employed to determine the reliability overall and the kappa coefficient. Values and the kappa coefficient that are near 1 denote simulation accuracy that is higher. The simulated performance of the model reaches a sufficient level in statistical significance whenever the kappa coefficient is higher than 0.75 [44]. The accuracy of kappa was confirmed to be 0.83.

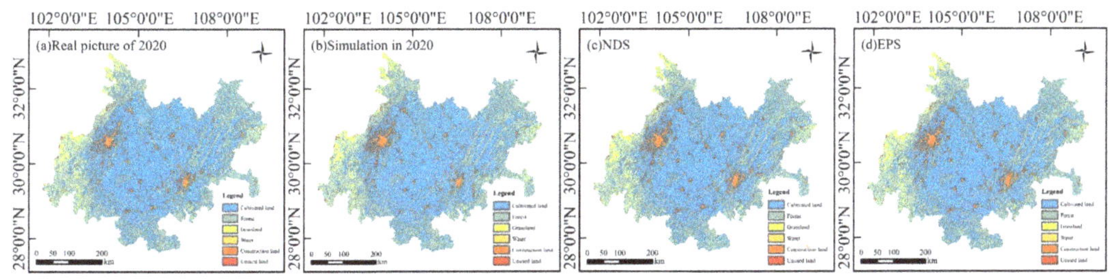


**Figure 4.** Comparison of simulation for 2020 and predictions of two scenarios in 2030.

According to the historical development trend between 2000 and 2020, predictions were made for the contribution ranking of the influencing factors of various land-use probabilities over the next decade (Figure 5). It is obvious from the figure that DEM has the greatest impact on cultivated land expansion, whereas the degree of contribution of other factors is not significantly different. When it comes to the contribution of forest land, the slope factor ranks highest among the fifteen selected driving factors. In addition, DEM has a strong contribution to grassland and water area, indicating that natural environmental factors play an important role. The degrees of contribution of various factors in construction land showed a ladder type. Population factors and DEM factors had the least influence on the unutilized land protrusion.

### 3.4. Changes of Carbon Storage between 2000 and 2030

The InVEST model was used to calculate carbon storage in the Chengdu-Chongqing urban agglomeration for 2000, 2010, and 2020. In order to simulate and forecast land use outcomes in 2030 and to forecast carbon storage capacity within gradual progression and ecological preservation scenarios, this was integrated with the PLUS model. In 2000, 2010, and 2020, the carbon storage of the Chengdu-Chongqing urban agglomeration was 5648.610 $\times$ $10^6$ t, 5669.267 $\times$ $10^6$ t, and 5673.100 $\times$ $10^6$ t, respectively, showing a continuous upward trend with an overall increase of 24.490 $\times$ $10^6$ t and an average annual increase of 1.225 $\times$ $10^6$ t. From 2000 to 2010, carbon storage in the urban agglomeration increased significantly, with an added value of 20.657 $\times$ $10^6$ t, which is an increase of 0.37%. In contrast, from 2010 to 2020, carbon storage in the urban agglomeration increased slightly, with an increment of 3.833 $\times$ $10^6$ t, which is an increase of 0.07%.

In the case of natural development, the carbon storage of the agglomeration in 2030 is predicted to be 5623.099 $\times$ $10^6$ t, a decrease of 50.001 $\times$ $10^6$ t compared with 2020 and reflecting an average annual decrease of 5.0001 $\times$ $10^6$ t. In contrast, under the ecological protection scenario, the carbon storage of the agglomeration in 2030 is predicted to be 5623.347 $\times$ $10^6$ t. The corresponding average annual decrease of 4.9753 $\times$ $10^6$ t indicates that the carbon storage deceleration is small. Under the ecological protection policy, which improves the effectiveness of regional ecological protection and carbon sequestration effects achieved in the Chengdu-Chongqing urban agglomeration. During 2020–2030, compared with the two typical development scenarios, the carbon storage of urban agglomerations under ecological protection measures that restrict the transfer of forest land and grassland to other land types tends to be more stable, avoiding a rapid decline.

Regarding the spatial distribution and evolution of carbon storage (Figure 6), carbon storage in the northwest region with Chengdu as the core of the urban agglomeration decreased slightly between 2000 and 2020 by 0.095 $\times$ $10^6$ t. On the contrary, the southeast region, with Chongqing as the core, exhibits a large increase in carbon storage by 22.722 $\times$ $10^6$ t. From 2020 to 2030, carbon storage of all cities in the Chengdu-Chongqing urban agglomeration decreased under the natural development scenario. Compared with the previous two decades, the northwest region with Chengdu as the core is still the region with the largest reduction in carbon storage, decreasing by 27.923 $\times$ $10^6$ t, accounting for 55.84% of the total reduction. Secondly, the carbon storage in the southeast region with Chongqing as the core exhibited a smaller decreased 22.078 $\times$ $10^6$ t, accounting for 44.16% of the total reduction. Under the ecological protection scenario, the northwest region with Chengdu as the core is still the city with the largest reduction of carbon storage at 27.840 $\times$ $10^6$ t, which corresponds to 99.70% of the natural development scenario, reflecting the effectiveness of ecological protection. The carbon storage of the southeast region with Chongqing as the core also decreased slightly, further reflecting the necessity of ecological protection.

### 3.5. Characteristics of Change in Carbon Storage Caused by Land Type Conversion

Due to area transfer and carbon density differences among different land types, the corresponding effects of change and transformation on carbon storage are different. Due to the change from a single land type to several land types between 2000 and 2020, the Chengdu-Chongqing urban agglomeration lost 25.447 $\times$ $10^6$ t of carbon storage. Quantitative conversion of cultivated land to construction land together with the conversion of forest land and grassland to cultivated land and construction land leads to decreased carbon storage in soil and vegetation. Because water area has a lower carbon density than other land types, converting it to another type of land can help create a carbon sink, which increases the amount of carbon that can be stored overall in the urban agglomeration. During the past 20 years, the transfer of cultivated land to other land types resulted in a reduction of carbon storage of 114.940 $\times$ $10^6$ t, with cultivated land mainly being converted into forest land, grassland, and construction land. Increasing conversion of forest land to other land types also increased, and the corresponding added value was 148.844 $\times$ $10^6$ t. The carbon storage of grassland correlated with the area change observed over the last

20 years; i.e., with a decrease in area, its carbon storage also decreased, and the conversion between different land types decreased by 57.241 × $10^6$ t. Because the carbon density of water areas is low and because its area does not change greatly, the carbon storage value of water areas does not change significantly between land type conversions. Due to its strong expansion, construction land increased continuously during the past 20 years and was mainly converted to arable land, forest land, and grassland. The unused land showed an overall trend of fluctuating growth, resulting in a small increase of carbon storage between land conversion, with an added value of 0.360 × $10^6$ t.

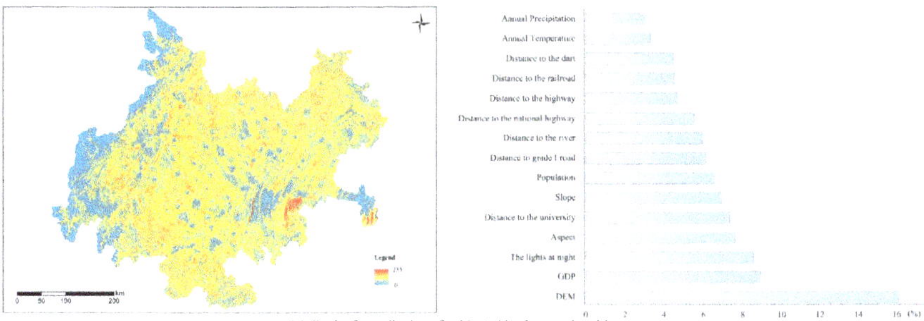


(a). Rank of contribution of cultivated land expansion drivers

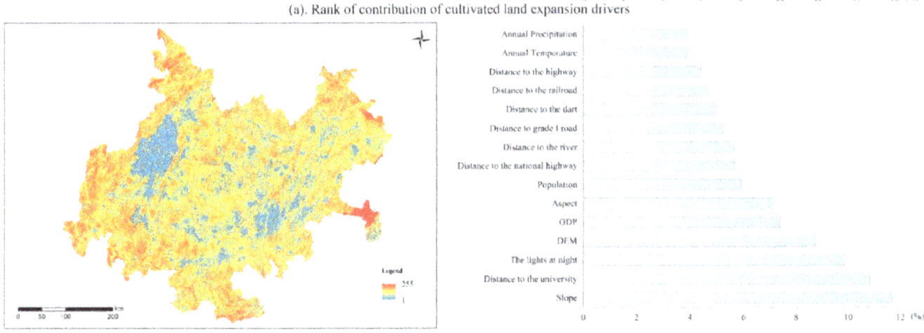


(b). Rank of contribution of forest land expansion drivers

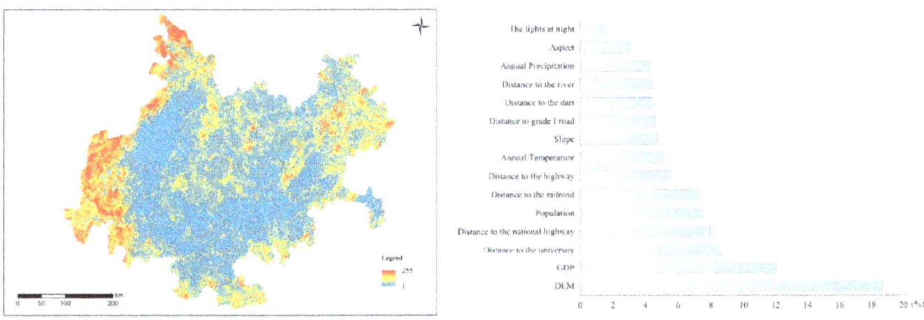


(c). Rank of contribution of grass land expansion drivers

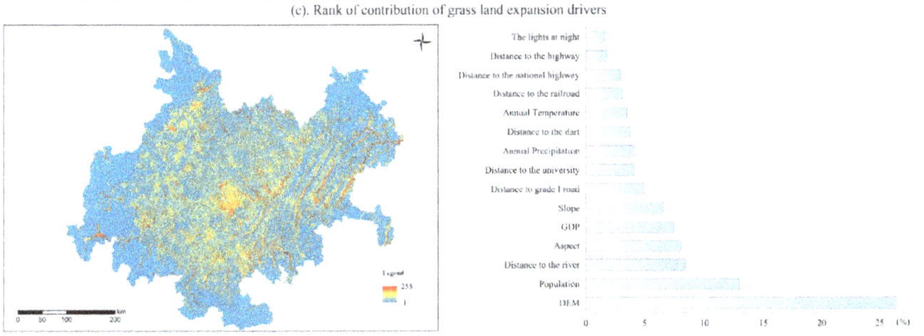


(d). Rank of contribution of water land expansion drivers

**Figure 5.** *Cont.*

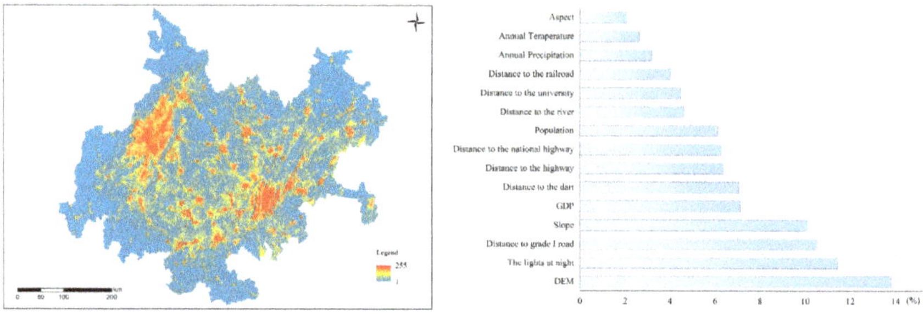


(e). Rank of contribution of construction land expansion drivers

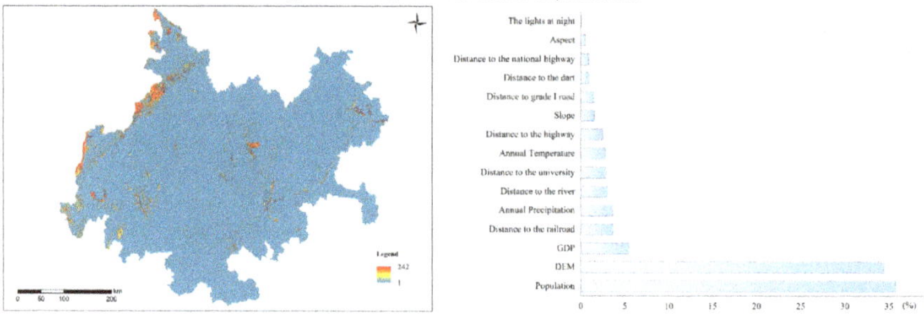


(f). Rank of contribution of unutilized land expansion drivers

**Figure 5.** Ranking of various land-use probabilities and their driving factors.

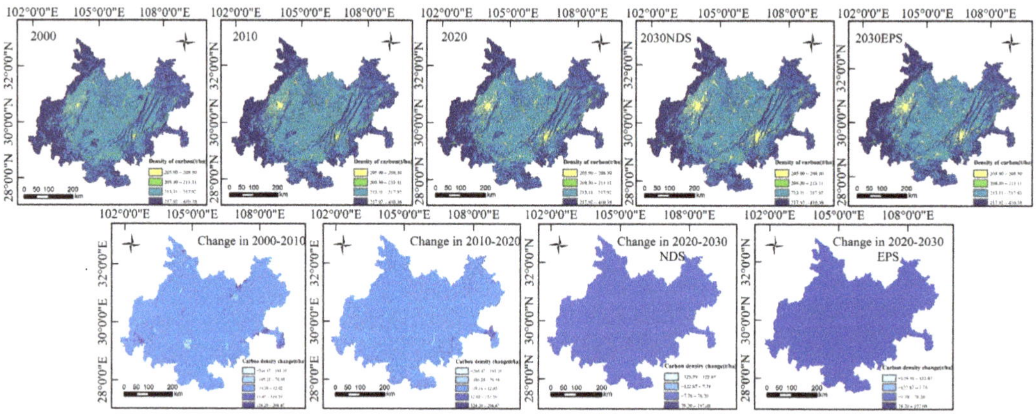


**Figure 6.** Carbon storage and its changes over different periods.

Compared with 2020 (Table 7), the carbon storage of the Chengdu-Chongqing urban agglomeration decreased by $2.955 \times 10^6$ t under the natural scenario by 2030 and significantly increased by $393.057 \times 10^6$ t under the ecological protection scenario. The main reason for this is the different transfer probability of cultivated land, forest land, grassland, and water area. In addition, guided by ecological protection, the conversion of other land types is restricted, and the transfer area to construction land decreases, resulting in increased carbon storage. Under the two tested scenarios, the carbon storage of forest land is the most significant. Despite the fact that each scenario indicated an upward trend, the carbon storage of forestland increased significantly under ecological protection scenario. First, there is a decline in the transformation of forested areas to agricultural land. Second, forestland controls the transfer of construction land and unused land, achieving the goal of regional carbon stability and reflecting the effectiveness and necessity of ecological protection policies. The change of carbon storage in grassland and water area was not obvious. The ability to store more carbon is significantly increased by converting water areas, building sites, and unused land into agricultural land, forest areas, and grassland. Overall, the conversion of various land types mainly results in increased construction land, which will prevent the metropolitan agglomeration from growing its carbon store in the foreseeable.

**Table 7.** Under scenarios of natural progression and ecological protection in 2020–2030, land type conversion will influence the amount of carbon stored in the atmosphere.

| Land Use Type | | Area (km$^2$) | | Change in Carbon Stock ($\times10^6$ t) | | Total ($\times10^6$ t) | |
|---|---|---|---|---|---|---|---|
| Converted from | Converted to | NDS Natural Development Scenario | EPS Ecological Protection Scenario | NDS Natural Development Scenario | EPS Ecological Protection Scenario | NDS Natural Development Scenario | EPS Ecological Protection Scenario |
| Cultivated land | Forest | 411.76 | 380.44 | −8.115 | −7.498 | −6.970 | −6.341 |
| | Grassland | 32.08 | 26.89 | −0.016 | −0.013 | | |
| | water | 22.48 | 12.75 | 0.011 | 0.006 | | |
| | Construction land | 1557.73 | 1576.12 | 1.153 | 1.166 | | |
| | Unused land | 0.23 | 0.24 | −0.003 | −0.003 | | |
| Forest | Cultivated land | 215.28 | 203.13 | 4.243 | 400.321 | 5.395 | 401.655 |
| | Grassland | 50.17 | 59.49 | 0.964 | 1.143 | | |
| | water | 1.16 | 1.09 | 0.023 | 0.022 | | |
| | Construction land | 7.68 | 7.87 | 0.157 | 0.161 | | |
| | Unused land | 1.00 | 1.01 | 0.008 | 0.008 | | |
| Grassland | Cultivated land | 42.01 | 28.14 | 0.021 | 0.014 | −1.362 | −2.230 |
| | Forest | 88.56 | 119.67 | −1.702 | −2.300 | | |
| | water | 270.12 | 0.70 | 0.262 | 0.001 | | |
| | Construction land | 54.92 | 53.23 | 0.068 | 0.065 | | |
| | Unused land | 0.92 | 0.87 | −0.010 | −0.010 | | |
| Water | Cultivated land | 23.16 | 21.05 | −0.011 | −0.010 | 0.071 | 0.067 |
| | Forest | 1.49 | 1.49 | −0.030 | −0.030 | | |
| | Grassland | 1.00 | 0.71 | −0.001 | −0.001 | | |
| | Construction land | 458.90 | 442.21 | 0.119 | 0.115 | | |
| | Unused land | 0.53 | 0.54 | −0.006 | −0.007 | | |
| Construction land | Cultivated land | 72.96 | 72.88 | −0.054 | −0.054 | −0.118 | −0.121 |
| | Forest | 3.02 | 3.16 | −0.062 | −0.065 | | |
| | Grassland | 0.76 | 0.75 | −0.001 | −0.001 | | |
| | water | 2.38 | 2.45 | −0.001 | −0.001 | | |
| | Unused land | 0.06 | 0.04 | −0.001 | 0.000 | | |
| Unused land | Cultivated land | 0.19 | 0.21 | 0.002 | 0.002 | 0.030 | 0.026 |
| | Forest | 1.81 | 2.08 | −0.014 | −0.016 | | |
| | Grassland | 0.89 | 0.85 | 0.010 | 0.010 | | |
| | water | 1.66 | 1.46 | 0.020 | 0.018 | | |
| | Construction land | 0.89 | 0.99 | 0.011 | 0.012 | | |
| | | Total ($\times10^6$ t) | | | | −2.955 | 393.057 |

## 4. Discussion

### 4.1. PLUS Analysis of Model Uncertainty

Currently, the majority of research on LUCC-related alterations to ecosystem carbon cycles is based on model simulations. Due to its complexity, LUCC can affect the energy flow in the ecosystem, but the existing models are hindered by uncertainties in simulating changes in the ecosystem carbon cycle caused by LUCC [34,45]. The reliability of prospective land-use change modeling scenarios largely determines the accuracy of modeling findings.

The driving factors selected for the PLUS model simulation used in this paper are terrain conditions, climate environment, social economy, and transportation accessibility, including fifteen factors, such as DEM, slope, aspect, temperature, precipitation, GDP, population, railway networks, and national road networks. These factors result in an accurate simulation of various land types and have different contributions to different types of land (Figure 4). Nonetheless, the PLUS model still has some limitations. First, in addition to natural factors, cultural factors feature many complex choices, such as cultural concepts, industrial output value, and POI; however, because it is challenging to quantify these elements inside the PLUS model, we do not incorporate them in our simulation method. Secondly, national policy orientation plays an important role in LUCC. Policy factors such as ecological protection red line, permanent basic farmland protection red line, and urban development red line—which all play a very important role in China's territorial space change—are difficult to assign specific values in the simulation process due to their complexity. Therefore, in order to better adapt to real scenarios of future land-use change, it is necessary to consider introducing more influencing factors in subsequent studies.

### 4.2. InVEST Model Uncertainty Analysis

The InVEST model can intuitively determine the effects of different types of conversion on carbon storage. Its results clearly reflect spatial and temporal variations of carbon storage in urban agglomerations and highlight the relationships between different land types, which can provide new ideas for regional development in terms of coordinating economic and ecological aspects. Nonetheless, it is important to note that the InVEST model makes more estimates for large-scale land changes based on established available carbon density values. In the carbon module, the change of carbon storage value due to vegetation growth and the internal structure of land use are ignored, resulting in errors in the change of spatial patterns of carbon storage and leading to uncertainty in the results [27]. In addition, although the carbon density values obtained from existing studies are close to those in the study area, these values may be influenced by human activities and changes in the natural environment. Therefore, the carbon density value also has a certain degree of uncertainty. Finally, while the carbon module considers differences in carbon density between different land-use types, it ignores differences in carbon sinks related to land-use types and the age organization of vegetation, which hinders the simulation of the estimation of the spatial pattern of carbon storage services. Therefore, in the study of future urban agglomeration, it is necessary to strengthen and verify the timeliness of data acquisition of carbon density values, carry out localized calibration, conduct field measurements of core indicators, and accurately estimate changes in regional carbon storage and based the on scientific and reasonable assumptions in order to better maintain the carbon balance of the regional ecosystem.

### 4.3. Advantages and Limitations of the Linkage Model

The Link PLUS and InVEST models have broad applications for guiding ecosystem services. The PLUS model's LEAS module extracts the growth of different types of property between the two steps for land-use change, which collects samples from the growing section. In order to investigate the variables of development probability related to each land-use type and assess the impact of each factor driving on land-use type expansion, the advancement and pushing factors of each land-use type are then paired using the random forest method [46]. This allows land-use change simulations combined with the InVEST model to be used as a means for studying future changes in regional carbon storage spatial patterns within urban agglomerations.

Although linkage models can effectively simulate the effects of ecosystems on carbon storage over short time scales, their application on longer time scales faces several limitations. The regional climate of the Chengdu-Chongqing urban agglomeration is humid all year, causing vegetation and soil carbon density to change constantly [47]. Consequently, the consequences of climate change could be disregarded whenever relational models are utilized for long-term projections. In addition, the original spatial resolution of LUCC data used was 30 m × 30 m. To ensure consistency, all spatial data were resampled to a grid of 100 m × 100 m. In future studies, data accuracy could be further improved to ensure the validity of simulation sampling and carbon storage measurements.

### 4.4. Spatial Structure of Urban Agglomerations and Carbon Storage

Urban agglomeration is a highly integrated urban complex with compact spatial organization and close economic ties formed by different levels of cities relying on transportation and communication and other infrastructure networks in a specific geographical area [48,49]. Agglomerations of urban space are dependent on land as a space carrier for social, economic, and ecological activities [50]. The land is also one of the most essential spatial attributes of urban development. Further, land-use type and its cover change represent the concrete expression of land as well as the main manifestation of urban agglomeration structure. Therefore, LUCC, which is crucial to the carbon cycle in terrestrial ecosystems, is at the heart of the urban agglomeration-carbon storage relationship.

In this study, we analyze the changes in carbon reserves caused by LUCC in Chengdu-Chongqing urban agglomerations, based on the relationship between urban spatial structure and carbon reserves. There was a significant change in cultivated land from 2000 to 2020. By converting this land type to another, the total area decreases, resulting in the largest decrease in carbon reserves. By contrast, forest land increases its carbon reserves as its transfer area increases. Other land types are similarly affected. In accordance with the historical development, the relationship between the spatial structure and carbon reserves of the natural scenario will remain the same in ten years. The total amount of carbon stored also increases as forest land is converted to grassland in the ecological scenario. There is no doubt that changes in land use will affect carbon reserves over time. There is a direct correlation between urban agglomeration's spatial structure and carbon storage. This relationship is traceable. According to Nicodemus Nyamari [51] and Cai [7], carbon reserves have changed in Kenya and China's Yangtze River Delta due to LUCC. In general, it can be observed that urban agglomerations and carbon reserves have a close relationship, and any changes in one will inevitably affect the other.

### 4.5. Development Strategy for Urban Agglomeration and Carbon Storage

In recent years, with the continuous development of urban agglomerations, contradictions of land use caused by urban expansion have become increasingly common. Continuously changing the land-use type can have a negative impact on the carbon sink of terrestrial ecosystems [52]. Consequently, China's territorial space planning must advance in order to achieve regional economic development goals while ensuring ecological protection in urban agglomerations. First, it is necessary to strictly abide by the "three red lines for protection" guided by national policies and appropriately control the increase of land used for construction. An example is Chengdu-Chongqing's urban agglomeration. Construction land in Chengdu and Chongqing as well as their surrounding large cities should be developed at a reduced pace. It is anticipated that small- and medium-sized cities will grow moderately because they are not occupying arable land. Further, the city needs to renovate the new construction space in order to tap into the potential of low-efficiency land and increase urban public and ecological space. Increasing construction land intensity is crucial to achieving limited growth and spatial transfer incentives in small towns and villages within urban agglomerations. Second, the ratio of forestland to grassland area should be increased to strengthen the ecological protection of high carbon density regions [53,54]. Therefore, it is imperative that the Chengdu-Chongqing urban agglomeration not only increase forest cover through afforestation in large areas but also optimize the urban vegetation structure. The objective is to guide the sustainable development of urban forest land and grasslands and to realize a harmonious coexistence pattern of life ecology. Enhance the total carbon storage capacity of the urban agglomeration and create an urban ecosystem that is healthy and stable as well as a harmonious living environment. Finally, it is imperative to focus on the complex function of land uses in order to complete the transformation from single land type to a production-life-ecological complex function. It is necessary to increase participation in ecological preservation, promote ideal land layouts, minimize carbon dioxide emissions, and enhance ecological efficiency thru the utilization of resources and large-scale land management. For urban agglomerations and metropolitan areas in other regions, this has certain reference value. Building an ecological security pattern and promoting an effective carbon cycle are the meanings of national development from the perspective of urban agglomeration.

### 4.6. Contribution to Research

Several contributions make this study different from others. On the one hand, this article quotes the latest simulation prediction model, PLUS, which has characteristics that the previous forecast model lacks. Various types of land can be better understood through this method. In addition, it contains a new multi-type seed growth mechanism that can simulate changes in plaque-level and land-level changes for a variety of land types. Moreover, it is coupled with a variety of target algorithms, enabling better planning decisions to be made. A FLUS model was used in previous research by Zuo et al. [55] to simulate land-use changes in 2020 in Chongqing. Additionally, Zhang et al. [56] simulated mainland China's ecosystem value using the FLUS model. In their research, they only needed to extract the first phase of the land in order to use the data for training, based on the probability of emergence and land competition. There is a lack of time concepts in this method as well as the ability to dig the changes in land use compared to the PLUS model.

On the other hand, Chengdu-Chongqing urban agglomeration is the largest urban group in southwestern China. In national strategic development, the Chengdu-Chongqing urban agglomeration plays a significant role due to their geographical location and historical significance. Carbon reserves in this area not only meet the needs of local ecological development but also get closer to the national dual carbon strategic planning goals. Data integration is also performed from a macro perspective, and different models can be used to better position Chengdu-Chongqing urban development. For the remaining urban agglomerations, this is of more guiding significance. Previous simulation prediction research using the PLUS model focused mostly on middle and small regions, similar to cities or specific places. In their analyses of the Hanzhong ecosystem and the southeast coastal protective forest ecosystem, Yang [57] and Bao [58] used the PLUS models. Study area is small and does not have universality for urban areas, urban agglomerations, or other regions. The main contributions of this study are therefore the two aspects above. The study presents multiple suggestions to improve the reference value of national strategic planning and regional development based on the results of the scenario simulation.

## 5. Conclusions

This study made projections for the carbon storage of something like the Chengdu-Chongqing metropolitan agglomeration in 2030 using the PLUS and InVEST models. These are its conclusions:

(1) Land use in the Chengdu-Chongqing urban agglomeration has changed significantly between 2000 and 2020, primarily due to a continuous increase of forest land area, water area, construction land area, and unused land area, together with a decrease of cropland and grassland areas. The driving force behind this change mainly comes from urbanization and the implementation of the "returning farmland to forest" policy. Carbon storage in the urban agglomeration has increased by $24.490 \times 10^6$ t in the past 20 years.

(2) In comparison, the accuracy of kappa is 0.83. According to the historical development trends from 2000 to 2020, the contribution of the probability impact factors of regional expansion have been calculated and ranked. The DEM exerts a significant influence, but other factors also contribute differently in specific situations.

(3) From 2020 to 2030, the cultivated lands, forests, grasslands, water areas, and unused lands in Chengdu-Chongqing will decline continuously under the natural development scenario. The area of construction land will continue to grow. The urban agglomeration's carbon storage will decrease from $5673.100 \times 10^6$ t in 2020 to $5623.099 \times 10^6$ t in 2030, i.e., a total decrease of $50.001 \times 10^6$ t.

(4) In the scenario of ecological preservation, crop land, water area, and unoccupied land will all decrease, while woods, grassland, and building land would all continue to grow. In this scenario, the urban agglomeration's carbon storage in 2020 will decrease from $5673.100 \times 10^6$ t to $5623.347 \times 10^6$ t in 2030, i.e., a total decrease of $49.753 \times 10^6$ t.
(5) Carbon storage under the ecological protection scenario can be reduced by $0.248 \times 10^6$ t relative to the natural development model. This slower reduction rate is conducive to the stabilization of carbon sinks. Under the ecological protection scenario, carbon storage in northwest China with Chengdu as its core decreased by $27.840 \times 10^6$ t, i.e., 99.70% of the natural development scenario. Carbon storage in southeast China, with Chongqing as its core, also declined slightly.

**Author Contributions:** Conceptualization, C.W. (Chaoyue Wang) and X.G.; methodology, C.W. (Chaoyue Wang), X.G. and T.L.; validation, C.W. (Chaoyue Wang), X.G. and T.L.; investigation, L.X., C.L. and C.W. (Chunbo Wang); writing—original draft, C.W. (Chaoyue Wang); writing—review and editing, C.W. (Chaoyue Wang), X.G. and T.L.; software, C.W. (Chaoyue Wang) All authors have read and agreed to the published version of the manuscript.

**Funding:** The Chongqing Education Commission's Humanities and Sociology Research Program (No. 21SKGH432) and the National Sociology Foundation of China (No. 21BMZ141) provided funding for this work. Project supported by China National Scholarship Foundation.

**Institutional Review Board Statement:** Not applicable.

**Informed Consent Statement:** Not applicable.

**Data Availability Statement:** Publicly available datasets were analyzed in this study. This data can be found here: Geospatial Data Cloud Website (http://www.gscloud.cn, accessed on 6 June 2021); RESDC (http://www.resdc.cn/, accessed on 12 June 2021); OpenStreetMap (https://www.openstreetmap.org/, accessed on 2 July 2021); World Clim (https://www.worldclim.org/, accessed on 16 August 2021).

**Acknowledgments:** We value the reviewers' critical and helpful criticism and recommendations, which boosted this manuscript's quality.

**Conflicts of Interest:** The authors say they have no competing interests. The study's design, data collection, analysis, or interpretation; the preparation of the paper; or the choice to publish the findings were all made independently of the funding sponsors.

## References

1. Schimel, D.S.; House, J.I.; Hibbard, K.A.; Bousquet, P.; Ciais, P.; Peylin, P.; Braswell, B.H.; Apps, M.J.; Baker, D.; Bondeau, A.; et al. Recent patterns and mechanisms of carbon exchange by terrestrial ecosystems. *Nature* **2001**, *414*, 169–172. [CrossRef] [PubMed]
2. Newbold, T.; Hudson, L.N.; Hill, S.L.L.; Contu, S.; Lysenko, I.; Senior, R.A.; Börger, L.; Bennett, D.J.; Choimes, A.; Collen, B.; et al. Global effects of land use on local terrestrial biodiversity. *Nature* **2015**, *520*, 45–50. [CrossRef] [PubMed]
3. Zhang, M.; Huang, X.; Chuai, X.; Yang, H.; Lai, L.; Tan, J. Impact of land use type conversion on carbon storage in terrestrial ecosystems of China: A spatial-temporal perspective. *Sci. Rep. UK* **2015**, *5*, 10233. [CrossRef]
4. Cantarello, E.; Newton, A.C.; Hill, R.A. Potential effects of future land-use change on regional carbon stocks in the UK. *Environ. Sci. Policy* **2011**, *14*, 40–52. [CrossRef]
5. Liu, W.; Zhan, J.; Zhao, F.; Yan, H.; Zhang, F.; Wei, X. Impacts of urbanization-induced land-use changes on ecosystem services: A case study of the Pearl River Delta Metropolitan Region, China. *Ecol. Indic.* **2019**, *98*, 228–238. [CrossRef]
6. Su, M.; Guo, R.; Hong, W. Institutional transition and implementation path for cultivated land protection in highly urbanized regions: A case study of Shenzhen, China. *Land Use Policy* **2019**, *81*, 493–501. [CrossRef]
7. Cai, W.; Peng, W. Exploring spatiotemporal variation of carbon storage driven by land use policy in the Yangtze river delta region. *Land* **2021**, *10*, 1120. [CrossRef]
8. Erik, N.; Heather, S.; Peter, H.; Marc, C.; Driss, E.; Stacie, W.; Steven, M.; Stephen, P.; Maya, M.A. Projecting Global land-use change and its effect on ecosystem service provision and biodiversity with simple models. *PLoS ONE* **2010**, *5*, e14327.
9. Xu, Z.; Fan, W.; Wei, H.; Zhang, P.; Ren, J.; Gao, Z.; Ulgiati, S.; Kong, W.; Dong, X. Evaluation and simulation of the impact of land use change on ecosystem services based on a carbon flow model: A case study of the Manas river basin of Xinjiang, China. *Sci. Total Environ.* **2019**, *652*, 117–133. [CrossRef]
10. Han, J.; Meng, X.; Zhou, X.; Yi, B.; Liu, M.; Xiang, W. A long-term analysis of urbanization process, landscape change, and carbon sources and sinks: A case study in China's yangtze river delta region. *J. Clean. Prod.* **2017**, *141*, 1040–1050. [CrossRef]

11. Li, C.; Zhao, J.; Thinh, N.; Xi, Y. Assessment of the effects of urban expansion on terrestrial carbon storage: A case study in Xuzhou City, China. *Sustainability* **2018**, *10*, 647. [CrossRef]
12. Wang, J.; Zhang, Q.; Gou, T.; Mo, J.; Wang, Z.; Gao, M. Spatial-temporal changes of urban areas and terrestrial carbon storage in the Three Gorges Reservoir in China. *Ecol. Indic.* **2018**, *95*, 343–352. [CrossRef]
13. Brown, D.G.; Verburg, P.H.; Pontius, R.G.; Lange, M.D. Opportunities to improve impact, integration, and evaluation of land change models. *Curr. Opin. Environ. Sustain.* **2013**, *5*, 452–457. [CrossRef]
14. Anputhas, M.; Janmaat, J.J.A.; Nichol, C.F.; Wei, X.A. Modelling spatial association in pattern based land use simulation models. *J. Environ. Manag.* **2016**, *181*, 465–476. [CrossRef]
15. He, C.; Zhang, D.; Huang, Q.; Zhao, Y. Assessing the potential impacts of urban expansion on regional carbon storage by linking the LUSD-urban and InVEST models. *Environ. Modell. Softw.* **2016**, *75*, 44–58. [CrossRef]
16. Aburas, M.M.; Ho, Y.M.; Ramli, M.F.; Ash Aari, Z.H. Improving the capability of an integrated CA-Markov model to simulate spatio-temporal urban growth trends using an analytical hierarchy process and frequency ratio. *Int. J. Appl. Earth Obs.* **2017**, *59*, 65–78. [CrossRef]
17. Etemadi, H.; Smoak, J.M.; Karami, J. Land use change assessment in coastal mangrove forests of Iran utilizing satellite imagery and CA–Markov algorithms to monitor and predict future change. *Environ. Earth Sci.* **2018**, *77*, 208. [CrossRef]
18. Liang, Y.; Hashimoto, S.; Liu, L. Integrated assessment of land-use/land-cover dynamics on carbon storage services in the Loess Plateau of China from 1995 to 2050. *Ecol. Indic.* **2021**, *120*, 106939. [CrossRef]
19. Sadat, M.; Zoghi, M.; Malekmohammadi, B. Spatiotemporal modeling of urban land cover changes and carbon storage ecosystem services: Case study in Qaem Shahr County, Iran. *Environ. Dev. Sustain.* **2020**, *22*, 8135–8158. [CrossRef]
20. Liu, X.; Liang, X.; Li, X.; Xu, X.; Ou, J.; Chen, Y.; Li, S.; Wang, S.; Pei, F. A future land use simulation model (FLUS) for simulating multiple land use scenarios by coupling human and natural effects. *Landsc. Urban Plan.* **2017**, *168*, 94–116. [CrossRef]
21. Deng, Y.; Yao, S.; Hou, M.; Zhang, T.; Lu, Y.; Gong, Z.; Wang, Y. Assessing the effects of the green for grain program on ecosystem carbon storage service by linking the InVEST and FLUS models: A case study of Zichang county in hilly and gully region of Loess Plateau. *Nat. Resour.* **2020**, *35*, 826–844.
22. Liu, X.; Wang, S.; Wu, P.; Feng, K.; Hubacek, K.; Li, X.; Sun, L. Impacts of urban expansion on terrestrial carbon storage in china. *Environ. Sci. Technol.* **2019**, *53*, 6834–6844. [CrossRef]
23. Gao, J.; Wang, L. Embedding spatiotemporal changes in carbon storage into urban agglomeration ecosystem management: A case study of the Yangtze River Delta, China. *J. Clean. Prod.* **2019**, *237*, 117764. [CrossRef]
24. Liang, X.; Guan, Q.; Clarke, K.C.; Liu, S.; Wang, B.; Yao, Y. Understanding the drivers of sustainable land expansion using a patch-generating land use simulation (PLUS) model: A case study in Wuhan, China. *Comput Environ Urban Syst.* **2021**, *85*, 101569. [CrossRef]
25. Xu, L.; Liu, X.; Tong, D.; Liu, Z.; Yin, L.; Zheng, W. Forecasting urban land use change based on cellular automata and the PLUS model. *Land* **2022**, *11*, 652. [CrossRef]
26. Maanan, M.; Maanan, M.; Karim, M.; Ait Kacem, H.; Ajrhough, S.; Rueff, H.; Snoussi, M.; Rhinane, H. Modelling the potential impacts of land use/cover change on terrestrial carbon stocks in north-west Morocco. *Int. J. Sust. Dev. World.* **2019**, *26*, 560–570. [CrossRef]
27. Etemadi, N.; Rickard, J.; Anderton, H.; Spall, S.; Hall, C.; Vaux, D.; Nachbur, U.; Silke, J. Modeling multiple ecosystem services, biodiversity conservation, commodity production, and tradeoffs at landscape scales. *Front. Ecol. Environ.* **2009**, *7*, 4–11. [CrossRef]
28. Chen, W.; Zhao, H.; Li, J.; Zhu, L.; Wang, Z.; Zeng, J. Land use transitions and the associated impacts on ecosystem services in the middle reaches of the Yangtze river economic belt in China based on the geo-informatic Tupu method. *Sci. Total Environ.* **2020**, *701*, 134690. [CrossRef] [PubMed]
29. Yang, J.; Gong, J.; Tang, W.; Liu, C. Patch-based cellular automata model of urban growth simulation: Integrating feedback between quantitative composition and spatial configuration. *Comput. Environ. Urban Syst.* **2020**, *79*, 101402. [CrossRef]
30. Clerici, N.; Cote-Navarro, F.; Escobedo, F.J.; Rubiano, K.; Villegas, J.C. Spatio-temporal and cumulative effects of land use-land cover and climate change on two ecosystem services in the Colombian Andes. *Sci. Total Environ.* **2019**, *685*, 1181–1192. [CrossRef] [PubMed]
31. Leh, M.D.K.; Matlock, M.D.; Cummings, E.C.; Nalley, L.L. Quantifying and mapping multiple ecosystem services change in West Africa. *Agric. Ecosyst. Environ.* **2013**, *165*, 6–18. [CrossRef]
32. Rodríguez-Echeverry, J.; Echeverría, C.; Oyarzún, C.; Morales, L. Impact of land-use change on biodiversity and ecosystem services in the Chilean temperate forests. *Landsc. Ecol.* **2018**, *33*, 439–453. [CrossRef]
33. Verburg, P.H.; Overmars, K.P. Combining top-down and bottom-up dynamics in land use modeling: Exploring the future of abandoned farmlands in Europe with the Dyna-CLUE model. *Landsc. Ecol.* **2009**, *24*, 1167. [CrossRef]
34. Zhao, M.; He, Z.; Du, J.; Chen, L.; Lin, P.; Fang, S. Assessing the effects of ecological engineering on carbon storage by linking the CA-Markov and InVEST models. *Ecol. Indic.* **2019**, *98*, 29–38. [CrossRef]
35. Polasky, S.; Nelson, E.; Pennington, D.; Johnson, K.A. The Impact of Land-use change on ecosystem services, biodiversity and returns to landowners: A case study in the state of Minnesota. *Environ. Resour. Econ.* **2011**, *48*, 219–242. [CrossRef]
36. Chen, T.; Peng, L.; Wang, Q. Scenario decision of ecological security based on the trade-off among ecosystem services. *China Environ. Sci.* **2021**, *41*, 3956–3968.

37. Nie, X.; Lu, B.; Chen, Z.; Yang, Y.; Chen, S.; Chen, Z.; Wang, H. Increase or decrease? Integrating the CLUMondo and InVEST models to assess the impact of the implementation of the major function oriented zone planning on carbon storage. *Ecol. Indic.* **2020**, *118*, 106708. [CrossRef]
38. Xie, X.L.; Sun, B.; Zhou, H.Z.; Li, Z.P.; Li, A.B. Organic carbon density and storage in soils of China and spatial analysis. *Acta Ecol. Sin.* **2004**, *41*, 35–43.
39. Li, K.; Wang, S.; Cao, M. Vegetation and soil carbon storage in China. *Sci. China* **2004**, *47*, 49–57. [CrossRef]
40. Huang, M.; Ji, J.; Cao, M.; Li, K. Modeling study of vegetation shoot and root biomass in China. *Acta Ecol. Sin.* **2006**, *26*, 4156–4163.
41. Li, W.; Zhang, C.; Li, S. Forest carbon storage in Guangxi Province estimated by 8th forest inventory data. *Southwest For. Univ. (Nat. Sci. Ed.).* **2017**, *37*, 127–133.
42. Zhang, M.; Lai, L.; Huang, X.; Chuai, X.; Tan, J. The carbon emission intensity of land use conversion in different regions of China. *Resour. Sci.* **2013**, *35*, 792–7999.
43. Chen, L.; Liu, G.; Li, H. Estimating net primary productivity of terrestrial vegetation in China using remote sensing. *Remote Sens.* **2002**, *2*, 129–135.
44. Pontius, R.G.; Boersma, W.; Castella, J.; Clarke, K.; de Nijs, T.; Dietzel, C.; Duan, Z.; Fotsing, E.; Goldstein, N.; Kok, K.; et al. Comparing the input, output, and validation maps for several models of land change. *Ann. Reg. Sci.* **2008**, *42*, 11–37. [CrossRef]
45. Houghton, R.A. Revised estimates of the annual net flux of carbon to the atmosphere from changes in land use and land management 1850–2000. *Tellus B Chem. Phys. Meteorol.* **2003**, *55*, 378–390.
46. Wang, Z.; Zeng, J.; Chen, W. Impact of urban expansion on carbon storage under multi-scenario simulations in Wuhan, China. *Environ. Sci. Pollut. Res. Int.* **2022**, *29*, 45507–45526. [CrossRef]
47. Pliscoff, P.; Luebert, F.; Hilger, H.H.; Guisan, A. Effects of alternative sets of climatic predictors on species distribution models and associated estimates of extinction risk: A test with plants in an arid environment. *Ecol. Model.* **2014**, *288*, 166–177. [CrossRef]
48. Fang, C. Progress and the future direction of research into urban agglomeration in China. *ACTA Geogr. Sin.* **2014**, *69*, 1130–1144.
49. Zhu, Z.; Zhu, X.; Li, S. Evolution process and characteristics of spatial structure of urban agglomeration in the middle reaches of the Yangtze River. *ACTA Geogr. Sin.* **2021**, *76*, 799–817.
50. Long, H. Land use transition and land management. *Geogr. Res.* **2015**, *34*, 1607–1618.
51. Nyamari, N.; Cabral, P. Impact of land cover changes on carbon stock trends in Kenya for spatial implementation of REDD+ policy. *Appl. Geogr.* **2021**, *133*, 102479. [CrossRef]
52. Wang, Y.; Meng, J.; Qi, Y.; Peng, F. Review of ecosystem management based on the InVEST model. *Chin. J. Ecol.* **2015**, *34*, 3526–3532.
53. Zhu, L.; Li, L.; Liu, S.; Li, Y. The evolution of village land-use function in the metropolitan suburbs and its in spiration to rural revitalization: A case study of Jiangjiayan Village in Chengdu City. *Geogr. Res.* **2019**, *38*, 535–549.
54. Zhu, W.; Zhang, J.; Cui, Y.; Zheng, H.; Zhu, L. Assessment of territorial ecosystem carbon storage based on land use change scenario: A case study in Qihe River Basin. *Acta Geogr. Sin.* **2019**, *74*, 446–459.
55. Zuo, Y.; Cheng, J.; Fu, M. Analysis of Land Use Change and the Role of Policy Dimensions in Ecologically Complex Areas: A Case Study in Chongqing. *Land* **2022**, *11*, 627. [CrossRef]
56. Zhang, J.; Li, X.; Zhang, C.; Yu, L.; Wang, J.; Wu, X.; Hu, Z.; Zhai, Z.; Li, Q.; Wu, G. Assessing spatiotemporal variations and predicting changes in ecosystem service values in the Guangdong–Hong Kong–Macao Greater Bay Area. *Gisci. Remote Sens.* **2022**, *59*, 184–199. [CrossRef]
57. Yang, S.; Su, H.; Zhao, G. Multi-scenario simulation of urban ecosystem service value based on PLUS model: A case study of Hanzhong city. *J. Arid. Land Resour. Environ.* **2022**, *36*, 86–95.
58. Bao, S.; Yang, F. Spatio-Temporal Dynamic of the Land Use/Cover Change and Scenario Simulation in the Southeast Coastal Shelterbelt System Construction Project Region of China. *Sustainability* **2022**, *14*, 8952. [CrossRef]

*land*

Article

# Dynamics between Direct Industrial Real Estate and the Macroeconomy: An Empirical Study of Hong Kong

**Daniel Lo [1,*], Yung Yau [2], Michael McCord [1] and Martin Haran [1]**

1 Faculty of Computing, Engineering and the Built Environment, Ulster University, Newtownabbey BT37 0QB, UK
2 Department of Sociology and Social Policy, School of Graduate Studies, Lingnan University, Hong Kong
* Correspondence: d.lo@ulster.ac.uk

**Abstract:** Pricing of direct industrial real estate (DIRE) has long been under-researched due to the paucity of analysable data. Compared to other types of real estate, DIRE has often been regarded as more inefficient because of information asymmetry amongst market players stemming from a lack of market transparency. Therefore, pricing of DIRE usually does not follow a random walk and should be more predictable than other types of real estate. Along this line of reasoning, this study empirically investigates the causal relationships between the price-to-rent ratio of DIRE and macroeconomic attributes using cointegration and causality techniques. More specifically, we employ data on the market of Hong Kong to investigate the lead-lag relationships between the price-to-rent ratio of DIRE and a wide spectrum of macroeconomic and financial indicators, including inflation, money supply, national income, exchange rates, performance of housing market and other economic indicators specific to the industrial sector. The results of our statistical tests reveal significant evidence that DIRE is generally moving in syncs with other segments of the economy over time in terms of long-term cointegration. Further, DIRE tends to lag behind the overall macroeconomy in terms of Granger causation with the price-to-rent ratio exhibiting varying lengths of time lag with the macroeconomic determinants. The findings of the study carry important implications for informing property valuation practices and industrial land policy, particularly in designing urban revitalization programmes aimed at optimising industrial land use.

**Keywords:** price-to-rent; industrial real estate; macroeconomics; market efficiency; Granger causality; Hong Kong

**Citation:** Lo, D.; Yau, Y.; McCord, M.; Haran, M. Dynamics between Direct Industrial Real Estate and the Macroeconomy: An Empirical Study of Hong Kong. *Land* **2022**, *11*, 1675. https://doi.org/10.3390/land11101675

Academic Editors: Bindong Sun, Tinglin Zhang, Wan Li, Chun Yin and Honghuan Gu

Received: 2 September 2022
Accepted: 26 September 2022
Published: 28 September 2022



## 1. Introduction

Direct industrial real estate (DIRE) is often characterised by low liquidity and transparency, and high physical heterogeneity and information asymmetry between market participants. Undertaking the valuation of DIRE is therefore a decidedly challenging exercise, not least when the market is sluggish with a paucity of transactions, or when special-purpose properties of rare structural features and functionality are concerned. Nonetheless, numerous studies have been conducted to explore the determinants of pricing with respect to the price and rent of DIRE at the property level, with the majority of research endeavours centred around how they can be explained by property attributes such as building age and structural design [1,2], proximity to labour market, accessibility or distance to infrastructure and economic centres [3–5], and industrial agglomeration [1,3,6]. As far as property valuation is concerned, the price-to-rent ratio has frequently been employed by real estate practitioners, traders, and policy makers to assess whether a given property market is overheated by, for instance, comparing the ratios cross-sectionally and contemporaneously with other similar markets, or temporally with the historical trends of the subject market [7] In the context of residential real estate, a significant deviation of the price-to-rent ratio from its long-term historical trends usually signals a decline in housing affordability or excessive speculative activity within the market, and a consequent mean reversion to its equilibrium value would likely occur. For commercial real estate, including DIRE, the price-to-rent ratio could indeed reveal useful information about the behaviours of market players and the interaction between the property market and the macroeconomy, including the legal and regulatory environment: a high price-to-rent ratio may imply market players anticipating a shortage of new supply in the foreseeable future, whilst a low ratio value may be indicative of a surging demand for rental properties resulting from, for example, a stamp duty increase.

An extensive body of literature has been directed at examining the dynamics between the ratio and the macroeconomy, with the majority of them focusing primarily on residential real estate and, to a lesser extent, the commercial office sector. Of particular societal relevance and importance are the branches of studies that delve into topics in relation to "buy versus rent" [8], generation rent [9,10], affordability [11,12] and the role of macroeconomics in the determination of price/rent of office real estate [13] However, relatively few scholarly works have given a thorough empirical account of the price-to-rent ratio of DIRE and its relationships with its underpinning macroeconomic determinants that shape the property market landscape. A more integrative and nuanced conceptualization of the dynamics between DIRE and the wider economy is of paramount importance in formulating industrial policy and guiding economic transformation on one hand and informing valuation practices of DIRE on the other. Against this backdrop, this study attempts to explore and dissect the macroeconomic and financial determinants of the price-to-rent ratio of DIRE using data from the Hong Kong market. More specifically, it is positioned to examine the long-term cointegration and causal relationships between the ratio and a wide spectrum of macroeconomic indicators, including attributes related to the general economic conditions, forex market, manufacturing sector, and other external economic factors with a purpose to enhance the empirical understanding of the market drivers of DIRE. Previous studies [14–16] in residential and commercial real estate (including the retail and warehouse subsectors) have illuminated that macroeconomic attributes can be used to explain the pricing behaviours of traders within a property market with the price-rent dynamics critically dependent on the liquidity, transparency, and/or efficiency of an economy. For example, Duca and Ling [14] observe that the capitalisation rate behaves differentially during periods of market boom and bust, with capital availability/liquidity being one of the key factors affecting the pricing of the market across the real estate industries of retail, industrial warehouse, apartment, and commercial office. Given the more inefficient nature of DIRE relative to other property sectors, we contend that the price-to-rent dynamics should be more pronounced and hence more noticeable with respect to the general macroeconomic conditions. In addition, since increasing the supply of DIRE is, in many circumstances, a technically more cumbersome and time-consuming process in view of, for example, town planning procedures, we therefore further posit that the price/rent interaction of DIRE could be less instantaneously responsive to changes in the market fundamentals of the economy.

Indeed, the primary motivation of the current study lies in the rather unique vicissitude of the Hong Kong manufacturing market and its industrial property sector over the past century, which makes the city an interesting subject upon which to conduct research. Since the "Reform and Openness" initiatives of China, designed and launched by Deng Xiaoping in 1979, the industrial sector of Hong Kong as a whole has witnessed a process of gradual, but inevitable, de-industrialisation with a constant outflow of manufacturing jobs, facilities, and investments to the mainland, particularly to the Pearl River Delta region, due to the imbalances and disparity of land supplies, production costs, and supplies of labour between the two places. The rapid and large-scale modernisation across China has concomitantly marked the end of the era of "Made in Hong Kong" when the city was once hailed as the Pearl of the Orient under the effective and efficacious governance and forward-looking leadership of the British colonial government based on the principles of laissez-faire capitalism and rule of law. During the colonial time, Hong Kong was, astoundingly, one of the world's leading exporters in many industries such as clothing and textile, plastics, toys, and electronics, despite the city's relatively small population and lack of natural resources. With the de-industrialisation of the city came the debilitation of its industrial real estate sector. The significance of the sector has shrunk substantially both in terms of output and labour force over the past four decades. Vacant, dilapidated, under-utilised, and poorly maintained decades-old industrial buildings scatter across the city, with some being converted—lawfully or unlawfully—to structures for residential and/or non-industrial uses. In 2010, the government of Hong Kong launched a scheme known as the "Revitalisation of Industrial Buildings", with an objective to optimise land use in order to meet the incessantly changing needs of the society. To facilitate the growth and development of new industries such as testing and certification services, cultural and creative industries, and environmental industries as promoted in the Policy Address in 2009–2010 (See: https://www.policyaddress.gov.hk/09-10/eng/index.html, accessed on 1 August 2022), the government introduced a number of measures to increase industrial land supply by, for instance, putting forward a tailor-made lease modification system, which provided favourable terms and conditions to land users/owners to redevelop their existing properties or apply for a wholesale conversion of the buildings with a land premium determined by their most suitable use. Hundreds of applications under the scheme have been approved, with millions of square metres of industrialised land being revitalised or created.

The remainder of the paper is organised as follows. Section 2 discusses the literature on the relationships between the price-to-rent ratio and macroeconomic attributes, covering not only the industrial property sector, but also the residential and commercial property markets. Section 3 presents the research methods of the study, providing an in-depth discussion on the Johansen cointegration test and the Granger causality model. Section 4 presents the key research findings, followed by an analysis of the results. The last section concludes the study.

## 2. Literature Review

From the perspective of price discovery, the dynamics between property price and rent as well as their cross-sectional and temporal associations with other macroeconomic determinants can indeed be explained by the DiPasquale and Wheaton model (or the DW model) [17]. Pertinently, property such as other asset classes can be viewed as an investment, with its price equal to the summation of all future discounted incomes that can be derived. The demand for real estate and the total housing stock jointly determines the level of rent. When the demand increases due to an exogenous shock, for instance, rent should increase as a result, given that the amount of housing stock remains largely the same in the short term. The surging rent should, on average, translate into a higher price of property through a process known as capitalisation. The rate of capitalisation, commonly termed "investment yield", indicates the opportunity cost that investors require to invest in the assets. Four economic attributes are encapsulated in the DW model, which governs the fashion in which the yield is determined, namely, long-term interest rate, expected growth rate of rental income, the risk associated with the generation of rental income, and government policy.

A large volume of research in the housing literature has been devoted to uncovering and explaining the joint dynamics of real estate prices and rents by employing different statistical methods and datasets. Pertinently, most of the attention within the literature on the price-to-rent ratio has so far been drawn to the residential real estate sector given its more transparent nature with relatively more readily available data for undertaking empirical analysis. For example, Sommer et al. [18] explored the linkages between property rents and prices for the US market using a dynamic equilibrium stochastic model of housing tenure choice. The results indicated that during the period of 1995 and 2005, over half of the growth in the price-to-rent ratio was a consequence of a higher per capita income, more relaxed lending requirements, and low interest rates. In a subsequent study, Kishor and Morley [19] examined market fundamentals that could affect the price-to-rent ratios of eighteen cities in the US. Their investigation was based upon Campbell and Shiller's analytical framework [20], which was, in turn, premised on the decomposition of the price-to-rent ratio into two components: an unobserved component, which is determined by the expected real estate return and the growth rate of real rental income, and a residual component, which explains the non-stationary temporal deviation of the ratio from its present value. The results discovered that large cities tend to have a present value component greater than that implied by the statistical model. In other words, large cities generally have a higher average price level relative to its rent counterpart.

Ayuso and Restoy [7] corroborated, using a general intertemporal asset pricing model, that residential market prices tend to mean-reverse to their long-term equilibrium over a long time horizon. Using data on the UK, US, and Spanish markets, their research findings displayed that the past overvaluation of the property sectors as reflected by high price-to-rent ratios are attributable to the slow adjustment of rents and inelasticity of housing supply in the presence of demand shocks. From a global economic perspective, Beltratti and Morana [21] revealed strong and persistent interlinkages between the housing markets of the G7 nations with their price fluctuations governed by some common global macroeconomic factors such as investment flows, productivity growth, global stock prices, and oil prices. Importantly, bidirectional causal linkages are evident between house prices and macroeconomic developments, with investment exhibiting a greater impact upon house price shocks relative to consumption and output factors. In relation to the dynamics between house price and rent, some existing empirical evidence in the housing literature suggested that they react to macroeconomic attributes in a differential manner. For instance, compared to rent, house price is more responsive to changes in short-term interest rates and the general productivity of the economy. On the other hand, rent is a lagging indicator of price and short-term interest rates [22].

In the context of the UK, Bracke [23], by evaluating the micro-spatial property price structure of London, illuminated that residential neighbourhoods of higher economic standing tend to be associated with higher price-to-rent ratios. Clark and Lomax [24] also utilised the British housing market transaction and rental price data and established a strong positive empirical relationship linking the degree of physical desirability of a neighbourhood and the price-to-rent ratio. Further, the study showed that detached and semi-detached houses, on average, display higher price-to-rent ratios relative to terraced houses and apartments. More recently, McCord et al. [25] and Lo et al. [8] explored the price-to-rent characteristics of the UK market for different property types, confirming that the detached sector tends to Granger-cause other submarkets in terms of pricing. It is further evident that the price-to-rent ratios of the detached sector are the largest as well as the most volatile. In their subsequent studies [15,16], they empirically revealed that GDP, money supply, foreign exchange markets, and the performance of the equity market are important drivers of the price-to-rent ratio. Further, Lo et al. [26] demonstrated how real estate pricing is intimately correlated with the efficiency of the market by examining the spatial autocorrelation structure of house prices in the UK.

Despite the DIRE sector being relatively under-researched, some studies in the literature did attempt to provide evidence-based investigations that are empirically insightful, dissecting the price–rent dynamics and the market fundamentals that underpin them. For instance, Ambrose [27] investigated 57 industrial real estate listings in Georgia, USA, for the period of 1986–1987 and observed that the industrial property market was indeed priced fairly rationally by investors through valuations that were based on building-specific and locational attributes. Further, industrial properties that were listed for sale were over 200% as large and had smaller finished space for office than properties for lease. In other words, traders in the market priced property attributes differently when they were buying than when they were leasing.

Another strand of studies focused more exclusively on the efficiency of industrial property. Atteberry and Rutherford [28] utilised the hedonic valuation model to examine over 700 industrial property sales in Dallas, USA. They detected a significant time lag between past and current industrial real estate prices, seemingly suggesting that the industrial markets under investigation were not informationally efficient, since the current prices did not fully reflect the past price information. In a similar study, Chai [29] scrutinised the pricing of the warehouse sector and revealed that the general market conditions comove with income-generating industrial properties. More specifically, the demand for and supply of warehouses jointly determine the levels of vacancy rates, rents, and net operating income of the firms within the sector, which indeed echoed the propositions and assumptions of the DW model. Mueller [30] also reported similar empirical findings in line with the DW model, revealing that local demand and supply are key determinants affecting the rental growth rates of industrial real estate. On a different note, Wang et al. [31] observed that the value of an industrial property could be heavily influenced by the level of technology at both industry and national levels. They conjectured that technological progress could optimise production methods and, hence, improve energy efficiency at the property level, which would translate into higher property values. Further, new technological breakthroughs could stimulate more industrial development, creating even more demand for industrial land that could result in higher land prices.

Some studies have shed empirical light on the relationships between the price/rent performance of DIRE and macroeconomic attributes. Notably, Thompson and Tsolacos [32] probed into the British industrial property data and established that industrial property real rents or their growth rates are closely and positively related to the general productivity of the economy proxied by GDP, which determines the demand for manufacturing goods. Further, they found that real rents and construction costs collectively determine the supply. A later investigation by Jones and Orr [33], also utilising British data, further revealed that the supply attributes of industrial real estate are of higher elasticity than those of the retail sector. In subsequent studies, Benjamin et al. [1] and Jackson and White [34] both averred that price inflation and interest rates are two important determinants of real prices and rents of industrial property. In particular, the two factors would generally lead to a decline in the performance of industrial property in terms of price and rent, especially in times of economic slowdown.

Appositely, the interrelationships between price, rent, return, and macroeconomic determinants for income-generating property have been examined more rigorously within the commercial office and retail real estate sectors. Using global real estate data on the Asian, European, and American markets, de Wit and van Dijk [35] found that employment, vacancy rates, GDP, general price level, and stock are significant drivers of the returns on office space investments. A later study by Karakozova [36] further confirmed their results using the Finnish property market data, with findings pointing to GDP and the territory sector employment rates being significant drivers of direct commercial office returns, whilst pointing out that factors affecting industrial real estate returns have received little attention in the literature. In a similar vein, Lieser and Groth [37] explored demographic, social, economic, and institutional attributes that could have an impact on commercial property investment activity. The study is empirically insightful and comprehensive in that it covered forty-seven countries over a nine-year investigation period, showing within a panel data regression framework that economic factors (e.g., size of economy, GDP growth), depth and sophistication of capital markets (proxied by factors such as market liquidity, amount of initial public offering activity, and ease of accessibility to capital), political stability, and sociocultural characteristics (such as general human development levels and the control of bribery and corruption) all have a statistically significant impact on the financial performance of commercial real estate. In the context of retail real estate, Ho and Faishal bin Ibrahim [38] measured the degree of association between the sector and the macroeconomic environment, providing evidence that pro-growth policy and GDP growth tend to have a positive association with rents and returns.

Despite a plethora of quantitative investigations examining the dynamics between the pricing of commercial real estate and the macroeconomy in the literature, to the best of our knowledge, so far there has been no empirical study exclusively directed at exploring the relationships between the price-to-rent ratio of industrial real estate and sector-level and macroeconomic attributes. From a practical standpoint, it is crucial to understand how the interaction of industrial property price and rent responds to or influences different economic fundamentals within the sector and the wider economy. A better conceptualisation of how the pricing of DIRE with respect to key macroeconomic attributes is of significance, not only in terms of formulating company-level investment strategies, but also steering the direction of industrial real estate development within a city or nation at a macro policy design level through, for instance, introducing more pro-DIRE measures and regulations, encouraging more FDI and streamlining existing taxation policy to facilitate more efficient and effective industrial land use and development.

## 3. Methodology

Based on the economic and methodological rationales of previous empirical studies in the literature [14–16], we employ Johansen cointegration and Granger causality techniques to explore potential cointegration and causal relationships between the price-to-rent ratio of DIRE and an array of macroeconomic determinants deemed to display pricing dynamics with the real estate market of Hong Kong. The investigation period spans from 2010 Q1 to 2019 Q4, which was selected with a purpose to model a relatively stable market environment and investment climate, minimising noises caused by irrelevant exogeneous factors such as the global financial crisis during 2007/2008 and the recent outbreak of COVID-19 during 2019/2020.

We explore the short- and long-term relationships between the price-to-rent ratios of the industrial real estate market, and a basket of its economic determinants, which can be categorised into five groups of variables, as follows:

Group 1—General macroeconomic attributes

The first group of attributes comprises (i) inflation (I), (ii) employment rate (ER), (iii) GDP growth (GDP), and (iv) stock market performance (HSI). We posit that the industrial real estate market should, to certain extent, perform in tandem with the general economy. For example, industrial land price should possibly move in sync with the general price level of goods and services, as reflected by the inflation rate; the general productivity or overall strength of the economy as signalled by the employment rate, GDP growth, and the performance of the stock market should have a positive association with the industrial sector. However, their causal relationships, we surmise, might be statistically ambiguous, which would require further empirical inquiry.

Group 2—Liquidity-related attributes

The four attributes encompassed in this group are (i) the exchange rate of RMB/HKD (RMB), (ii) (inverse of) DXY (DXY), (iii) M3 money supply (M3), and (iv) foreign direct investment (FDI). According to the Quantity Theory of Money, asset prices increase with the amount of money or liquidity of capital being circulated within an economy, the holding velocity of money, and the level of real output constant. Accordingly, we hypothesise that (i) money supply, as represented by M3, should be positively associated with industrial land price in the long run; (ii) when there is an increase in the exchange rate between RMB and HKD, a higher volume of capital-chasing local assets will flow from China to Hong Kong as HKD-denominated assets will become more financially appealing from the perspective of Chinese investors, driving real estate asset prices up; (iii) along the same line of thought, when DXY, which measures the strength of the USD, appreciates, the HKD, through the linked exchange rate mechanism, should appreciate in the short term against other international currencies. Hence, when the inverse of DXY increases, we should expect a larger amount of international capital pouring into the Hong Kong economy, propelling the prices of its assets, including those of industrial real estate assets; (iv) likewise, when there is an increased amount of foreign direct investment within the economy, we should also expect industrial property prices to surge.

Group 3—Housing market attributes

We consider this category of attributes, consisting of (i) average residential property price (RHP) and (ii) residential price-to-rent ratio (PtR_R), to detect any temporal linkages between the housing market and the DIRE market in terms of pricing. Several studies (e.g., [39,40]) in the literature have documented temporal co-movement between the two sectors of real estate. The associations between them could be plausibly due to the underlying market fundamentals they share in common, causing their price/rent characteristics to trend in a similar fashion over time.

Group 4—Industrial sector-specific attributes

It is logical to assume that the price-to-rent ratio of the industrial property market should be correlated to the performance and characteristics of the manufacturing sector and the DIRE. Briefly, if the manufacturing sector is in expansion with growing return and profitability, the demand for industrial space should plausibly increase, impacting the pricing of industrial real estate. To account for the size, profitability, productivity, and revenue of the manufacturing sector of Hong Kong, we incorporate the following attributes specific to the manufacturing sector in our models: (i) value added (VA), (ii) number of industrial establishments (IE), (iii) total workforce (W), (iv) operating expenses (OE), and gross sales (S).

Group 5—External economic factors

Lastly, we posit that the steadfast growth of the manufacturing sector in mainland China over the past decades, particularly in provinces geographically proximate to Hong Kong such as Guangdong, should have a quantitative and spatial impact on the overall development of the city's industrial sector, consequentially shrinking its industrial real estate. Therefore, we encompass the rate of change in the volume of manufacturing output of Guangdong (GD) in our models to evaluate whether the Hong Kong industrial real estate sector is affected by its neighbouring region in terms of pricing. An inverse operation is performed on the variable given that it is expected to be inversely correlated to the DIRE in Hong Kong.

The abovementioned variables are measured on a quarterly basis. In addition, the price-to-rent time series are analysed for three administratively defined districts, namely, Hong Kong Island (HKI), Kowloon (K), and the New Territories (NT), to determine whether there is any spatial heterogeneity across the industrial real estate market with respect to the macroeconomic attributes. Table 1 below provides detailed descriptions of the variables examined in our analysis.

Empirical Models

We undertake stationarity tests on the time series prior to examining any potential cointegration and causal relationships between two variables. Failure to detect and account for the non-stationarity of a time series could produce spurious empirical results [41] In our study, we employ the Augmented Dickey Fuller (ADF) unit root test to detect whether a unit root is found within a given time series, with the general equation of the test given by:

$$\Delta Y_t = \alpha + \beta T + \varnothing Y_{t-1} + \sum_{i=1}^{k} \partial \Delta Y_{t-i} + \varepsilon_t \tag{1}$$

where $Y_t$ denotes the level of the time series; $\alpha$ is an intercept term; $T$ is a temporal trend; $k$ represents the number of time periods (i.e., the lag length) for achieving white noise governed by the Schwarz Information Criterion (SIC); and $\varepsilon_t$ is an error term, which is a mean of zero and finite variances.

Cointegration Tests

We utilised the Johansen cointegration test to detect any long-term cointegration associations between the price-to-rent time series and macroeconomic attributes. Statistically, the components of a vector $C$ ~CI ($i$, $j$) are said to be cointegrated of order $i$, $j$ if $C_t$ is stationary at the first difference, i.e., I(1), and we can observe a non-zero vector P such that $P'C_t$ follows $I(i–j)$ with $i \geq j > 0$.

**Table 1.** Descriptions of variables.

| Variable | Definition | Inverse Operator | Unit | Data Source |
|---|---|---|---|---|
| Price-to-rent ratios | Average industrial land price divided by average industrial land rent | No | Ratio | Rating and Valuation Department, Hong Kong |
| Inflation (I) | The year-on-year growth rate of consumer price index | No | Percentage | Census and Statistics Department, Hong Kong |
| Employment rate (ER) | The percentage of population aged 15 and over who have been at work for pay or profit during the 7 days before enumeration or have had formal job attachment | No | Percentage | Census and Statistics Department, Hong Kong |
| GDP growth (GDP) | The year-on-year GDP growth rate | No | Percentage | Census and Statistics Department, Hong Kong |
| Year-on-year stock market return (HSI) | The year-on-year return on the Hang Seng Index | No | Percentage | Yahoo Finance |
| RMB/HKD (RMB) | The exchange rate between the Chinese Yuan and Hong Kong Dollar | No | Ratio | Yahoo Finance |
| US Dollar Index (DXY) | The geometric weighted average of six international currencies (EUR, JPY, GBP, CAD, SEK, and CHF) against the USD | Yes | Percentage | Tradingview |
| Foreign direct investment (FDI) | The volume of direct foreign investment | No | HKD (M) | Census and Statistics Department, Hong Kong |
| M3 Money supply growth (ME) | Year-on-year growth of M3 money supply | No | HKD (M) | Hong Kong Monetary Authority |
| Residential house price (RHP) | Average house prices of Class C [1] residential real estate in the respective districts of Hong Kong Island, Kowloon, and New Territories | No | HKD | Rating and Valuation Department, Hong Kong |
| Residential price-to-rent (PtR_R) | Price-to-rent ratios of Class C residential real estate in the respective districts of Hong Kong Island, Kowloon, and New Territories | No | Ratio | Rating and Valuation Department, Hong Kong |
| Industrial value added (VA) | GDP attributable to the secondary (i.e., manufacturing) sector | No | HKD (M) | Census and Statistics Department, Hong Kong |
| Number of industrial establishments (IE) | Number of industrial establishments in the manufacturing sector | No | Number | Census and Statistics Department, Hong Kong |
| Industrial workforce (W) | Number of persons engaged in the manufacturing sector | No | Number | Census and Statistics Department, Hong Kong |
| Operating expense (OE) | Total amount of expenses for purchases of materials, supplies, and industrial work and services | No | HKD (M) | Census and Statistics Department, Hong Kong |
| Industrial sales (S) | Total sales of goods, industrial work, and services | No | HKD (M) | Census and Statistics Department, Hong Kong |
| Manufacturing output of Guangdong province (GD), China | Total amount of industrial output of Guangdong province | Yes | RMB (100 M) | Bureau of Statistics, Guangdong Province, China |

[1] The Rating and Valuation Department of the Hong Kong Government provides data on average prices and rents measured on a per $m^2$ basis for five classes of private residential property, which are defined by unit size: Class A (below 39.9 $m^2$), Class B (40–69.9 $m^2$), Class C (70–99.9 $m^2$), Class D (100–159.9 $m^2$), and Class E (over 160 $m^2$).

Mathematically, we can detect a cointegration relationship between two time series, $X_t$ and $Y_t$, by conducting a regression of $Y_t$ on $X_t$ as in Equation (2) below.

$$Y_t = \alpha + \beta X_t + u_t \tag{2}$$

The regression residuals $u_t$ are examined for stationarity by carrying out a unit root test. $X_t$ and $Y_t$ are said to be cointegrated if $u_t$ is a stationary time series. As pointed out by Dickey et al., the above method based on Engle and Granger [42] could be sensitive to the choice of dependent variable, potentially leading to statistical inconsistencies. Hence, a modified modelling approach based on Johansen [43,44] is employed, which considers the following equation:

$$\Delta Y_t = \eta Y_{t-1} + \sum_{i=1}^{k} \Gamma_i \Delta Y_{t-i} + BX_t + \varepsilon_t \tag{3}$$

where $\Delta Y_t$ is the change in $Y_t$, $\eta = \sum_{i=1}^{k} A_i - I$, and $\Gamma_i = -\sum_{j=1+1}^{k} A_i$. $Y_t$ denotes a $k$-vector of I(1), which is non-stationary. $X_t$, on the other hand, represents a $d$-vector of deterministic variables. $\eta$ depicts the rank of the coefficient matrix, i.e., the number of cointegrating vectors within the time series. The lag length is determined by the Schwarz Information Criterion. Trace test statistics can be obtained by undertaking the likelihood ratio test on the number of cointegrating vectors between the time series. If the two variables are cointegrated, then they should be integrated to order one.

Granger Causality Test in Error Correction Models (ECMs)

If two time series are cointegrated, their long-term temporal relationship should be determined within the framework of the error correction model (ECM) [45]. The equation of an ECM-based causality equation can be formulated as follows:

$$\Delta Y_t = \lambda + \sum_{i=1}^{u} \alpha_i \Delta Y_{t-i} + \sum_{j=1}^{v} \beta_j \Delta X_{t-j} + \phi z_{t-1} + \varepsilon_t \tag{4}$$

where $\lambda$ is an intercept term, and $z_{t-1}$ is the error correction (EC) term of the equation with a coefficient of $\phi$. The ECM equation contains information about both short- and long-term dynamics between the two time series, with $u$ and $v$ being the number of lags that are large enough to produce an error term that is white noise. It is further noted that all terms are I(0) in Equation (4). $\beta_j s$ measure the short-term influence of $Y$ to changes in $X$. In other words, $\beta_j s$ signal the short run elasticity of $Y$ with respect to $X$. On the other hand, the error correction term, $z_{t-1}$, represents the long-term dynamic between the two variables. Mathematically, $z_{t-1}$ is given by:

$$z_{t-1} = Y_{t-1} - w_0 - w_1 X_{t-1} + w_2 t \tag{5}$$

where $w_1$ is the coefficient on the lagged independent variable $X_{t-1}$, which indicates the degree of $Y$'s long-run elasticity with respect to $X$ [46]. The speed of adjustment of short-term disequilibria is captured by the coefficient of the EC term, $\phi$. The EC term should have a positive sign if changes in $Y$ are greater than its long-term average value. In other words, $\Delta Y_t$ should tend to decrease in value so as to follow the path of its long-term equilibrium. On the contrary, the EC term should be negatively signed if $\Delta Y_t$ is below its average value, which "pushes" $Y$ upward over time. If there is a long-term lead–lag relationship between the variables, the coefficient $\phi$ should be negative. Put differently, the null hypothesis of long-term non-causality should be rejected if $\phi$ is negative at the conventional statistically significant level.

Given that the coefficients of the lagged variables $\Delta X_{t-j}$ measure the short-term interaction between the time series, the Wald $X^2$ test can be employed to determine the short-term Granger causality by checking the coefficient restriction on the lagged first difference terms. If the coefficients are statistically different from zero, the null hypothesis of short-term non-causality should be falsified. By examining the data on the DIRE of Hong Kong using the cointegration and causality techniques, the study aims to explore whether the pricing of the property market is causally linked with the macroeconomic attributes in a Granger fashion in the long run.

## 4. Results and Discussion

### 4.1. Results of the ADF Test

We design the specification of each of the ADF equations based on an initial graphical analysis proposed by [47]: If a time series graphically shows a time trend, stochastic or deterministic, a linear trend and an intercept term should be encapsulated into the equation to minimise omitted variable biases. The summary of the results of the ADF tests is reported in Table 2 (see Appendix A for full results). The three price-to-rent ratio time series for the residential markets of Hong Kong, as well as those for the employment rate and stock market performance, are found to be stationary at level, implying that they tend not to exhibit statistical properties that depend on the sample period of investigation. The rest of the time series under investigation are non-stationary at level, but stationary at first difference. Accordingly, we apply differencing on the time series and transform the two sets of variables to I(0) and I(I) based on their stationarity [41].

**Table 2.** Summary of the results of the ADF tests.

| Time Series | Stationarity |
|---|---|
| PtR_R(HK), PtR_R(K), PtR_R(NT), ER and HSI | Stationary at level |
| PtR(HK), PtR(K), PtR(NT), I, RMB, FDI, ME, DXY, GDP, VA, IE, W, OE, IS, RHP(HK), RHP(K), RHP(NT) and GD | Stationary at first difference |

Note: The tests are conducted based on the 5% significance level. Full results are available upon request.

### 4.2. Results of the Cointegration and Granger Causality Tests

The results of the cointegration tests on the time series are depicted in Table 3. We perform the tests on the price-to-rent ratio time series for the three districts of Hong Kong with respect to each of the macroeconomic attributes. Trace statistics and eigenvalues are utilised to detect whether a cointegration relationship exists between a given pair of time series. It is evident that most pairs of time series are cointegrated in the long term at the 5% statistical significance level, implying that they tend to move in tandem over time. The only two pairs of time series that are not cointegrated are PtR(NT) vs. DXY and PtR(NT) vs. VA. Procedurally, the findings of the cointegration analysis determine the methodological approach that should be adopted to examine the lead–lag relationship between a pair of time series: if they are found to be non-cointegrated, the corresponding causality equation should be constructed within an ordinary vector autoregressive regression framework. Otherwise, the causality should be examined by adopting an error correction model approach. Table 4 presents the results of the Granger causality tests, detailing the chi-square statistics of the Wald test and the t-statistics of the error correction term in Equation (4). The $R^2$; adjacent $R^2$, The Akaike information criterion (AIC), Schwarz criterion (SC), Durbin–Watson (DW) value, F-statistic, and the coefficient on the EC term of each pair of the Granger models are also reported in the table.

Emanating from Table 4 are a number of noteworthy and interesting findings in relation to the causal dynamics between the DIRE and the macroeconomic factors. First, the general economic indicators, including inflation, employment, and stock market, appear to Granger-cause the DIRE market in the long run, but not the other way around. The time lags between the variables are typically two to three quarters. It is further noticeable that GDP growth displays a bidirectional causal linkage with the price-to-rent variables across the three districts. In other words, the growth of national income is both a cause and a consequence of the price-to-rent ratio, in a Granger sense.

**Table 3.** Results of cointegration tests.

| | | Y = Price-to-Rent Ratio of Industrial Real Estate | | | | | |
|---|---|---|---|---|---|---|---|
| | | Hong Kong Island | | Kowloon | | New Territories | |
| | X | None<br>Trace Stat. (Prob)<br>Eigenvalue<br>(Prob) | At most 1<br>Trace Stat. (Prob)<br>Eigenvalue<br>(Prob) | None<br>Trace Stat. (Prob)<br>Eigenvalue<br>(Prob) | At most 1<br>Trace Stat.<br>(Prob)<br>Eigenvalue<br>(Prob) | None<br>Trace Stat.<br>(Prob)<br>Eigenvalue<br>(Prob) | At most 1<br>Trace Stat.<br>(Prob)<br>Eigenvalue<br>(Prob) |
| Group 1 | I | 33.33664<br>(0.00) ***<br>0.491059<br>(0.00) *** | 9.021425<br>(0.00) ***<br>0.221663<br>(0.00) *** | 26.63423<br>(0.00) ***<br>0.402259<br>(0.03) ** | 8.108709<br>(0.00) ***<br>0.201677<br>(0.00) *** | 44.26229<br>(0.00) ***<br>0.561689<br>(0.00) *** | 14.56851<br>(0.00) ***<br>0.332810<br>(0.00) *** |
| | ER | 45.18061<br>(0.00) ***<br>0.564146<br>(0.00) *** | 14.45404<br>(0.00) ***<br>0.323383<br>(0.00) *** | 38.93924<br>(0.00) ***<br>0.480966<br>(0.00) *** | 14.67519<br>(0.00) ***<br>0.327415<br>(0.00) *** | 32.14486<br>(0.00) ***<br>0.409107<br>(0.02) ** | 12.67842<br>(0.00) ***<br>0.290121<br>(0.00) *** |
| | GDP | 43.89405<br>(0.00) ***<br>0.563733<br>(0.00) *** | 13.20252<br>(0.00) ***<br>0.300105<br>(0.00) *** | 37.07822<br>(0.00) ***<br>0.473791<br>(0.00) *** | 13.32214<br>(0.00) ***<br>0.302364<br>(0.00) *** | 29.48732<br>(0.00) ***<br>0.361579<br>(0.00) *** | 12.88332<br>(0.05) **<br>0.294041<br>(0.00) *** |
| | HSI | 42.77438<br>(0.00) ***<br>0.587911<br>(0.01) *** | 9.973314<br>(0.0016) ***<br>0.236276<br>(0.0016) *** | 34.32131<br>(0.00) ***<br>0.484660<br>(0.00) *** | 9.792978<br>(0.00) ***<br>0.232545<br>(0.00) *** | 24.12839<br>(0.00) ***<br>0.337069<br>(0.09) * | 8.918264<br>(0.00) ***<br>0.214185<br>(0.00) *** |
| Group 2 | RMB | 51.00004<br>(0.00) ***<br>20.11395<br>(0.00) *** | 0.566021<br>(0.00) ***<br>0.419358<br>(0.00) *** | 39.50740<br>(0.00) ***<br>0.505264<br>(0.00) *** | 13.46932<br>(0.00) ***<br>0.305134<br>(0.00) *** | 35.73811<br>(0.00) ***<br>0.457544<br>(0.00) *** | 13.10710<br>(0.00) ***<br>0.298298<br>(0.00) *** |
| | 1/DXY | 44.46733<br>(0.00) ***<br>0.563370<br>(0.00) *** | 13.80661<br>(0.00) ***<br>0.311439<br>(0.00) *** | 36.93921<br>(0.00) ***<br>0.467219<br>(0.00) *** | 13.64235<br>(0.00) ***<br>0.308375<br>(0.00) *** | 29.17684<br>(0.00) ***<br>0.339327<br>(0.00) *** | 13.84045<br>(0.09)*<br>0.312069<br>(0.00) *** |
| | FDI | 44.28514<br>(0.00) ***<br>0.569169<br>(0.00) *** | 13.12970<br>(0.00) ***<br>0.298726<br>(0.00) *** | 34.51324<br>(0.00) ***<br>0.446863<br>(0.00) *** | 12.60369<br>(0.00) ***<br>0.288685<br>(0.00) *** | 30.38143<br>(0.00) ***<br>0.407928<br>(0.02) ** | 10.98872<br>(0.00) ***<br>0.256950<br>(0.00) *** |
| | M3 | 45.98939<br>(0.00) ***<br>15.25011<br>(0.00) *** | 0.564296<br>(0.00) ***<br>0.337785<br>(0.00) *** | 33.33011<br>(0.00) ***<br>0.417475<br>(0.00) *** | 13.33592<br>(0.00) ***<br>0.302624<br>(0.00) *** | 30.78925<br>(0.00) ***<br>0.415742<br>(0.02)** | 10.90501<br>(0.00) ***<br>0.255267<br>(0.00) *** |
| Group 3 | RHP | 38.70646<br>(0.00) ***<br>0.555853<br>(0.00) *** | 8.677265<br>(0.00) ***<br>0.209050<br>(0.00) *** | 40.04824<br>(0.00) ***<br>0.535338<br>(0.00) *** | 11.68977<br>(0.00) ***<br>0.270897<br>(0.00) *** | 31.59580<br>(0.00) ***<br>0.419772<br>(0.02) ** | 11.45546<br>(0.00) ***<br>0.266265<br>(0.00) *** |
| | PtR_R | 65.04982<br>(0.00) ***<br>0.711587<br>(0.00) *** | 19.034549<br>(0.00) ***<br>0.402346<br>(0.00) *** | 37.28334<br>(0.00) ***<br>0.504795<br>(0.00) *** | 11.08849<br>(0.00) ***<br>0.258951<br>(0.00) *** | 48.69358<br>(0.00) ***<br>0.606138<br>(0.00) *** | 14.21862<br>(0.00) ***<br>0.319064<br>(0.00) *** |

**Table 3.** *Cont.*

| | | Y = Price-to-Rent Ratio of Industrial Real Estate | | | | | |
|---|---|---|---|---|---|---|---|
| | | Hong Kong Island | | Kowloon | | New Territories | |
| Group 4 | VA | 47.06448<br>(0.00) ***<br>0.579933<br>(0.00) *** | 15.84018<br>(0.00) ***<br>0.355967<br>(0.00) *** | 38.28487<br>(0.00) ***<br>0.453507<br>(0.00) *** | 16.53248<br>(0.00) ***<br>0.368234<br>(0.00) *** | 30.53394<br>(0.00) ***<br>0.359872<br>(0.07) * | 14.47482<br>(0.00) ***<br>0.331072<br>(0.00) *** |
| | IE | 45.03834<br>(0.00) ***<br>0.558500<br>(0.00) *** | 15.60557<br>(0.00) ***<br>0.351756<br>(0.00) *** | 38.46194<br>(0.00) ***<br>0.473172<br>(0.00) *** | 15.39018<br>(0.00) ***<br>0.347866<br>(0.00) *** | 33.03596<br>(0.00) ***<br>0.424461<br>(0.02) ** | 13.14782<br>(0.00) ***<br>0.305954<br>(0.00) *** |
| | W | 51.88273<br>(0.00) ***<br>0.645208<br>(0.00) *** | 14.57871<br>(0.00) ***<br>0.332999<br>(0.00) *** | 40.70381<br>(0.00) ***<br>0.484332<br>(0.00) *** | 16.86130<br>(0.00) ***<br>0.373978<br>(0.00) *** | 42.19879<br>(0.00) ***<br>0.522432<br>(0.00) *** | 15.59302<br>(0.00) ***<br>0.351530<br>(0.00) *** |
| | OE | 48.60125<br>(0.00) ***<br>0.627156<br>(0.00) *** | 13.08383<br>(0.00) ***<br>0.304719<br>(0.00) *** | 39.70403<br>(0.00) ***<br>0.502303<br>(0.00) *** | 14.58450<br>(0.00) ***<br>0.333107<br>(0.00) *** | 32.11027<br>(0.00) ***<br>0.393540<br>(0.04) ** | 14.10606<br>(0.00) ***<br>0.324184<br>(0.00) *** |
| | S | 49.24067<br>(0.00) ***<br>0.633734<br>(0.00) *** | 13.08244<br>(0.00) ***<br>0.304693<br>(0.00) *** | 40.09045<br>(0.00) ***<br>0.504138<br>(0.00) *** | 14.83795<br>(0.00) ***<br>0.337785<br>(0.00) *** | 32.34709<br>(0.00) ***<br>0.395053<br>(0.04) ** | 14.25298<br>(0.00) ***<br>0.326937<br>(0.00) *** |
| Group 5 | Inverse GD | 51.26316<br>(0.00) ***<br>0.613456<br>(0.00) *** | 16.09428<br>(0.00) ***<br>0.352723<br>(0.00) *** | 39.80090<br>(0.00) ***<br>0.445102<br>(0.00) *** | 18.00900<br>(0.00) ***<br>0.385367<br>(0.00) *** | 95.57611<br>(0.00) ***<br>0.886784<br>(0.00) *** | 14.97301<br>(0.00) ***<br>0.332807<br>(0.00) *** |

Note: *** denotes 1% statistical sig.; ** 5% sig.; * 10% sig. Full results are available upon request.

**Table 4.** Results of Granger causality tests.

| | Y = Price-to-Rent | Hong Kong Island | | | Kowloon | | | New Territories | | |
|---|---|---|---|---|---|---|---|---|---|---|
| | X | Short-term chi-sq (Prob) | Long-term t-statistic (Prob) | $R^2$, Adj $R^2$<br>AIC; SC<br>DW; F<br>EC<br>Lag based on SIC | Short-term chi-sq (Prob) | Long-term t-statistic (Prob) | $R^2$; Adj $R^2$<br>AIC; SC<br>DW; F<br>EC<br>Lag based on SIC | Short-term chi-sq (Prob) | Long-term t-statistic (Prob) | $R^2$; Adj $R^2$<br>AIC; SC<br>DW; F<br>EC<br>Lag based on SIC |
| Group 1 | I | | | | | | | | | |
| | X→Y | 1.958261<br>(0.3756) | −4.937419<br>(0.0000) *** | 0.739127;<br>0.683225<br>5.660394;<br>5.971464<br>1.933729;<br>13.22197<br>−2.133166<br>Lag = 2 | 6.664995<br>(0.0357) ** | −3.941771<br>(0.0005) *** | 0.503569;<br>0.397191<br>4.779536;<br>5.090606<br>2.064788;<br>4.733765<br>−0.967190<br>Lag = 2 | 9.719646<br>(0.0078) *** | −2.042917<br>(0.0506) ** | 0.482433;<br>0.371525<br>4.765590;<br>5.076660<br>2.172811;<br>4.349871<br>−0.338907<br>Lag = 2 |
| | Y→X | 1.632070<br>(0.6521) | 4.031771<br>(0.0005) *** | 0.592373;<br>0.461932<br>0.524383;<br>0.928420<br>2.022130;<br>4.541320<br>0.000102<br>Lag = 3 | 2.054872<br>(0.5611) | 4.291386<br>(0.0002) *** | 0.680403;<br>0.578131<br>0.281092;<br>0.685129<br>2.010390;<br>6.652928<br>0.061624<br>Lag = 3 | 20.56348<br>(0.0045) *** | −3.064139<br>(0.0091) *** | 0.822174;<br>0.603312<br>0.203525;<br>0.997537<br>2.224203;<br>3.756579<br>−0.124806<br>Lag = 7 |

**Table 4.** *Cont.*

| | Y = Price-to-Rent | | Hong Kong Island | | | Kowloon | | | New Territories | |
|---|---|---|---|---|---|---|---|---|---|---|
| Group 1 | ER | | | | | | | | | |
| | X→Y | 1.985117<br>(0.3706) | −4.612383<br>(0.0001) *** | 0.707950;<br>0.647526<br>5.734648;<br>6.042554<br>1.977709;<br>11.71635<br>−2.034770<br>Lag = 2 | 3.574402<br>(0.1674) | −4.701829<br>(0.0001) *** | 0.541435;<br>0.446559<br>4.668355;<br>4.976262<br>1.916656;<br>5.706784<br>−1.610318<br>Lag = 2 | 12.99918<br>(0.0430) ** | −2.766420<br>(0.0132) ** | 0.811747;<br>0.656716<br>4.346942;<br>5.034005<br>1.932279;<br>5.236011<br>−2.138274<br>Lag = 6 |
| | Y→X | 23.15500<br>(0.0003)<br>*** | 1.958936<br>(0.0642) * | 0.639065;<br>0.422504<br>−3.281432;<br>−2.691898<br>2.207866;<br>2.950969<br>0.007809<br>Lag = 2 | 2.666108<br>(0.4460) | 3.768261<br>(0.0009) *** | 0.820853;<br>0.765731<br>−3.686986;<br>−3.287039<br>2.060798;<br>14.89154<br>0.008717<br>Lag = 3 | 1.069561<br>(0.5858) | 3.861961<br>(0.0006) *** | 0.840052;<br>0.806959<br>−3.760328;<br>−3.452421<br>2.086704;<br>25.38473<br>0.001029<br>Lag = 2 |
| | GDP | | | | | | | | | |
| | X→Y | 3.413240<br>(0.1815) | −4.797933<br>(0.0000) *** | 0.731594;<br>0.676061<br>5.650226;<br>5.958132<br>1.986572;<br>13.17418<br>−1.997423<br>Lag = 2 | 6.523643<br>(0.0383) ** | −4.751474<br>(0.0001) *** | 0.592315;<br>0.507966<br>4.550747;<br>4.858654<br>1.951795;<br>7.022226<br>−1.431508<br>Lag = 2 | 18.89957<br>(0.0003) *** | −2.863480<br>(0.0082) *** | 0.594807;<br>0.470133<br>4.635098;<br>5.035045<br>2.210271;<br>4.770878<br>−0.747543<br>Lag = 3 |
| | Y→X | 11.77424<br>(0.0082)<br>*** | −3.213163<br>(0.0035) *** | 0.479295;<br>0.319078<br>3.923093;<br>4.323040<br>1.957848;<br>2.991535<br>−0.389145<br>Lag = 3 | 8.278265<br>(0.0406) ** | −3.677872<br>(0.0011) *** | 0.502512;<br>0.349439<br>3.877480;<br>4.277426<br>1.879596;<br>3.282827<br>−0.512787<br>Lag = 3 | 10.31308<br>(0.0161) ** | −4.332135<br>(0.0002) *** | 0.580910;<br>0.451959<br>3.705996;<br>4.105943<br>1.828528;<br>4.504890<br>−0.710709<br>Lag = 3 |
| | HSI | | | | | | | | | |
| | X→Y | 2.554783<br>(0.2788) | −5.642554<br>(0.0000) *** | 0.769634;<br>0.721972<br>5.497393;<br>5.805299<br>1.957942;<br>16.14777<br>−2.374291<br>Lag = 2 | 5.441938<br>(0.0658) * | −5.572355<br>(0.0000) *** | 0.630735;<br>0.554335<br>4.451768;<br>4.759674<br>1.833642;<br>8.255722<br>−1.743983<br>Lag = 2 | 7.561901<br>(0.0228) ** | −2.962609<br>(0.0060) *** | 0.543689;<br>0.449279<br>4.614976;<br>4.922882<br>1.947396;<br>5.758848<br>−0.841496<br>Lag = 2 |
| | Y→X | 14.78769<br>(0.0634) * | 0.671763<br>(0.5145) | 0.807270;<br>0.534237<br>−1.681588;<br>−0.840869<br>1.504834;<br>2.956668<br>0.030857<br>Lag = 8 | 17.34044<br>(0.0153) ** | 3.685796<br>(0.0022) *** | 0.769862;<br>0.539723<br>−1.704402;<br>−0.964279<br>1.916224;<br>3.345210<br>0.035582<br>Lag = 7 | 4.086430<br>(0.2523) | −4.200321<br>(0.0003) *** | 0.468145;<br>0.304498<br>−1.305758;<br>−0.905811<br>1.927231;<br>2.860693<br>−0.044620<br>Lag = 3 |

**Table 4.** *Cont.*

| | Y = Price-to-Rent | | Hong Kong Island | | | Kowloon | | | New Territories | | |
|---|---|---|---|---|---|---|---|---|---|---|---|
| Group 2 | RMB | | | | | | | | | | |
| | | X→Y | 42.70838<br>(0.0000)<br>*** | −5.056897<br>(0.0001) *** | 0.903347;<br>0.845354<br>5.110120;<br>5.699653<br>1.974470;<br>15.57707<br>−3.761158<br>Lag = 5 | 1.105385<br>(0.5754) | −4.250169<br>(0.0002) *** | 0.606239;<br>0.524771<br>4.515997;<br>4.823904<br>1.897391;<br>7.441443<br>−1.183168<br>Lag = 2 | 37.82489<br>(0.0000) *** | −3.776939<br>(0.0010) *** | 0.774242;<br>0.676087<br>4.211121;<br>4.704944<br>4.379529;<br>1.948474<br>−1.309565<br>Lag = 4 |
| | | Y→X | 35.37248<br>(0.0000)<br>*** | 3.099452<br>(0.0065) *** | 0.829431;<br>0.688962<br>−4.685261;<br>−3.998198<br>1.847607;<br>5.904733<br>0.026114<br>Lag = 6 | 15.67224<br>(0.0004)<br>*** | 3.746013<br>(0.0008) *** | 0.536986;<br>0.441190<br>−4.336964;<br>−4.029058<br>2.041578;<br>5.605523<br>0.012468<br>Lag = 2 | 6.505884<br>(0.0387) ** | 4.159254<br>(0.0003) *** | 0.548309;<br>0.454856<br>−4.361723;<br>−4.053816<br>1.931055;<br>5.867199<br>0.009320<br>Lag = 2 |
| | M3 | | | | | | | | | | |
| | | X→Y | 1.086964<br>(0.5807) | −5.056468<br>(0.0000) *** | 0.735879;<br>0.681233<br>5.634130;<br>5.942037<br>1.952700;<br>13.46637<br>−2.238344<br>Lag = 2 | 17.90539<br>(0.0001)<br>*** | −4.973773<br>(0.0000) *** | 0.613794;<br>0.533889<br>4.496623;<br>4.804530<br>2.029943;<br>7.681579<br>−1.406375<br>Lag = 2 | 10.38446<br>(0.0156) ** | −3.868645<br>(0.0007) *** | 0.585580;<br>0.458066<br>4.657617;<br>5.057563<br>2.039515;<br>4.592277<br>−1.340024<br>Lag = 3 |
| | | Y→X | 2.142361<br>(0.3426) | 0.197897<br>(0.8445) | 0.302903;<br>0.158676<br>24.22203;<br>24.52994<br>2.221385;<br>2.100185<br>952.4243<br>Lag = 2 | 0.656998<br>(0.7200) | −1.475277<br>(0.1509) | 0.338300;<br>0.201396<br>24.16992;<br>24.47782<br>2.112508;<br>2.471084<br>−7803.518<br>Lag = 2 | 3.568198<br>(0.1679) | −1.705136<br>(0.0989) * | 0.343396;<br>0.207547<br>24.16219;<br>24.47009<br>2.099994;<br>2.527773<br>−8115.359<br>Lag = 2 |
| | Inverse DXY | | | | | | | | | | |
| | | X→Y | 2.078931<br>(0.3536) | −4.874538<br>(0.0000) *** | 0.725904;<br>0.669195<br>5.671200;<br>5.979107<br>2.083470;<br>12.80042<br>−2.065161<br>Lag = 2 | 4.330951<br>(0.1147) | −4.534927<br>(0.0001) *** | 0.555297;<br>0.463290<br>4.637659;<br>4.945565<br>1.842611;<br>6.035346<br>−1.386748<br>Lag = 2 | 0.123896<br>(0.9399) | NA | 0.067120;<br>−0.0317451<br>4.649881;<br>4.867573<br>1.909190;<br>0.054042 |
| | | Y→X | 3.126074<br>(0.2095) | 1.210780<br>(0.2358) | 0.389267;<br>0.262908<br>−12.31526;<br>−12.00736<br>2.055909;<br>3.080652<br>$6.37 \times 10^{-5}$<br>Lag = 2 | 7.542668<br>(0.0230) ** | 1.948857<br>(0.0610) * | 0.458534;<br>0.346507<br>−12.43564;<br>−12.12774<br>1.797401;<br>4.093056<br>$6.00 \times 10^{-5}$<br>Lag = 2 | 2.727393<br>(0.2557) | NA | 0.082326;<br>−0.032384<br>−12.72465;<br>−12.50696<br>2.094587;<br>0.717690 |

**Table 4.** *Cont.*

| | Y = Price-to-Rent | Hong Kong Island | | | Kowloon | | | New Territories | | |
|---|---|---|---|---|---|---|---|---|---|---|
| Group 2 | FDI | | | | | | | | | |
| | X→Y | 19.61473 (0.0001) *** | −6.131878 (0.0000) *** | 0.781301; 0.736053 5.445418; 5.753324 1.836672; 17.26710 −2.22298 Lag = 2 | 3.528794 (0.1713) | −4.775521 (0.0000) *** | 0.580850; 0.494130 4.578481; 4.886387 1.897027; 6.697946 −1.583678 Lag = 2 | 2.479107 (0.2895) | −2.820380 (0.0086) *** | 0.461262; 0.349799 4.781030; 5.088936 1.923781; 4.138250 −0.890324 Lag = 2 |
| | Y→X | 3.560104 (0.1686) | −1.933148 (0.0630) * | 0.503902; 0.401261 26.59458; 26.90248 2.216759; 4.909359 −27276.01 Lag = 2 | 0.957831 (0.6195) | 0.412721 (0.6828) | 0.448994; 0.334993 26.69955; 27.00745 2.187116; 3.938505 8706.255 Lag = 2 | 1.277016 (0.5281) | 1.931720 (0.0632) * | 0.503010; 0.400184 26.59637; 26.90428 2.151742; 4.891873 33290.95 Lag = 2 |
| Group 3 | RHP | | | | | | | | | |
| | X→Y | 9.867830 (0.0072) *** | −6.369505 (0.0000) *** | 0.813884; 0.775377 5.284093; 5.592000 1.975044; 21.13615 −2.607543 Lag = 2 | 11.17467 (0.0037) *** | −6.425298 (0.0000) *** | 0.674943; 0.607690 4.324252; 4.632159 1.697076; 10.03586 −1.963624 Lag = 2 | 9.339842 (0.0094) *** | −3.521734 (0.0014) *** | 0.605306; 0.523645 4.469911; 4.777817 1.920124; 7.412441 −1.211647 Lag = 2 |
| | Y→X | 1.728108 (0.4214) | 1.540901 (0.1342) | 0.187383; 0.019256 21.08895; 21.39686 2.078472; 1.114531 44.22952 Lag = 2 | 0.668911 (0.7157) | 0.489285 (0.6283) | 0.181072; 0.011638 20.82714; 21.13504 2.081673; 1.068691 573.1691 Lag = 2 | 1.422560 (0.4910) | 1.138375 (0.2643) | 0.319341; 0.178515 19.36244; 19.67035 1.772581; 2.267628 671.0844 Lag = 2 |
| | PTR_R | | | | | | | | | |
| | X→Y | 3.723506 (0.1554) | −6.472288 (0.0000) *** | 0.811471; 0.772465 5.296975; 5.604881 2.187910; 20.80376 −2.253099 Lag = 2 | 2.822869 (0.2438) | −4.968101 (0.0000) *** | 0.577532; 0.490125 4.586366; 4.894273 1.820342; 6.607370 −1.708182 Lag = 2 | 3.727361 (0.2924) | −3.778072 (0.0008) *** | 0.634194; 0.521638 4.532838; 4.932785 2.328615; 5.634494 −1.095677 Lag = 3 |
| | Y→X | 1.383438 (0.5007) | 1.431040 (0.1631) | 0.204225; 0.039582 4.455241; 4.763148 2.069618; 1.240412 0.850319 Lag = 2 | 0.218645 (0.8964) | 0.420556 (0.6772) | 0.186884; 0.018653 5.215656; 5.523562 2.109472; 1.110877 0.1980069 Lag = 2 | 1.056220 (0.5897) | 1.294459 (0.2057) | 0.233955; 0.075463 3.367587; 3.675493 2.208741; 1.476133 0.222972 Lag = 2 |

**Table 4.** *Cont.*

| Y = Price-to-Rent | | Hong Kong Island | | | Kowloon | | | New Territories | | |
|---|---|---|---|---|---|---|---|---|---|---|
| Group 4 | VA | | | | | | | | | |
| | X→Y | 1.326520<br>(0.5152) | −5.309510<br>(0.0000) *** | 0.746388;<br>0.693916<br>5.593529;<br>5.901436<br>1.969655;<br>14.22464<br>−2.021836<br>Lag = 2 | 2.067229<br>(0.3557) | −5.024951<br>(0.0000) *** | 0.568995;<br>0.479821<br>4.606372;<br>4.914279<br>1.884626;<br>6.380763<br>−1.672703<br>Lag = 2 | 0.561103<br>(0.7554) | NA | 0.020047;<br>−0.102447<br>4.636363;<br>4.854055<br>1.963868;<br>0.163658 |
| | Y→X | 1.975198<br>(0.5776) | 0.372068<br>(0.7129) | 0.715701;<br>0.628224<br>9.877112;<br>10.27706<br>1.855581;<br>8.181613<br>1.994489<br>Lag = 3 | 22.44526<br>(0.0021)<br>*** | 3.715394<br>(0.0023) *** | 0.931559;<br>0.853341<br>9.154412;<br>9.940792<br>2.163976;<br>11.90972<br>47.56728<br>Lag = 7 | 1.671632<br>(0.4335) | NA | 0.971420;<br>0.967847<br>9.973481;<br>10.19117<br>2.247396;<br>271.9114 |
| | IE | | | | | | | | | |
| | X→Y | 1.916654<br>(0.3835) | −5.286917<br>(0.0000) *** | 0.745101;<br>0.692363<br>5.598590;<br>5.906497<br>1.996232;<br>14.12843<br>−2.131167<br>Lag = 2 | 3.124283<br>(0.2097) | −5.162852<br>(0.0000) *** | 0.602126;<br>0.519807<br>4.526389;<br>4.834295<br>1.926339;<br>7.314551<br>−1.510751<br>Lag = 2 | 1.934065<br>(0.3802) | −3.923758<br>(0.0005) *** | 0.552655;<br>0.460101<br>4.595130;<br>4.903037<br>2.035659;<br>5.971155<br>−0.855768<br>Lag = 2 |
| | Y→X | 2.011355<br>(0.3658) | 0.702251<br>(0.4881) | 0.795408;<br>0.753079<br>8.151648;<br>8.459554<br>1.903162;<br>18.79097<br>1.434385<br>Lag = 2 | 3.970888<br>(0.1373) | 1.557829<br>(0.1301) | 0.806344;<br>0.766278<br>8.096713;<br>8.404620<br>2.019635;<br>20.12507<br>2.658745<br>Lag = 2 | 3.311911<br>(0.1909) | 1.287965<br>(0.2079) | 0.812432;<br>0.773625<br>8.064771;<br>8.372678<br>1.888818;<br>20.93516<br>5.286974<br>Lag = 2 |
| | W | | | | | | | | | |
| | X→Y | 3.405656<br>(0.1822) * | −5.378885<br>(0.0000) *** | 0.750835;<br>0.699283<br>5.575839;<br>5.883746<br>2.187224;<br>14.56478<br>−2.372866<br>Lag = 2 | 6.066062<br>(0.0482) ** | −5.137415<br>(0.0000) *** | 0.621716;<br>0.543450<br>4.475897;<br>4.783804<br>1.837567;<br>7.943667<br>−1.534625<br>Lag = 2 | 5.188325<br>(0.0747) * | −3.161005<br>(0.0037) *** | 0.552116;<br>0.459450<br>4.596335;<br>4.904241<br>1.900366;<br>5.958146<br>−0.174312<br>Lag = 2 |
| | Y→X | 1.863957<br>(0.3938) | 1.357643<br>(0.1850) | 0.750386;<br>0.698742<br>13.13369;<br>13.44160<br>2.022517;<br>14.52992<br>29.89178<br>Lag = 2 | 2.111863<br>(0.3479) | 1.158494<br>(0.2561) | 0.751795;<br>0.700442<br>13.12804;<br>13.43594<br>2.009519;<br>14.63982<br>33.89436<br>Lag = 2 | 2.581034<br>(0.2751) | 2.632681<br>(0.0134) ** | 0.792251;<br>0.749269<br>12.95011;<br>13.25802<br>2.064789;<br>18.43198<br>98.47509<br>Lag = 2 |

**Table 4.** *Cont.*

| | Y = Price-to-Rent | Hong Kong Island | | | Kowloon | | | New Territories | | |
|---|---|---|---|---|---|---|---|---|---|---|
| Group 4 | OE | | | | | | | | | |
| | X→Y | 0.474635<br>(0.7887) | −5.694094<br>(0.0000) *** | 0.763015;<br>0.713984<br>5.525719;<br>5.833625<br>2.038555;<br>15.56180<br>−2.306712<br>Lag = 2 | 0.928861<br>(0.6285) | −5.472595<br>(0.0000) *** | 0.609229;<br>0.528380<br>4.508375;<br>4.816281<br>1.946536;<br>7.535369<br>−1.535889<br>Lag = 2 | 14.87873<br>(0.0050) *** | −6.234057<br>(0.0000) *** | 0.764768;<br>0.662493<br>4.252232;<br>4.746054<br>1.888514;<br>7.477583<br>−3.376131<br>Lag = 4 |
| | Y→X | 31.32345<br>(0.0001)<br>*** | 1.745425<br>(0.1087) | 0.959275;<br>0.892633<br>14.30348;<br>15.19091<br>1.831260;<br>14.39452<br>163.5588<br>Lag = 8 | 22.19343<br>(0.0024)<br>*** | 3.367591<br>(0.0046) *** | 0.925420;<br>0.840186<br>14.75712;<br>15.54350<br>2.086138;<br>10.85742<br>1099.633<br>Lag = 7 | 20.79339<br>(0.0077) ** | 2.851538<br>(0.0158) ** | 0.936674;<br>0.833050<br>14.74493;<br>15.63235<br>2.304097;<br>9.039124<br>161.2888<br>Lag = 8 |
| | S | | | | | | | | | |
| | X→Y | 0.436888<br>(0.8038) | −5.696917<br>(0.0000) *** | 0.763285;<br>0.714309<br>5.524582;<br>5.832489<br>2.044707;<br>15.58499<br>−2.317066<br>Lag = 2 | 0.948955<br>(0.6222) | −5.463812<br>(0.0000) *** | 0.608442;<br>0.527430<br>4.510386;<br>4.818292<br>1.944401;<br>7.510521<br>−1.536109<br>Lag = 2 | 22.60465<br>(0.0002) *** | −6.082147<br>(0.0000) *** | 0.757631;<br>0.652253<br>4.282121;<br>4.775943<br>1.871033;<br>7.189665<br>−3.352625<br>Lag = 4 |
| | Y→X | 38.78412<br>(0.0000)<br>*** | 2.158196<br>(0.0539) * | 0.962532;<br>0.901219<br>14.25850;<br>15.14592<br>1.856848;<br>15.69889<br>165.1394<br>Lag = 8 | 31.14026<br>(0.0001)<br>*** | 2.960016<br>(0.0130) ** | 0.947979;<br>0.862854<br>14.58664;<br>15.47407<br>2.217405;<br>11.13635<br>1083.299<br>Lag = 8 | 17.33499<br>(0.0154) ** | 3.373241<br>(0.0045) *** | 0.914213;<br>0.816171<br>14.93146;<br>15.71784<br>2.626648;<br>9.324673<br>652.6142<br>Lag = 7 |
| Group 5 | Inverse GD | | | | | | | | | |
| | X→Y | 1.847946<br>(0.3969) | −4.079670<br>(0.0003) *** | 0.792944;<br>0.750105<br>5.390711;<br>5.698618<br>1.966966;<br>18.50982<br>−1.867908<br>Lag = 2 | 0.931476<br>(0.6277) | −4.692798<br>(0.0001) *** | 0.544498;<br>0.450256<br>4.661653;<br>4.969559<br>1.841403;<br>5.777667<br>−1.588144<br>Lag = 2 | 8.810966<br>(0.0660) * | −5.214551<br>(0.0000) *** | 0.737014;<br>0.622672<br>4.363760;<br>4.857583<br>1.879728;<br>6.445714<br>−2.286376<br>Lag = 4 |
| | Y→X | 8.347398<br>(0.0154)<br>** | −0.431743<br>(0.6691) | 0.487439;<br>0.381392<br>0.381392;<br>−18.43094<br>2.090009;<br>4.596438<br>$-1.14 \times 10^{-6}$<br>Lag = 2 | 1.322985<br>(0.5161) | −0.615597<br>(0.5430) | 0.336976;<br>0.199799<br>−18.48146;<br>−18.17355<br>2.169379;<br>2.456497<br>$-1.96 \times 10^{-6}$<br>Lag = 2 | 3.459276<br>(0.1773) | 2.976093<br>(0.0058) *** | 0.484575;<br>0.377935<br>−18.73328;<br>−18.42537<br>1.914375;<br>4.544041<br>$5.88 \times 10^{-6}$<br>Lag = 2 |

Note: *** denotes 1% statistical sig.; ** 5% sig.; * 10% sig. Full results are available upon request.

Second, the four liquidity-related variables, namely, RMB, M3, DXY, and his, are long-term leading indicators for the price-to-rent ratio in a Granger fashion. They tend to lead the DIRE market by two quarters, with an exemption of the exchange rate of RMB/HKD being ahead of the Hong Kong Island and New Territories industrial real estate market by five and four quarters, respectively. Third, it is statistically evident that the residential market and the industrial real estate market are causally correlated with the causal pathway running from the former to the latter across the three regions in Hong Kong. The average house prices, as well as the price-to-rent ratios, Granger-cause the pricing of the DIRE market.

Fourth, the results seem to suggest that the industrial sector specific attributes, including value-added, number of industrial establishments, size of workforce, operation expense, and sales, are prone to leading the DIRE market in terms of property pricing over a long time horizon, with these attributes showing unidirectional Granger causation with the price-to-rent variables. Lastly, Table 4 further reveals that the rate of the industrialisation process in China, proxied by the total amount of manufacturing output in the province of Guangdong, has a statistically significant negative impact on the price-to-rent ratio of the DIRE of Hong Kong. The inverse of GD Granger-causes the three price-to-rent variables in the long run, statistically significant at the 5% level.

## 5. Discussion

The findings stemming from the above cointegration and causality analyses reveal several empirically important observations pertaining to the market nature and fundamentals of the DIRE of Hong Kong. First and foremost, the DIRE market is highly cointegrated with other segments of the economy in terms of pricing, despite its shrinking economic significance over the past decades. For example, the general price level, employment level, and the year-on-year performance of the Hang Seng Index seem to be moving synchronously with the price-to-rent ratios of the DIRE over time, as the results of the cointegration analysis suggest. More crucially, the DIRE market appears to causally lag behind the general economy by circa two quarters for most economic indicators. From the perspective of price discovery, this could be attributed to the inefficient nature of the DIRE market, which is highly heterogenous and informationally untransparent. In the context of Hong Kong, there is no well-established industrial real estate agency that is specialised in collating and centralising market data and information of the DIRE the same manner we can observe for the residential property sector, which partially explains why the DIRE is not informationally transparent. In addition, the GDP variable showing a bidirectional Granger causal connection with the DIRE may indicate that the industrial sector and the general economy are inextricably linked. A growing economy could naturally, on one hand, result in a swelling demand for industrial goods through the effect of wealth accumulation, which in turn drives the prices of DIRE up. On the other hand, an expansion of the industrial sector could induce more demand for other economic activities such as accounting, legal services, and construction through a feedback loop, which explains the causality from DIRE to GDP. Indeed, recent developments in Hong Kong seem to have proven themselves to be a revelation of the observed bidirectional causal link between the industrial real estate sector and the general economy. A decade of strong and persistent economic growth of the city has induced more government-led high-end industrial property development programmes focusing chiefly on scientific innovation and technological advancements, with hectares of land such as Hong Kong Science Park designated for these specific land uses. The city has also witnessed a rapid expansion in the sectors of computer cloud servicing and data processing and storage, which has simultaneously fuelled demand for special-purpose industrial properties for data centres and warehouses, for instance.

Second, the findings of our Granger causality analysis seem to confirm the Quantity Theory of Money. The four liquidity variables, namely, M3, FDI, RMB, and 1/DXY, all exhibit a unidirectional lead–lag relationship with the price-to-rent variables. When the RMB appreciates against the HKD, or the DXY becomes weaker relative to other international currencies, HKD-denominated assets, including DIRE, would become more financially appealing to non-local buyers, causing the prices of DIRE to escalate. Along a similar line of reasoning, when the money supply (M3) and FDI increase within the economy of Hong Kong, prices of DIRE assets, which are denominated by HKD, would be buoyed by the influx of new liquidity. Consequently, the price-to-rent ratios of DIRE should also increase.

Third, the housing market and the industrial real estate market of Hong Kong are intertwined in terms of cointegration. Further, we observe a Granger causation running from the two housing market variables to the DIRE price-to-rent ratios across the three districts. One possible explanation for such a causal observation is that land supply for residential development is extremely limited in Hong Kong. The scarcity of developable land more often than not drives residential property prices up to an exorbitant level, making the remaining undeveloped land even more scarce and valuable. Therefore, some developers in Hong Kong would turn to converting existing industrial or commercial properties into residential housing units through the rezoning of land and/or lease modifications, resulting in swelling the prices of DIRE. Upon closer examination of the data, the price-to-rent ratios of industrial properties have increased by circa 460% in the New Territories over the sample period, seemingly confirming the persistently strong demand for this category of real estate.

Another important revelation of our empirical results is that we can indeed use certain information that is specific to the industrial/manufacturing sector to predict the movement of the price-to-rent ratios of the DIRE. The five industry variables, namely, value-added, number of industrial establishments, total industrial workforce, operating expenses, and sales, are all causally correlated with the DIRE variables in a Granger fashion. In other words, the growth of the industrial sector, which is, generally or by definition, accompanied by a higher level of output, a more intense formation of establishments, larger numbers of workers, and more capital expenditure for investment and revenue, could lead to an elevation of the price-to-rent ratio, but not the other way around.

Last but not least, the Granger causality analysis on the variable GD unequivocally confirms the commonly held view that the process of industrialisation in mainland China has a long-lasting negative impact upon the pricing of industrial real estate in Hong Kong. The larger the output of the manufacturing sector in Guangdong, the more the prices of DIRE in Hong Kong are depressed relative to rents. As aforementioned, this could be attributable to the great disparity of land and property prices between the two places, resulting in a clear yet sophisticated division of labour in production between the two places, with the mainland specialising in the actual hand-on manufacturing whilst Hong Kong plays a managerial role by providing tertiary-level services along the production line to support the industrial operations in China. Indeed, the average industrial land prices of Guangdong have skyrocketed over the past ten to fifteen years. Take Shenzhen, a neighbour city of Hong Kong, as an example: the average industrial land prices of the special economic zone of China were circa RMB 500 at the beginning of the market cycle in 2008, with prices topped out at circa RMB 4500 in 2019, posting an eight-fold increase (Source: CEIC Data (see https://www.ceicdata.com/en/china/land-price-city-industrial/cn-land-price-industrial-shenzhen, accessed on 1 August 2022), whilst the Hong Kong industrial land market observed a comparatively more modest growth of circa 400% during the same time period.

## 6. Conclusions

This study has made at least two contributions to the literature of industrial real estate. First, it dissects the pricing of the industrial real estate market using the price-to-rent ratio, which has not yet been thoroughly explored in the existing literature. The affordability/overvaluation of real estate, which is commonly measured by the price-to-rent ratio, is generally examined in the context of residential property, and to a lesser extent, commercial office property. Relatively little is known about the dynamics between price and rent in the field of DIRE. Second, it empirically examines the cointegration and lead–lag relationships between the price-to-rent ratio and a large array of macroeconomic and financial determinants in a holistic and systematic fashion using the property and economic data of the Hong Kong property market. The results reveal that the industrial property market in Hong Kong is generally informationally inefficient, which can, to a large degree, be explained or predicted by sector-level and economy-wide fundamentals using Granger causality techniques. Specifically, macroeconomic attributes such as employment rates, money supply, FDI, inflation, equity market, and industrial sector-specific factors including sales and value-added are observed to Granger-cause the price-to-rent movements of the DIRE. Further, GDP and the pricing of the DIRE are interlinked by bidirectional Granger causal links.

The findings emanating from this study carry significant and potentially far-reaching practical implications for investors and policy makers. For example, our causality analysis suggests that the price-to-rent ratio of DIRE is a lagging indicator of a number of macroeconomic attributes including housing prices and the stock market index, providing valuable insights to traders or investors keen to develop arbitrage investment strategies to exploit the market. In addition, from a policy-making point of view, a more thorough conceptualisation of the determinants of the industrial land market should help policy makers design and formulate policies and regulations that are in the interest of the long-term sustainable development of the industrial sector and its stakeholders. For instance, our analysis demonstrates that the performance of DIRE is closely causally linked to the amount of foreign direct investment. Policy makers intent on revitalising the industrial real estate market in Hong Kong could perhaps consider introducing more pro-FDI policy measures within the DIRE sector through offering tax incentives and/or providing a more level playing field amongst local and foreign-market players. In addition, our findings appear to suggest that the performance of the industrial sector is a leading indicator of the DIRE, signalling that if the government is keen to reinvigorate the traditional industrial land market, an indirect way would be to provide financial support to the industrialists and/or facilitate the development of the manufacturing sector, whose expansion would, in turn, reinforce the growth of the DIRE in the long run in terms of pricing.

Whilst the paper empirically reveals significant lead–lag relationships between the pricing of DIRE and its macroeconomic attributes, shedding new light on issues around appraisals and market forecast of industrial real estate, we believe future research efforts in the area of research could be devoted to exploring the micro-spatial dynamics between a given industrial market and its economic, social, and/or demographic determinants using geo-referenced or spatially granular information. The classification of the three districts in the current study is based on a set of politically-imposed geographical boundaries, not on the actual underlying economic working or fundamentals of the markets. With the use of GIS techniques, the effect of macroeconomic and other factors affecting the DIRE could be identified and measured in a more spatially explicit manner. Lastly, the cointegration and causal relationships examined in this study could be re-evaluated by employing a multivariate modelling approach in future studies when data of a sufficiently long time span are available. Given the relatively small sample size and the associated statistical complications arising from issues such as degree of freedom, a bivariate analytical framework is chosen in this study, which could potentially ignore indirect channels through which cointegration linkages and causal associations amongst the variables could be formed (we thank a reviewer of our article for highlighting the issue concerning the bi-variability of our models) [48,49].

**Author Contributions:** Conceptualization, D.L. and Y.Y.; methodology, D.L. and M.M.; software, D.L.; validation, D.L., Y.Y., M.M. and M.H.; formal analysis, D.L. and M.M.; investigation, D.L., Y.Y., M.M. and M.H.; resources, D.L.; data curation, D.L.; writing—original draft preparation, D.L.; writing—review and editing, Y.Y., M.M. and M.H.; visualization, Y.Y., M.M. and M.H.; supervision, D.L.; project administration, D.L.; funding acquisition, D.L. All authors have read and agreed to the published version of the manuscript.

**Funding:** This research received no external funding.

**Institutional Review Board Statement:** Not applicable.

**Informed Consent Statement:** Not applicable.

**Data Availability Statement:** The data and information utilised in this research are publicly available on the websites of the Rating and Valuation Department and Census and Statistics Department of the Hong Kong Government. Please refer to Table 1 above.

**Conflicts of Interest:** The authors declare no conflict of interest.

## Appendix A

**Table A1.** Full results of ADF tests.

| Variable | t-statistic (level) Prob (level) t-statistic (1st diff) Prob (1st diff) | Variable | t-statistics (level) Prob (level) t-statistic (1st diff) Prob (1st diff) | Variable | t-statistic (level) Prob (level) t-statistic (1st diff) Prob (1st diff) |
|---|---|---|---|---|---|
| PR(HK) | −2.756358<br>(0.0740) *<br>−8.830476<br>(0.0000) *** | 1/DXY | −1.189171<br>(0.6695)<br>−6.773987<br>(0.0000) *** | PtR_R(NT) | −6.381109<br>(0.0000) ***<br>−8.379050<br>(0.0000) *** |
| PR(K) | −2.776863<br>(0.0711) *<br>−5.448885<br>(0.0001) *** | FDI | −2.636348<br>(0.0682) *<br>−6.559269<br>(0.0000) *** | VA | −2.731652<br>(0.0781) *<br>−6.041304<br>(0.0000) *** |
| PR(NT) | −2.375346<br>(0.1551)<br>−5.433777<br>(0.0001) *** | M3 | −0.892727<br>(0.7800)<br>−4.934122<br>(0.0003) *** | IE | −1.235322<br>(0.6490)<br>−6.047171<br>(0.0000) *** |
| I | −1.210615<br>(0.6599)<br>−6.027234<br>(0.0000) *** | RHP(HK) | −1.035658<br>(0.7309)<br>−5.990680<br>(0.0000) *** | W | −1.163165<br>(0.6802)<br>−6.062103<br>(0.0000) *** |
| ER | −3.024789<br>(0.0415) **<br>−5.027312<br>(0.0044) *** | PHP(K) | −1.063254<br>(0.7202)<br>−1.063254<br>(0.7202) | OE | −2.793364<br>(0.0687) *<br>−5.993349<br>(0.0000) *** |
| GDP | −1.534898<br>(0.5057)<br>−7.305895<br>(0.0000) *** | PHP(NT) | −0.730942<br>(0.8264)<br>−4.812043<br>(0.0004) *** | S | −2.786345<br>(0.0697) *<br>−6.006693<br>(0.0000) *** |
| HSI | −3.626772<br>(0.0100) ***<br>−7.468470<br>(0.0000) *** | PtR_R(HK) | −7.176993<br>(0.0000) ***<br>−9.832806<br>(0.0000) *** | 1/GD | −1.986720<br>(0.2912)<br>−5.963282<br>(0.0000) *** |
| RMB | −1.347538<br>(0.5976)<br>−5.334136<br>(0.0001) *** | PtR_R(K) | −3.049188<br>(0.0391) **<br>−8.880541<br>(0.0000) *** | | |

Note: *** denotes 1% statistical sig.; ** 5% sig.; * 10% sig. with lag length determined by SIC.

## References


1. Benjamin, J.; Zietz, E.; Sirmans, S. The environment and performance of industrial real estate. *J. Real Estate Lit.* **2003**, *11*, 279–324. [CrossRef]
2. Henneberry, J. Conflict in the industrial property market. *Town Plan. Rev.* **1988**, *59*, 241. [CrossRef]
3. Lin, S.W.; Ben, T.M. Impact of government and industrial agglomeration on industrial land prices: A Taiwanese case study. *Habitat Int.* **2009**, *33*, 412–418. [CrossRef]
4. Arauzo Carod, J.M. Determinants of industrial location: An application for Catalan municipalities. *Pap. Reg. Sci.* **2005**, *84*, 105–120. [CrossRef]
5. Martin, P.; Rogers, C.A. Industrial location and public infrastructure. *J. Int. Econ.* **1995**, *39*, 335–351. [CrossRef]
6. Hsia, M.; Green, M. Briefing: European industrial property location. *Prop. Manag.* **1991**, *9*, 51–65. [CrossRef]
7. Ayuso, J.; Restoy, F. House prices and rents: An equilibrium asset pricing approach. *J. Empir. Financ.* **2006**, *13*, 371–388. [CrossRef]
8. Lo, D.; McCord, M.; McCord, J.; Davis, P.; Haran, M. Rent or buy, what are the odds? Analysing the price-to-rent ratio for housing types within the Northern Ireland housing market. *Int. J. Hous. Mark. Anal.* **2021**, *14*, 1062–1091. [CrossRef]
9. Hoolachan, J.; McKee, K.; Moore, T.; Soaita, A.M. Generation rent and the ability to settle down: Economic and geographical variation in young people's housing transitions. *J. Youth Stud.* **2017**, *20*, 63–78. [CrossRef]
10. McKee, K.; Soaita, A.M.; Hoolachan, J. Generation rent and the emotions of private renting: Self-worth, status and insecurity amongst low-income renters. *Hous. Stud.* **2020**, *35*, 1468–1487. [CrossRef]
11. Anacker, K.B. Introduction: Housing affordability and affordable housing. *Int. J. Hous. Policy* **2019**, *19*, 1–16. [CrossRef]

12. Wetzstein, S. Assessing post-GFC housing affordability interventions: A qualitative exploration across five international cities. *Int. J. Hous. Policy* **2021**, *21*, 70–102. [CrossRef]
13. Akinsomi, O.; Mkhabela, N.; Taderera, M. The role of macro-economic indicators in explaining direct commercial real estate returns: Evidence from South Africa. *J. Prop. Res.* **2018**, *35*, 28–52. [CrossRef]
14. Duca, J.V.; Ling, D.C. The other (commercial) real estate boom and bust: The effect of risk premia and regulatory capital arbitrage. *J. Bank. Financ.* **2020**, *112*, 105317. [CrossRef]
15. Lo, D.; McCord, M.; Davis, P.T.; McCord, J.; Haran, M. Causal relationships between the price-to-rent ratio and macroeconomic factors: A UK perspective. *J. Prop. Investig. Financ.* 2022; *forthcoming*. [CrossRef]
16. Lo, D.; Yau, Y.; McCord, M.; Haran, M. Lead-lag relationship between the price-to-rent ratio and the macroeconomy: An empirical study of the residential market of Hong Kong. *Buildings* **2022**, *12*, 1345. [CrossRef]
17. DiPasquale, D.; Wheaton, W.C. The markets for real estate assets and space: A conceptual framework. *Real Estate Econ.* **1992**, *20*, 181–198. [CrossRef]
18. Sommer, K.; Sullivan, P.; Verbrugge, R. *Run-Up in the House Price-Rent Ratio: How Much Can Be Explained by Fundamentals?* US Department of Labor, Bureau of Labor Statistics, Office of Prices and Living Conditions: Washington, DC, USA, 2010.
19. Kishor, N.K.; Morley, J. What factors drive the price-rent ratio for the housing market? A modified present-value analysis. *J. Econ. Dyn. Control* **2015**, *58*, 235–249. [CrossRef]
20. Campbell, J.Y.; Shiller, R. The dividend price ratio and expectations of future dividends and discount factors. *Rev. Financ. Stud.* **1988**, *1*, 195–288. [CrossRef]
21. Beltratti, A.; Morana, C. International house prices and macroeconomic fluctuations. *J. Bank. Financ.* **2010**, *34*, 33–545. [CrossRef]
22. Manganelli, B.; Morano, P.; Tajani, F. House prices and rents. The Italian experience. *WSEAS Trans. Bus. Econ.* **2014**, *11*, 219–226.
23. Bracke, P. Housing prices and rents: Microevidence from a matched data set in central London. *Real Estate Econ.* **2015**, *43*, 403–431. [CrossRef]
24. Clark, S.; Lomax, N. Rent/price ratio for English housing sub-markets using matched sales and rental data. *Am. Real Estate Assoc.* **2020**, *52*, 136–147. [CrossRef]
25. McCord, M.; Lo, D.; McCord, J.; Davis, P.; Haran, M. Measuring the cointegration of housing types in Northern Ireland. *J. Prop. Res.* **2019**, *36*, 343–366. [CrossRef]
26. Lo, D.; Liu, N.; McCord, M.; Haran, M. Information transparency and pricing strategy in the Scottish housing market. *Int. J. Hous. Mark. Anal.* **2022**, *15*, 429–450. [CrossRef]
27. Ambrose, B. An analysis of the factors affecting light industrial property valuation. *J. Real Estate Res.* **1990**, *5*, 355–370. [CrossRef]
28. Atteberry, W.L.; Rutherford, R.C. Industrial real estate prices and market efficiency. *J. Real Estate Res.* **1993**, *8*, 377–385. [CrossRef]
29. Chai, Y.W. Determinants of NOI for warehouse properties. *Real Estate Financ.* **1997**, *14*, 48–54.
30. Mueller, G. Real estate rental growth rates at different points in the physical market cycle. *J. Real Estate Res.* **1999**, *18*, 131–150. [CrossRef]
31. Wang, J.M.; Shi, Y.F.; Zhang, J. Energy efficiency and influencing factors analysis on Beijing industrial sectors. *J. Clean. Prod.* **2017**, *167*, 653–664. [CrossRef]
32. Thompson, B.; Tsoolacos, S. Rent adjustments and forecasts in the industrial market. *J. Real Estate Res.* **1990**, *17*, 151–167. [CrossRef]
33. Jones, C.; Orr, A. Local commercial and industrial rental trends and property market constraints. *Urban Stud.* **1999**, *36*, 223–237. [CrossRef]
34. Jackson, C.; White, M. Inflation and rental change in industry property: A multi-level analysis. *J. Prop. Investig. Financ.* **2005**, *23*, 342–363. [CrossRef]
35. De Wit, I.; Van Dijik, R. The global determinants of direct office real estate returns. *J. Real Estate Financ. Econ.* **2003**, *26*, 24–45. [CrossRef]
36. Karakozova, O. Modelling and Forecasting Property Rents and Returns. Ph.D. Thesis, Swedish School of Economics and Business Administration, Helsinki, Finland, 2005.
37. Lieser, K.; Groh, P.A. The determinants of international commercial real estate investment. *J. Real Estate Financ. Econ.* **2014**, *48*, 611–659. [CrossRef]
38. Ho, K.H.; Faishal bin Ibrahim, M. Explaining the macro-economy and retail real estate sector dynamic interaction between prime and suburban retail real estate sectors. *J. Prop. Investig. Financ.* **2010**, *28*, 77–108.
39. Thibodeau, T.G. Estimating the effect of high-rise office buildings. *Land Econ.* **1990**, *66*, 402–408. [CrossRef]
40. Gyourko, J. Understanding commercial real estate: How different from housing is it? *J. Portf. Manag.* **2009**, *35*, 23–37. [CrossRef]
41. Granger, C.W.J.; Newbold, P. Spurious regressions in econometrics. *J. Econom.* **1974**, *2*, 111–120. [CrossRef]
42. Engle, R.F.; Granger, C.W.J. Cointegration and error correction: Representation, estimation and testing. *Econometrica* **1987**, *55*, 251–276. [CrossRef]
43. Johansen, S. Estimation and Hypotheses Testing of Co-integration Vectors in Gaussian Vector Autoregressive Models. *Econometrica* **1991**, *59*, 1551–1580. [CrossRef]
44. Johansen, S. *Likelihood-Based Inference in Cointegrated Vector Autoregressive Models*; Oxford University Press: Oxford, UK, 1995.
45. Granger, C.W.J. Some recent developments in a concept of causality. *J. Econom.* **1988**, *39*, 199–211. [CrossRef]
46. Thomas, R.L. *Modern Econometrics: An Introduction*; Addison-Wesley: Wokingham, UK, 1997.

47. Hamilton, J.D. *Time Series Analysis*; Princeton University: Princeton, NJ, USA, 1994.
48. Janakiramanan, S.; Lamba, A.S. An empirical examination of linkages between Pacific-Basin stock markets. *J. Int. Financ. Mark. Inst. Money* **1998**, *8*, 155–173. [CrossRef]
49. Yang, J.; Kolari, J.W.; Min, I. Stock market integration and financial crises: The case of Asia. *Appl. Financ. Econ.* **2003**, *13*, 477–486. [CrossRef]

*land*

MDPI

*Article*

# Exploring the Role of Transit Ridership as a Proxy for Regional Centrality in Moderating the Relationship between the 3Ds and Street-Level Pedestrian Volume: Evidence from Seoul, Korea

**Seung-Nam Kim [1,*], Juwon Chung [2] and Junseung Lee [3]**

1 Department of Urban Design and Studies (209-707), Chung-Ang University, 84 Heukseok-ro, Dongjak-gu, Seoul 06974, Korea
2 Department of Urban Planning and Design Research, The Seoul Institute, 57 Nambusunhwan-ro, 340-gil, Seocho-gu, Seoul 06756, Korea
3 Department of Urban Design and Studies (207-737), Chung-Ang University, 84 Heukseok-ro, Dongjak-gu, Seoul 06974, Korea
* Correspondence: snkim@cau.ac.kr; Tel.: +82-2-82-5377

**Abstract:** The preference for walking and the resulting pedestrian activities have been considered key success factors for streets, neighborhoods, and cities alike. Although micro- and meso-scale built environment factors that encourage walking have been investigated, the role of macroscopic factors such as regional centrality in explaining street-level pedestrian volume is often neglected. Against this backdrop, this study examines the relationship between built environments and street-level pedestrian volume using Smart Card and pedestrian volume survey data from Seoul after controlling for transport ridership as a proxy for regional centrality. As a preliminary study, we analyzed 36 regression models applying different sets of transit ridership variables and found that the combination of bus ridership within 400 m and subway ridership within 300 m best explained the variation in pedestrian volume on a street. Then, the effects of the 3D variables (density, diversity, and design) on pedestrian volume were compared before and after controlling for ridership within this spatial range. The results demonstrated that, after taking transit ridership into account, the influence of built environment variables is generally reduced, and the decrease is more pronounced among walkshed-level 3D variables than street-level variables. Particularly, while the effect of "design" (street connectivity) on pedestrian volume appeared to be negatively significant in the constrained model, it was found to be insignificant in the unconstrained model which controlled for transit ridership. This suggests that the degree of street connectivity is influenced by regional centrality, and accordingly, the coefficient of the "design" variable in our constrained model might be biased. Thus, to accurately understand the effect of the meso-scale 3D variables on pedestrian volume, both micro- and macro-scale built environmental factors should be controlled.

**Keywords:** walking; pedestrian volume; built environment; land use; transit ridership; regional centrality; Seoul

**Citation:** Kim, S.-N.; Chung, J.; Lee, J. Exploring the Role of Transit Ridership as a Proxy for Regional Centrality in Moderating the Relationship between the 3Ds and Street-Level Pedestrian Volume: Evidence from Seoul, Korea. *Land* **2022**, *11*, 1749. https://doi.org/10.3390/land11101749

Academic Editors: Bindong Sun, Tinglin Zhang, Wan Li, Chun Yin and Honghuan Gu

Received: 17 August 2022
Accepted: 1 October 2022
Published: 9 October 2022

**Publisher's Note:** MDPI stays neutral with regard to jurisdictional claims in published maps and institutional affiliations.



## 1. Introduction

Walking has received significant attention in urban and transportation planning as the most traditional, universal, affordable, sustainable, inclusive, and even irreplaceable mode of travel [1–3]. As the reckless spread of urban sprawl increases car dependencies and human-made greenhouse gas emissions accelerate climate change, researchers have further emphasized the importance of walking as a means of transportation [4–7]. Furthermore, walking determines various outcomes for individuals and cities. It affects individual health and social networks, and determines street safety and attractiveness, neighborhood livability and vitality, and even the prosperity of regional and state economies [8–16]. Preference for walking and the resulting pedestrian volume in public spaces is undoubtedly a crucial success factor for streets, neighborhoods, and cities [8,17].

Accordingly, urban design and transportation researchers have actively examined the physical conditions that encourage walking. For example, Jacobs [8] emphasized the importance of sidewalks, parks, land use mix, small urban blocks, and old buildings as prerequisites for walking and street vitalization. Ewing and Handy [18] suggested a conceptual framework explaining the link between the built environment and how an individual feels about the environment as a place to walk. From a more empirical point of view, studies have analyzed the relationship between the built environment and street-level pedestrian volume [19,20].

However, previous studies have seldom deliberated on the macroscopic mechanism that determines the pedestrian volume at a specific street segment and have mainly focused on the influences of the microscopic built environment alone. The pedestrian volume measured on a particular street includes the number of people that happen to be just passing by that point; the number is likely to be determined by the regional centrality and resulting background floating population before the microscopic built environment of that place. The effect of the built environment on the number of pedestrians may vary greatly depending on whether such regional centrality is controlled for, but previous studies have rarely considered this factor. This research gap has been derived from the absence of appropriate methods to control for regional centrality or the resulting background floating population. Here, the transit ridership data would be a reliable alternative to fill this gap in a city such as Seoul, Korea, where public transportation serves a wide range of metropolitan areas and almost everyone uses a Smart Card for payment on that service.

Against this backdrop, using Smart Card and pedestrian volume survey (PVS) data from Seoul, this study aims to analyze the relationship between the built environment and street-level pedestrian volume after controlling for transit ridership as a proxy for regional centrality. Specifically, this study compares the effect of traditional 3D variables (density, diversity, and design) on pedestrian volume between the constrained (excluding transit ridership variables) and unconstrained models (including transit ridership variables). Since the amount of walking on a street determines an individual's perception of the street environment as well as the outcomes of surrounding areas, the results of this study are expected to contribute not only to more accurately estimating the pedestrian volume but also to predicting the success of the street, neighborhood, and city.

## 2. Theoretical Background and Literature Review

### 2.1. Pathways from Built Environment to Street-Level Pedestrian Volume: A Conceptual Framework

Traditional travel behavior studies regarding mode choice and trip frequency have focused on the role of the built environment in enhancing active transportation. In particular, Cervero and Kockelman [21] suggested the 3D variables: density, diversity, and design. Following this, Ewing and Cervero [22] included two additional D variables: destination accessibility and distance to public transit. Based on these variables, several studies empirically investigated the effect of 3D to 5D variables after controlling for other factors [21–29]. Studies have also examined the role of microscopic streetscape elements such as architectural details and aesthetics, street furniture, pavement design, microclimate, and safety [30–35].

However, the theoretical framework of pedestrian volume studies is distinct from individual-level walking behavior studies. Contrary to urban design theorists' hopeful belief that more people visit and stay in better places [33,36,37], the number of pedestrians measured on a street (i.e., pedestrian volume) is determined by a wide variety of environmental factors including the built form, weather conditions, and even socio-cultural background [16,29,30,38]. From the viewpoint of transportation planning, to explain the pedestrian volume on a specific street, it is necessary to understand all the factors influencing trip generation, mode choice, and even the route choice of people around that place. Specifically, it is important to consider not only the impact of neighborhood-level walkability and urban design quality but also the effect of regional accessibility and centrality [39]. According to Chung et al. [16], the built environment's role appears differently depending on the activity type of pedestrians. Here, we expand this framework and examine the role of the built environment as a determinant of each behavioral type by scale (Figure 1).

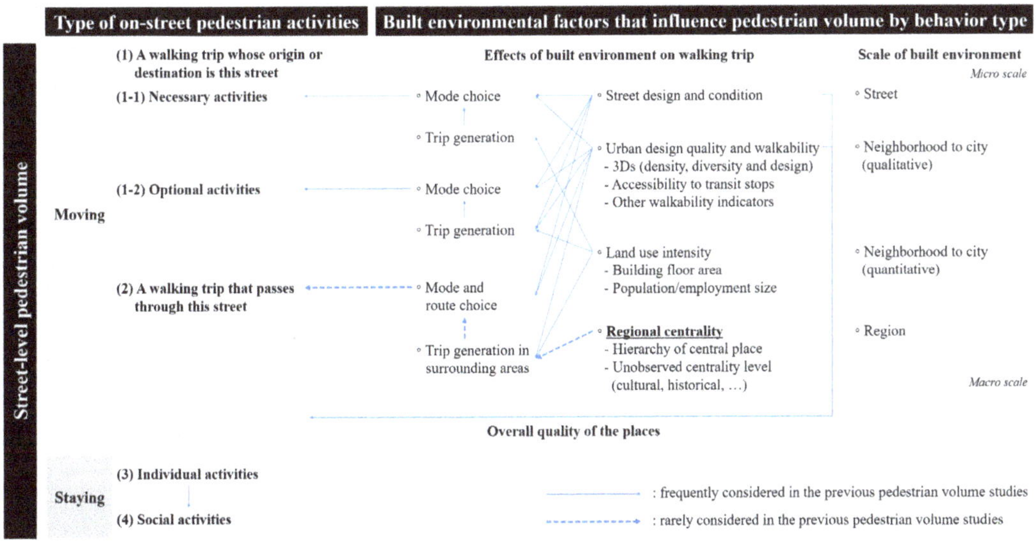


**Figure 1.** Pathways from built environment to street-level pedestrian volume: a conceptual framework.

Figure 1 illustrates a conceptual framework explaining the pathways from the built environment to street-level pedestrian volume. The pathways in this diagram are not only theoretically based, but also intuitively predictable. When studying pedestrian volume on a street, urban researchers tend to focus more on people who visit or stay at a place for a specific purpose [33,36,37,40]. However, when counting pedestrians on an actual street, a significant number of pedestrians that happen to be just passing by the place are also included regardless of the area's quality. Even though the number of people staying on the street is closely related to the quality of the place, many pedestrians may still pass by, in which case the connection to the built environment is relatively small [33]. Therefore, it is necessary to examine the determinants of walking in accordance with the type of pedestrian activities.

First, the pedestrian activities are divided into "moving" and "staying." For pedestrians on the "move", the activity can be classified broadly into two types: (1) the walking trip whose origin or destination is this street, and (2) the walking trip that passes through this street. Here, type (1) is classified again according to whether the purpose of the walking activity is necessary (1-1) or optional (1-2), and the built environmental factors affecting the activities also change accordingly. Necessary activities (1-1), such as commuting to work or school, are not greatly affected by the quality of the place [33]. Hence, trip generation of necessary activities is directly determined by land use intensity (i.e., employment size or building floor area) rather than the microscopic street environment [41]. However, the proportion of *walking* trips among the total necessary trips generated (mode choice) is affected by the street's design and condition and the urban design quality and walkability of surrounding areas. On the other hand, optional activities occurring in a desired place are more influenced by the built environment at the stage of trip generation (1-2) [33]. Land use intensity is a key driver, but the street- and neighborhood-level environment also matters. The factors determining the proportion of walking trips among the total generated trips are the same as those of necessary activities.

The number of walking trips passing through a specific street segment, type (2), is proportional to the amount of the floating population in the surrounding areas. We therefore need to consider the amount of trips generated in its surrounding areas rather than just that street segment [42]. The floating population is influenced not only by land use intensity and urban design quality, but also by more macroscopic physical aspects such as regional centrality by determining the total number of potential activities [39]. Even if the microscopic conditions of a specific area are the same, the number of trips generated in that area may vary depending on its historical, cultural, and economic centrality and status. Of course, how many of those trips are walking trips and how many of those walking trips will pass through the street (i.e., the internal capture of trips) are influenced by the urban design quality/walkability of surrounding areas and the design/condition of street segments [39].

The number of pedestrians performing "staying" activities (types 3 and 4), such as sitting on a bench or waiting for someone, is undoubtedly proportional to the total number of pedestrians moving in the area. Since only pedestrians on the move can stay, more moving activities cause more staying activities. Therefore, all the built environmental factors that generate moving activities indirectly generate staying activities at the same time. In addition, Gehl [33] argued that when there is a desirable place to stay, people who were passing by may also remain in that place; in a good built environment, these individual activities can develop into social activities, such as conversations and group activities. Thus, the conversion rate of moving activities to staying activities is closely related to the overall quality of the place [37,43].

Despite the above mechanism, the pedestrian volume on the street is just a single value, and the type of activity comprising that value cannot be known. Thereby, when explaining the variance in pedestrian volume using an econometric approach, all built environments ranging from the street (micro) to regional (macro) scales should be comprehensively considered. The next sub-section reviews the empirical evidence on this issue.

### 2.2. Empirical Evidence on the Relationship between Built Environment and Pedestrian Volume

Studies have empirically analyzed the relationship between the built environment and street-level pedestrian volume using data from diverse cities worldwide. Based on the conceptual framework explained in Figure 1, key findings of previous studies are summarized as follows.

Regarding the street-level built environment, Ewing et al. [25] examined the role of streetscape features to explain pedestrian volume using pedestrian count data from 588 blocks in New York City. They revealed that the design and use of adjacent buildings (i.e., proportion of first floor with windows, proportion of retail frontage, and proportion of active-use buildings) and street furniture encouraged pedestrian activities. In terms of specific street conditions, Rodríguez et al. [44] showed that sidewalk width was positively associated with pedestrian count in 338 street segments in Bogota. Using PVS data from Seoul, several studies also found that wider sidewalks and roadways, the existence of street furniture and crosswalks, and the absence of slopes and parked vehicles were crucial predictors of pedestrian volume [16,19,20,45–50].

The role of the 3D variables (density, diversity, and design) has also been the focus of previous studies. To measure those variables, studies have applied 50 to 500 m buffers from the survey spots, most frequently using 400 m (quarter mile). Although the specific type of measure is different (population density, job density, commercial density, floor area ratio, etc.), most studies confirmed that the higher the "density", the greater the pedestrian volume [25,26,44–46,50,51]. However, when two or more density variables were included in the model at the same time, as in Kang [20], conflicting signs of coefficients were also drawn. Here, when the spatial range for variable measurement is constant (e.g., 500 m buffer), both density and size (both population and employment) variables have the same meaning. Therefore, a couple of studies additionally controlled for the land use intensity using the building floor area variable [48]. Next, Lee et al. [45], Lee et al. [48], and Hajrasouliha and Yin [26] showed that a higher land use mix ("diversity") generates more walking activities. Lastly, studies also revealed a strong association between good street connectivity ("design") and pedestrian volume on the street by applying diverse forms of measures: road density [44], intersection density [26], intersection type (number of ways) [47,51], and average street length [51]. By contrast, the proportion of major arterials was found to be negatively associated with pedestrian volume [51].

In addition to the 3Ds, studies have also demonstrated that good public transit accessibility could be a key explanatory factor of pedestrian volume. Because people may walk a few blocks before boarding or alighting transport, pedestrian volume on the street is closely related to public transportation facilities in the surrounding areas [42]. While most studies have applied transit accessibility in the form of "existence/number of transit stops/stations within certain areas" or "distance to the nearest stops/stations" [20,25,45,47–49,51,52], Jang et al. [50] additionally considered the number of subway station entrances and subway/bus lines, daily service frequencies, and the distance decay effect.

Compared to the street- or neighborhood-level built environment, regional centrality has been rarely considered. Using PVS data from Seoul, Kang [20] investigated the effects of land-use accessibility and centrality on pedestrian count. His series of multilevel models showed that residential-, commercial-, and office-use accessibility and centrality, which were measured in terms of betweenness, straightness, and a gravity index, were positively associated with both weekday and Saturday pedestrian volume. However, regional centrality indicators using built environmental attributes are not a suitable variable to explain the variance in pedestrian volume as measured at different times. This is because the former is static, whereas the latter is day-dependent.

Thus, the surrounding area's transit ridership corresponding to the day when the PVS was taken has been considered as an alternative way to control for regional centrality. For instance, Rodríguez et al. [44] used the bus rapid transit (BRT) ridership to control for the BRT station demand itself in a model that explained the variation in pedestrian volume around the BRT stops. Using Smart Card data from Seoul, Chung et al. [16] attempted to control for the regional centrality of PVS spots by employing the bus and subway ridership of the surrounding areas to explain pedestrian volume. In both studies, transit ridership was identified as a robust estimator of walking activity. However, neither study tested how much spatial range should be considered when measuring the variable. Despite their different influences, Chung et al. [16] used a 400 m (1/4 mile) buffer for bus and subway en bloc. Unlike this, the present study firstly identifies the size of the influential area of transit stops/stations and then measures ridership within those identified areas as a proxy for regional centrality and the resulting floating population. Afterward, the study compares the effects of the traditional built environment factors (i.e., 3D variables) on pedestrian volume before and after controlling for transit ridership.

## 3. Empirical Setting

### 3.1. Study Area

The study area covers Seoul, the capital of South Korea, with a population of approximately 10 million (approximately 19.4% of the national population) over an area of 605 km$^2$ (0.6% of the national territory) as of 2019 [53]. As per the *2030 Seoul Master Plan* [54], Seoul pursues a polycentric spatial structure consisting of three central business districts (CBDs), seven regional centers, and twelve local centers. However, most employment is concentrated in the triangle-shaped area connecting the three CBDs, as can be illustrated through a regional job accessibility (RA) measure (Figure 2). (The RA measure was conceptually proposed by Hansen [55]. Since the RA of a given Transportation Analysis Zone (TAZ) is defined as the sum of jobs inversely weighted by their distance from the TAZ, TAZs with higher RAs represent the employment centers of the region [56]. To measure the RAs of 424 TAZs in Seoul, we used employment data from 1,138 TAZs not only in Seoul but also in its surrounding areas (i.e., Incheon and Gyeonggi Province) in 2015. Term $RA_i = \sum_j E_j \times e^{(\beta d_{ij})}$, where $RA_i$ is regional job accessibility in TAZ $i$; $E_j$ is total employees of TAZ $j$, $\beta$ is the distance resistance coefficient = 0.280 [57], and $d_{ij}$ is the distance between TAZ $i$ and $j$ [56].)

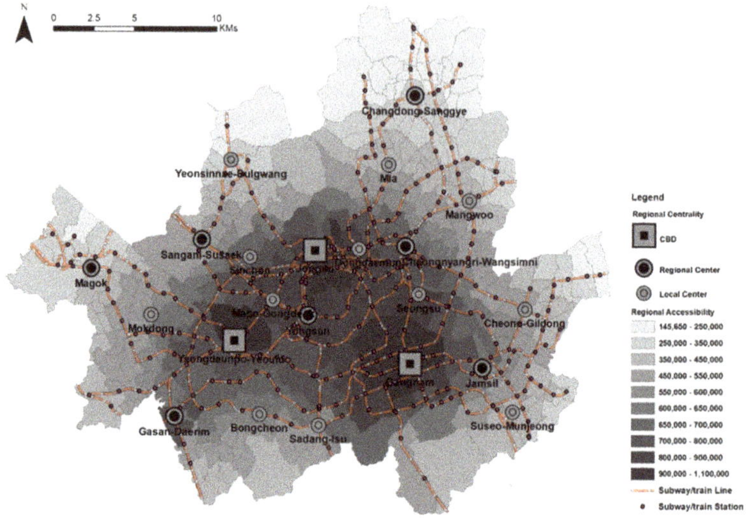


**Figure 2.** Map of Seoul: Location of central places suggested in the *2030 Seoul Master Plan* [54], regional job accessibility of 424 *Dong* (literally means a neighborhood in Korean, and it is the smallest administrative unit in Seoul. This spatial unit, *Dong*, is used as the TAZ in Korea) in 2015, and location of subway lines and stations in 2015.

Seoul's sizable population density (17,013 people/km$^2$) makes public transportation feasible [15,54]. In 2019, Seoul had 10 subway lines supporting 329 stations and 354 bus lines with 6254 stops [58–60] (Figure 2). The collective commuting modal share of public transport, walking, and cycling was relatively high at 77.8% in 2019 [61]. Seoul also has a very convenient and advanced payment system called "Smart Card" [48,62]. According to Lee [63], as of 2014, 99.02% of transit users paid for their fares with a Smart Card. Thus, transit ridership data collected through Smart Card usage information can be one of the most suitable proxies for the floating population and regional centrality in Seoul.

In addition, from 2009 to 2015, the Seoul Metropolitan Government investigated and provided quality data to the public for pedestrian volume studies. Since the data (i.e., PVS data) contain quite large samples covering almost all built-up areas of Seoul, which is one of the largest metropolitan cities, it has a variety of advantages over other research data worldwide [16]. Accordingly, the data have been widely used for pedestrian volume studies [16,19,20,45–50]. We believe that Seoul, which has a quality Smart Card system as well as transit ridership and PVS data, is a most suitable study area for this study.

### 3.2. Data and Variables

To explore the relationship between the built environment and street-level pedestrian volume, this study used the following data from Seoul, 2015: (1) PVS data, (2) transit ridership data, and (3) various spatial data for measuring the built environments. Specific measurements were selected based on the literature review previously provided (Section 2; Figure 1). The following sub-sections describe the study's data, measurements, and analysis method and model specifications.

#### 3.2.1. Pedestrian Volume (Dependent Variable)

To measure street-level pedestrian volume (the dependent variable), we used the 2015 PVS data from Seoul. The Seoul City Government and National Information Society Agency had jointly launched this survey in 2009, and it was conducted annually until 2015. Namely, the data used in this study were the most recent. The survey was conducted by trained investigators who employed manual counting in the field. As Chung et al. [16] argued, this approach has various advantages over self-report surveys or automated counting using GPS, sensors, and computer vision techniques.

In the 2015 survey, 1223 spots (street segments) were investigated. The spots were officially selected from an initial 10,000 representative spots selected by the Seoul Metropolitan Government in 2009, including (1) arterial roads, (2) streets in CBDs, regional centers, and local centers, (3) streets in low-rise residential and mixed-use areas, and (4) roads along the Han River [64]. Using cluster analysis, the 10,000 spots were clustered so that 5 to 20 spots were included in one cluster based on land use and location (longitude and latitude coordinates) [65]. Then, in each annual survey, at least one survey spot candidate was randomly selected from each cluster, and the spots were finalized after confirming both the possibility of securing a view and the presence of a place to avoid sunlight through a preliminary field survey [64,65]. As a result, the survey spots covered a considerable portion of built-up urban areas in Seoul, as shown in Figure 3. Each street segment was investigated three times a week (either Tuesday or Thursday; Friday; and Saturday) from 2–31 October 31. On each day, every pedestrian was counted from 07:30 to 19:30. Because the survey was not conducted on a single day, it is necessary to consider factors that might vary depending on the survey date in the model specification process.

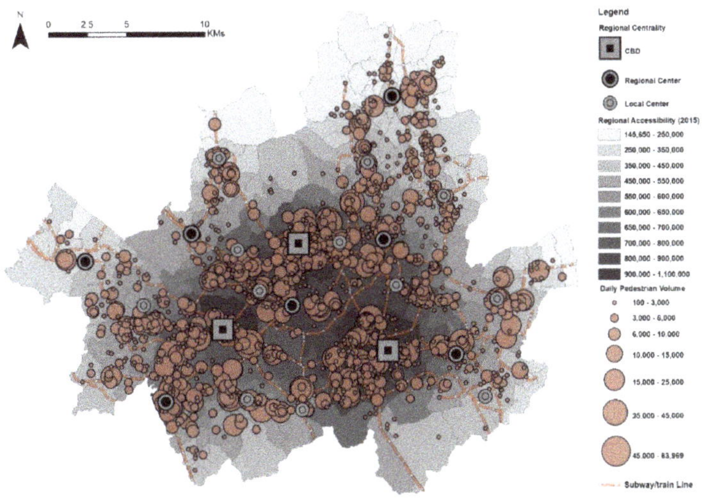


**Figure 3.** Central places in 2014 [54], regional job accessibility in 2015 (authors' calculation), and 2015 PVS spots and their results (2015 PVS data).

The data cleaning procedures followed those of Chung et al. [16]. First, we excluded the data gathered on Thursdays as the survey was conducted on only two days (15–22 October) as a supplementary survey for the Tuesday survey, and the number of spots included was quite small (one and three spots, respectively). Second, we also excluded the data surveyed from 28–31 October, which were not provided by the Seoul Open Data Plaza due to data processing errors. Finally, to render the distribution closer to normality, we used the log-transformed daily pedestrian volume of 2889 observations counted at 1162 survey spots as a dependent variable. Since the data contained the exact X- and Y-coordinates and various physical conditions of the survey spots, we were able to analyze the effect of the specific built environment on street-level pedestrian volume.

### 3.2.2. Transit Ridership (Proxy for Regional Centrality)

As discussed in Section 2, using Smart Card data, we controlled for daily transit ridership of the PVS days as a proxy for regional centrality. The transit ridership variable is expected to contribute in two major ways to explaining the variation in street-level pedestrian volume. First, because people may walk before boarding and after getting off a transit [42], transit riders of surrounding areas naturally/directly become pedestrians on the street. Second, transit ridership can be a proxy variable representing the level of centrality of that area. In other words, in Seoul, a large number of transit users in an area means that there is a larger floating population in the surrounding areas. Either way, transit ridership is inevitably closely related to the floating population, and as explained in Figure 1, the floating population can potentially be connected to the amount of walking trips that pass through a certain street, alongside the consequent staying activities.

Figure 4 shows that subway stations and their ridership are concentrated in the CBD triangle area. Bus stops are relatively more distributed, but major stops with the highest ridership are populated in major central places, particularly Gangnam CBD area (Figure 5). These support our premise that transit ridership can be a useful proxy for regional centrality. In addition, given the characteristic of the dependent variable being composed of non-single survey days, transit ridership, which is a time-dependent variable, is more suitable for controlling for regional centrality than other static variables such as regional job accessibility and central place hierarchy, and thereby addresses the limitations of prior studies using PVS data.

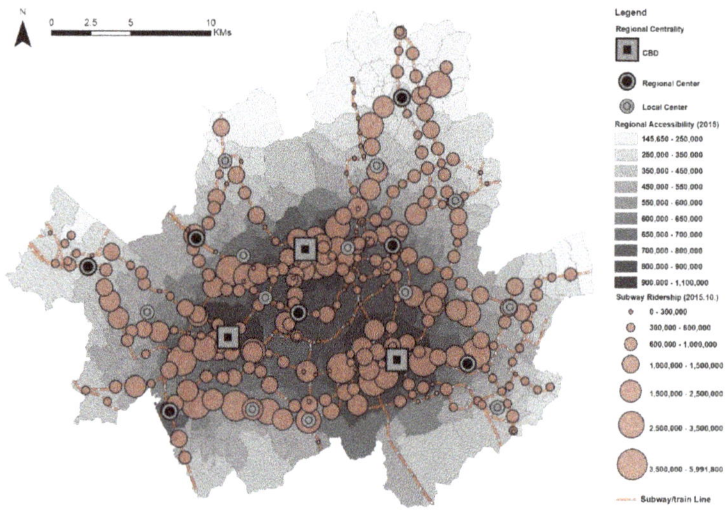


**Figure 4.** Central places in 2014 [54], regional job accessibility in 2015 (authors' calculation), and total subway ridership in October 2015 (2015 subway ridership data).

Specifically, the log-transformed daily total number of passengers boarding and alighting at all bus stops and subway stations within certain areas of the survey spot on the survey day was measured as an independent variable in the model. Here, it is important to determine the transit stops' influential area from the PVS spot. In contrast to the conventionally used 400 m to 500 m walkshed in previous walkability studies when measuring the built environment, there is little empirical guidance regarding the spatial extent of transit ridership that best explains the pedestrian volume on a street. Accordingly, in Section 4.1, we identify the size of the influential areas of bus stops and subway stations, respectively, by successively applying 100 to 600 m buffers at 100 m intervals for measuring each transit's ridership, and finding the ridership set with the highest explanatory power for a preliminary regression model among the 36 possible combinations.

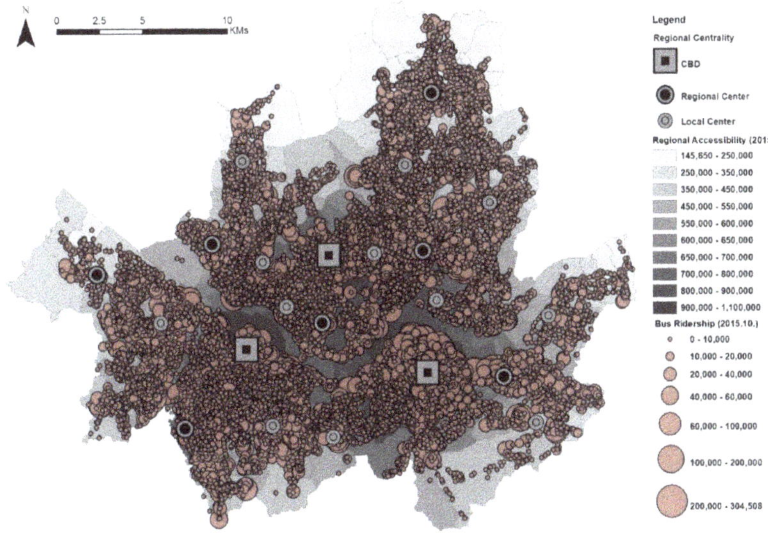


**Figure 5.** Central places in 2014 [54], regional job accessibility in 2015 (authors' calculation), and total bus ridership in October 2015 (2015 bus ridership data, only 12 PVS days included).

### 3.2.3. Walkshed-Level Built Environment (3D) Variables

As for walkshed-level built environment variables, we measured and employed the 3D variables within 500 m of the survey spot. Unlike the size of the transit stations' influential areas, the size of the walkshed is generally agreed in previous studies to be approximately a 400 m to 500 m buffer [20,66,67].

First, for the "density" measure, we controlled for both population and job density (the number of residents and workers in the 500 m walkshed). Since the area of the walkshed is constant, this measure has the same meaning as controlling a fixed population (i.e., population and employment size). As explained in Section 2, density variables are applied in almost all previous studies (e.g., [15]). They are mainly expected to affect trip generation and mode choice, thereby explaining the variation in pedestrian volume.

Second, for the "diversity" measure, while previous studies mostly applied the extent of land use mix such as the entropy index [68], we employed the facility accessibility index, borrowing the core concept of WalkScore™ [69]. This covers not only land use diversity but also proximity to pedestrian-friendly facilities. As Frank et al. [70] revealed, the higher the mixed land use, the closer the convenience facilities such as grocery stores, shops, and restaurants, and the more walking activities occur in the specified areas. Lee and Moudon [27] also argued that distance to routine daily destinations can be simple and effective alternatives to a complicated land use mix index. Using data from the Ministry of Security and Public Administration, Korea, this study categorized pedestrian-friendly facilities into (a) restaurants and cafes, (b) shopping facilities, (c) cultural and leisure facilities, (d) educational facilities, and (e) public transportation facilities (Table 1). We then estimated the weight of each category using the analytic hierarchy process (AHP) method developed by Saaty [71] by surveying 73 experts in the urban and transportation fields via email. Of the 53 experts who responded (53/73 = 72.6%), only responses of 43 experts (43/53 = 81.1%) whose consistency index was less than 0.2 were used for analysis. They averaged 13.6 years of professional experience, and 82% were either professors or researchers with either a master's or doctoral degree. Table 1 shows the result of AHP analysis. Finally, we calculated each PVS spot's facility accessibility index considering whether each facility was located within walkable distance from the spot. Specifically, we awarded 2 points if the facility was located within 250 m and 1 point for 500 m, similar to those of WalkScore™. Then, we summed all the points after multiplying their weights.

**Table 1.** Type and weight of pedestrian-friendly facilities.

| Category | | Specific Type of Facility | Weight |
|---|---|---|---|
| (a) | Restaurants and cafés | General restaurant, café | 0.164 |
| (b) | Shopping facilities | Market, department store, shopping center, outlet, mall, general store | 0.161 |
| (c) | Cultural and leisure facilities | Theater, exhibition hall, museum, auditorium, concert hall, zoo, botanical garden, gym, swimming pool | 0.141 |
| (c) | Educational facilities | Library, kindergarten, elementary/middle/high school, university | 0.177 |
| (e) | Public transportation facilities | Bus stop, bus terminal, subway station, train station | 0.357 |

Third, regarding the "design" measure, we applied the connectivity index (i.e., link/node ratio), which is a widely used indicator in previous studies and urban design guidelines [72–74]. As Jacobs [8] emphasized the importance of small blocks, the more connected the pedestrian path, the more walking activities people complete. The link/node ratio reflects at least three characteristics of street networks: whether the block size is small, whether the pedestrian path is properly connected, and whether the networks provide shorter and more varied routes. We measured the ratio using the pedestrian path network map in Seoul provided by the National Geographic Information Institute in Korea and following Steiner and Butler's [72] guideline. Along with the pedestrian volume and transit ridership variables, the 3D variables explained so far were also log-transformed to convert their distribution closer to normality. Resultantly, we could interpret the coefficients of key built environment variables as the elasticity of pedestrian volume.

#### 3.2.4. Street-Level Built Environment Variables

We also controlled for street-level built environment variables including adjacent land use, street type, and specific street conditions of the survey spot. Because the information was included in the PVS data, most previous papers have also controlled for them [16,19,20,49]. As Gehl [33] suggested, the quality of the microscopic urban spaces can also affect the number of people on a street.

First, to control for land use of the surrounding area, two land use dummy variables—commercial and residential—were used, with other uses as the reference group. In general, we can expect that people would be more likely to walk and stay in commercial streets with higher diversity and vitality [33,37,50]. Second, we employed the street type of the survey spot. This was defined as dummy variables for "street with a sidewalk" and "street without a sidewalk (i.e., shared street)", where "street with a sidewalk, but shared with bicycles" functioned as the reference group. Third, regarding specific street conditions, we controlled for the sidewalk width, number of traffic lanes, and various dummy variables assessing the presences of road centerlines, sloping road segments, pedestrian protection fences, nearby crosswalks, and obstacles to walking. Here, note that these dummy variables were measured based on their presence within 50 m from the survey spot, and the presence of a sloping road segment was entirely determined by the trained investigator's perception [16]. Definitions and descriptive statistics of all variables explained so far are provided in Table 2.

**Table 2.** Definitions and descriptive statistics.

| Variables | Definition (Unit) | Mean | S.D. | Min. | Max. |
|---|---|---|---|---|---|
| *Dependent variable* | | | | | |
| Pedestrian volume | Total daily pedestrian counts | 7121 | 9615 | 57 | 103,437 |
| | Ln (pedestrian volume) | 8.247 | 1.164 | 4.043 | 11.547 |
| *Transit ridership (proxy for regional centrality): macro-scale variables* | | | | | |
| Bus ridership | Daily bus ridership within 400 m buffer | 23,494 | 21,326 | 0 | 127,236 |
| | Ln (bus ridership + 1) | 9.696 | 0.953 | 0.000 | 11.750 |
| Subway ridership | Daily subway ridership within 300 m buffer | 27,020 | 43,941 | 0 | 247,209 |
| | Ln (subway ridership + 1) | 4.512 | 5.350 | 0.000 | 12.420 |
| *Walkshed-level variables (within 500 m buffer): meso-scale variables* | | | | | |
| Density (population) | Population density of census tracks within 500 m (capita/m$^2$) | 0.016 | 0.010 | >0.001 | 0.041 |
| | Ln (population density) | −4.467 | 0.936 | −9.520 | −3.190 |
| Density (job) | Job density of census tracks within 500 m (capita/m$^2$) | 0.019 | 0.022 | >0.001 | 0.124 |
| | Ln (job density) | −4.570 | 1.133 | −9.440 | −2.090 |
| Diversity | Facility accessibility index (see Section 3.2.3) | 0.406 | 0.118 | 0.050 | 0.710 |
| | Ln (facility accessibility index) | 0.277 | 0.170 | −0.528 | 0.837 |
| Design | Connectivity index (link/node ratio) | 1.338 | 0.221 | 0.590 | 2.310 |
| | Ln (connectivity index) | −0.951 | 0.334 | −2.996 | −0.342 |
| *Street-level variables (mostly within 50 m buffer): micro-scale variables* | | | | | |
| *Land use* | | | | | |
| Residential | Residential use (yes = 1) | 0.694 | 0.461 | | |
| Commercial | Commercial use (yes = 1) | 0.222 | 0.415 | | |
| Other land use (ref.) | Other land use (yes = 1) | 0.085 | 0.218 | | |
| *Street type* | | | | | |
| With sidewalk | Street with a sidewalk (yes = 1) | 0.711 | 0.454 | | |
| Without sidewalk | Street without a sidewalk shared with pedestrians and vehicles (yes = 1) | 0.084 | 0.278 | | |
| With shared sidewalk (ref.) | Sidewalk is shared with pedestrian and bicycle (yes = 1) | 0.205 | 0.404 | | |
| *Street condition* | | | | | |
| Sidewalk width | Width of sidewalk or fringe of the road for pedestrian passage (m) | 4.198 | 2.221 | 1.000 | 24.000 |
| # of traffic lanes | Number of traffic lanes (count) | 4.182 | 2.731 | 1.000 | 18.000 |
| Presence of centerline | Dummy (yes = 1, *within 50 m buffer*) | 0.730 | 0.444 | | |
| Presence of sloping road | Dummy (yes = 1, *within 50 m buffer*) | 0.213 | 0.409 | | |
| Presence of fence | Dummy (yes = 1, *within 50 m buffer*) | 0.225 | 0.417 | | |
| Presence of crosswalk | Dummy (yes = 1, *within 50 m buffer*) | 0.616 | 0.486 | | |
| Presence of obstacle | Dummy (yes = 1, *within 50 m buffer*) | 0.917 | 0.276 | | |

Number of observations = 2889.

### 3.3. Analysis Method and Model Specification

We employed a multiple regression approach to test the effects of the built environment variables explained above. Although we conceptually categorized them into three levels (street-, walkshed-, and regional-level), the variables had no actual hierarchical structure in the data, so there was no need to apply multi-level analysis. However, we applied spatial regression analysis methods to consider a potential spatial autocorrelation issue.

The analysis consisted of two steps. As a preliminary analysis, we identified the size of the influential areas of bus stops and subway stations that best explain the variation in pedestrian volume. We first ran 36 ordinary least squares (OLS) regression models by applying different sets of bus and subway transit riderships that were measured based on different sizes of influential areas (100 m to 600 m at 100 m intervals). Then, we conducted a spatial regression analysis including the transit riderships within the specific influential areas identified above and identified a more suitable model between the spatial lag and spatial error models. Lastly, we ran 36 spatial lag or spatial error models again by applying different sets of transit riderships and identified the final size of the bus stops' and subway stations' influential areas.

In the second step, we conducted two spatial regression analyses: constrained and unconstrained models. While the constrained model excluded the transit ridership variables, the unconstrained model included transit ridership measured within the specific influential areas identified in the first step. Then, we compared the results of the two differentiated models depending on whether or not transit ridership was controlled for. Except transit ridership, the same set of independent variables were applied in all regression models explained so far.

## 4. Results of Analysis

### 4.1. Preliminary Analysis: Determing the Size of the Transit Stops' Influential Area

Through the process described in Section 3.3., we found that the spatial error model had a better goodness of fit than the OLS and spatial lag models (see Appendices A and B). Then, we ran 36 spatial error models applying different sets of bus and subway transit riderships and found that the ridership variables in all 36 models had statistically and positively significant associations with street-level pedestrian volume at the 1% probability level (Table 3). Therefore, we determined the size of the transit stops' influential areas that best explain the variation in pedestrian volume based on the model fit measures. Table 3 illustrates that the model applying a bus ridership within 400 m and a subway ridership within 300 m shows the highest explanatory power (Pseudo $R^2$ = 0.5232). Log-likelihood statistics also show exactly the same pattern (see Appendix C).

**Table 3.** The explanatory power (Pseudo-$R^2$) of the spatial error models by the transit stop/station's influential area size.

| | | Subway Ridership Within | | | | | |
|---|---|---|---|---|---|---|---|
| | | **100 m** | **200 m** | **300 m** | **400 m** | **500 m** | **600 m** |
| **Bus ridership within** | 100 m | 0.492 | 0.509 | 0.518 | 0.513 | 0.511 | 0.495 |
| | 200 m | 0.493 | 0.509 | 0.516 | 0.511 | 0.510 | 0.497 |
| | 300 m | 0.504 | 0.518 | 0.5227 | 0.516 | 0.516 | 0.506 |
| | 400 m | 0.504 | 0.519 | **0.5232** | 0.514 | 0.512 | 0.503 |
| | 500 m | 0.496 | 0.515 | 0.519 | 0.509 | 0.504 | 0.494 |
| | 600 m | 0.480 | 0.500 | 0.508 | 0.500 | 0.496 | 0.481 |

Note: All models' bus and subway riderships had statistically positive associations with street-level pedestrian volume at $p = 0.01$. Other built environmental factors suggested in Table 2 were applied as control variables in all regression models.

These spatial ranges are not substantially different from other similar empirical studies ([75], light-rail station 326 m; [66], BART (Bay Area Rapid Transit) station 548 m; [76], transit stops 494 m) or the general walking distance (400 m to 500 m) conventionally adopted in previous studies [16,20,66,67]. This reflects the general tendency that people are willing to walk before boarding or after alighting from transit [42]. This result also supports the persisting claim that values the link between walking and public transportation [76–78]. Reflecting this result, we applied both the 400 m buffer bus and 300 m buffer subway transit riderships in the unconstrained model as a proxy for regional centrality.

### 4.2. Impact of Built Environment on Pedestrian Volume: Transit Ridership Controlled vs. Not Controlled

Appendix A (constrained models) and Appendix B (unconstrained models) present the results of OLS, spatial lag, and spatial error models of the log-transformed street-level pedestrian volume. In all models, the variance inflation factor (VIF) values were quite small; accordingly, no multicollinearity was found. Moran's I, Rho (ρ), and Lambda (λ) values in the models were statistically significant, indicating the presence of spatial autocorrelation. The results of the Lagrange Multiplier (LM) lag and error test demonstrated that the spatial error model had better goodness of fit than the spatial lag model. The improvement of fit measured by Log likelihood, Akaike Info Criterion (AIC), and Schwarz Criterion (SC) was also greater in the spatial error model than in the spatial lag model.

Table 4 summarizes the results of both the constrained (not controlled for transit ridership) and unconstrained (controlled for transit ridership) spatial error models. As expected, due to the strong significance of both ridership variables (coefficient: 0.267, 0.043; Z: 10.760, 11.775), the unconstrained model showed higher explanatory power (Pseudo-$R^2$: 0.523) than the constrained model (Pseudo-$R^2$: 0.475) and those of previous studies using the same dataset [19,20,45–50,79], although $R^2$ and Pseudo-$R^2$ are not statistically comparable. This implies that the area-wide floating population captured by transit ridership, which represents regional centrality, can explain street-level pedestrian volume well. Conversely, it would mean that, as previously explained, it is difficult to accurately grasp the impact of the built environments on street-level pedestrian volume when the macroscopic centrality and resulting floating population are not controlled for. The unconstrained model produces less biased coefficients than the constrained model. Thus, we interpret the following results by focusing on how the coefficients of the built environment variables change after controlling for transit ridership.

**Table 4.** Spatial regression (spatial error) models of log-transformed street-level pedestrian volume.

| Variables | Constrained Model | | Unconstrained Model | |
|---|---|---|---|---|
| | Coef. | z | Coef. | z |
| *Lambda* ($\lambda$) | 0.744 *** | 32.652 | 0.706 *** | 28.370 |
| *Constant* | 8.409 *** | 35.374 | 4.860 *** | 13.465 |
| *Transit ridership (proxy for regional centrality)* | | | | |
| log_bus ridership (400 m buff.) | | | 0.267 *** | 10.760 |
| log_subway ridership (300 m buff.) | | | 0.043 *** | 11.775 |
| *Walkshed-level 3D variables* | | | | |
| Density (log_population density) | −0.025 | −0.682 | −0.069 ** | −2.074 |
| Density (log_job density) | 0.191 *** | 5.623 | 0.126 *** | 4.028 |
| Diversity (log_facility accessibility index) | 0.495 *** | 7.292 | 0.200 *** | 3.009 |
| Design (log_connectivity index) | −0.343 *** | −2.783 | −0.144 | −1.229 |
| *Street-level variables* | | | | |
| *Land use* | | | | |
| Residential | 0.376 *** | 4.464 | 0.296 *** | 3.702 |
| Commercial | 0.665 *** | 7.208 | 0.433 *** | 4.889 |
| Other land use (ref.) | | | | |
| *Street type* | | | | |
| Street with a sidewalk | 0.663 *** | 10.936 | 0.647 *** | 11.199 |
| Street without a sidewalk | 0.550 *** | 6.896 | 0.558 *** | 7.342 |
| Street with a shared sidewalk (ref.) | | | | |
| *Street condition* | | | | |
| Sidewalk width | 0.080 *** | 9.562 | 0.065 *** | 8.180 |
| Number of traffic lanes | 0.022 ** | 2.414 | 0.016 | 1.836 |
| Presence of centerline | −0.242 *** | −3.986 | −0.169 *** | −2.910 |
| Presence of sloping road | −0.369 *** | −8.584 | −0.317 *** | −7.734 |
| Presence of fence | 0.127 *** | 3.078 | 0.144 *** | 3.647 |
| Presence of crosswalk | 0.185 *** | 4.599 | 0.192 *** | 5.025 |
| Presence of obstacle | −0.026 | −0.427 | 0.024 | 0.414 |
| *Summary Statistics* | | | | |
| N | 2,889 | | 2,889 | |
| *Pseudo-$R^2$* | 0.475 | | 0.523 | |
| *Moran's I* | 41.182 *** | | 36.238 *** | |
| *Robust LM (error)* | 388.289 *** | | 398.931 *** | |

** $p < 0.05$, *** $p < 0.01$.

First, with regard to walkshed-level 3D variables, the coefficient size and significance level showed quite substantial differences between both models. Of the two density variables, only job density was significant in the constrained model, while the unconstrained model showed significant but conflicting signs of coefficients for both variables (coefficient: −0.069, 0.126; Z: −2.074, 4.028). Although this appears to be a somewhat contradictory result, a reasonable interpretation is possible if both density variables are considered at the same time. As Chung et al. [16] asserted, areas with high job density would represent an inner-city area in Seoul, and more people walk in those areas than in the populated, outlying residential areas. Kang [20], who demonstrated the same results as our study, argued that previous studies showing a positive coefficient for the population density variable did not include job density in the model.

Next, the results demonstrated that "diversity" (facility accessibility index) was positively associated with pedestrian volume in both models. However, the coefficient size was dramatically reduced after controlling for transit ridership (coefficient: 0.495 to 0.200; Z: 7.292 to 3.009). This implies that the coefficient of "diversity" in the constrained model may be a combination of the direct influence of the variable on street-level pedestrian volume and the indirect influence via the generation of a floating population (transit ridership) in the surrounding area, as explained in Figure 1.

The results regarding "design" were distinctive between the two models. While the "design" variable (connectivity index) was significantly and negatively associated with street-level pedestrian volume in the constrained model (coefficient: −0.343; Z: −2.783), the relationship was not statistically significant in the unconstrained model (coefficient: −0.144; Z: −1.229). This suggests that there may be an unobserved relationship between regional centrality and street network design. In Section 5.1, we discuss this possibility through comparison with other empirical studies.

Lastly, the coefficient size and significance level of street-level built environment variables also tended to decrease after taking transit ridership into account. However, the tendency was not noticeable compared to the walkshed-level variables. This is considered to be so because the street-level environment has a weaker association with regional centrality than the neighborhood-level environment. In other words, microscopic street elements can affect walking independently of macroscopic factors. In the unconstrained model, significant variables generally showed the same results as previous studies [19,20,49,50,79] as well as the constrained model. The only variable whose results have changed is the number of traffic lanes. While this was positively associated with pedestrian volume in the constrained model (coefficient: 0.022; Z: 2.414), it appeared insignificant in the unconstrained model (coefficient: 0.016; Z: 1.836). This seems to be derived from the intervention of the regional centrality (transit ridership) variable in the "design" (street connectivity) variable. In fact, it is more likely that roads with more lanes are formed in places with greater regional centrality (where there are many transit users) [45].

## 5. Discussion

In Section 4, the analysis results were interpreted and discussed focusing on the effects of the 3D variables. Section 5 extends our discussion, focusing on the following two derivative issues.

### *5.1. Does "Design" Matter?*

The constrained model demonstrated that the street connectivity ("design") variable had a negative and significant effect on pedestrian volume, in contrast to general urban design theory [72]. This relationship was also identified in Kang's study [20] using the same data set as our study. In the study, he revealed that street intersection density, which is another representative variable measuring "design," had a negative relationship with pedestrian volume [20]. This result may be interpreted as follows: in the case of Seoul, where old and complex urban fabric and street network pattern extensively remains throughout the city, the higher the street connectivity, the smaller the pedestrian volume, unlike in Western cities.

However, this conclusion is still arguable. Although the specific measurements and study areas were different from the above studies, some studies have shown contradictory results. Using path analysis, Hajrasouliha and Yin [26] demonstrated that street network connectivity (intersection density variable) not only had a direct positive effect but also an indirect positive effect via the job density variable on pedestrian volume. Miranda-Moreno et al. [51] and Lee et al. [46] also showed that the number of roads at an intersection was positively associated with pedestrian volume.

Moreover, the relationship was insignificant in the unconstrained model, which controlled for regional centrality. This tendency also appeared in other studies. In Lee and Koo's [45] analysis model that controlled for regional centrality by employing street-level integration measured with space syntax, the association between the block-level street network density (street length/block area) and pedestrian volume was insignificant. What both studies had in common was that they controlled for regional centrality in any form. This suggests that regional centrality and the street network design variables are strongly associated. Hajrasouliha and Yin [26] also argued that street connectivity might have an indirect impact on pedestrian volume by influencing other built environment characteristics. In addition, in reality, a higher street network connectivity can be expected in urban areas with higher centrality than in peripheral areas. Therefore, we can interpret that, in the models where the required built environment variables (such as regional centrality) are not controlled for, street-connectivity-related variables function as proxy variables, having no conclusive effect on pedestrian volume.

This offers two more implications. First, in a model that does not control for regional centrality, the effect (coefficient value) of the street connectivity variable on pedestrian volume may not actually be a net effect of "design." Second, to reveal the net effect of "design" (whether "design" matters), regional centrality needs to be controlled for in the model as in the present study. In conclusion, after examining our results alone, unlike previous studies, "design" may not directly matter in terms of street-level pedestrian volume.

### *5.2. Which Is a Better Explanatory Variable, Accessibility to Transit or Transit Ridership?*

Walking and public transportation usage are complementary activities [76–78]. Accordingly, studies that analyzed the determinants of pedestrian volume reflected the factors indicating the degree of public transit use in their econometric models. Here, the methods fall into two main approaches. While the first approach measures and reflects "accessibility to public transit" using variables such as the distance to the stop/station, the existence of the stop/station, or stop/station density [20,49,50,52], the second approach reflects it more directly through utilizing "transit ridership" itself [16,44]. Although both approaches empirically revealed that the variables were significantly associated with pedestrian volume [16,20,25,47,49], the meaning of the results was different.

Studies that used accessibility to transit stop/station generally intended to test hypotheses that suggest: (1) more pedestrian traffic may occur in places with more transit stops/stations that generally have good accessibility to various urban services; and (2) if there is a transit stop/station nearby, some of the transit users flowing out may flow into the PVS point. Most studies used the accessibility variables to test the first hypothesis. However, rather than simple accessibility variables (distance to stop/station or density), "how many people *actually* use public transportation" can more accurately indicate the centrality of an area. This is because transit ridership is determined by regional centrality rather than the micro-scale built environment. At the same time, without controlling for regional centrality or the floating population, it is difficult to ascertain the net effect of the micro-level transit accessibility variable on pedestrian volume. Moreover, when testing the second hypothesis, the transit accessibility variable was only a proxy. It is reasonable to use a more straightforward variable, the number of transit users, which directly explains the variation of pedestrian volume.

Thus, researchers can choose from either form of variables according to the purpose of the analysis. If the goal of modeling is to better reflect the determinants of pedestrian volume, it is desirable to directly control for the transit ridership as in the present study. If the goal is to derive policy implications for public transportation infrastructure (e.g., optimal transit stop density or interval), applying accessibility-type variables is more desirable. In addition, the desired form of a variable may vary depending on the type or availability of research data. When using cross-sectional data, it is desirable to employ an accessibility variable to represent the static built environment, and when using time series data, it is desirable to apply a day-dependent ridership variable. However, at this point, whether reliable transit ridership data are well established, as in Seoul, can also be an important consideration. In the case of Seoul, where almost all citizens use a Smart Card [63], transit ridership can be a good proxy for regional centrality and the resultant floating population. This is highly accurate and easily obtainable data, which greatly improves the explanatory power of the model for predicting pedestrian volume on a specific street, while street vitality and commercial performance can be determined accordingly. Being an obvious benefit of smart technology, it can be a more accurate method than the previous approach that utilized the distance from or existence of transit stops/stations.

## 6. Conclusions

This study examined the relationship between the built environment and street-level pedestrian volume after controlling for regional centrality (transit ridership) using 2015 PVS and Smart Card data from Seoul. As a preliminary study, we analyzed 36 spatial regression models by applying different sets of bus and subway transit riderships and found that the combination of a bus ridership within 400 m and a subway ridership within 300 m best explained the variation in street-level pedestrian volume. These are not substantially different from other empirical studies [66,75,76] or a general walking distance of 400 m to 500 m conventionally adopted in previous studies [16,20,66,67].

After controlling for both ridership variables as a proxy for regional centrality, we examined the effect of the 3D variables on street-level pedestrian volume and compared this with the result of the model that did not control for ridership. Key findings are as follows. First, both transit ridership variables explained the variance in pedestrian volume well, greatly improving the explanatory power of the models. Specifically, we found that if daily ridership at bus stops (subway stations) located within 400 m (300 m) from a specific street point increased by 1%, the daily pedestrian volume at that street point increased by 0.267% (0.043%). Second, after taking transit ridership into account, the influence of built environment variables was generally reduced, and the decrease was more pronounced in the walkshed-level variables (i.e., 3D variables) than in the street-level variables. In particular, the influence of the "design" variable (street connectivity index) was found to be insignificant in the unconstrained model. This means that the degree of street connectivity is influenced by regional centrality, and accordingly, the coefficient of the "design" variable in our constrained model (or even in other previous studies' pedestrian volume estimation models) might be biased. Thus, to accurately understand the effect of the meso-scale 3D variables on pedestrian volume, both micro- and macro-scale built environmental factors must be controlled.

As explained throughout, this study provided more precise empirical evidence on the effects of traditional 3D variables on pedestrian volume by controlling for regional centrality. This analytical framework and these analysis results can be transferred to many metropolitan cities around the world. Since a large number of pedestrians on the streets contribute to the attractiveness of public spaces and vitality of the urban economy in most cities around the world, with the exception of some cities with poor security [80], many urban researchers have been and will continue to examine the determinants of pedestrian volume. Thus, a more accurate analysis method as to which built environment attributes may encourage pedestrian activities would be meaningful not only in Seoul but also for all cities on earth.

In addition, even though this study focused on the theoretical and analytical frameworks to identify the physical factors influencing street-level pedestrian volume more accurately, it also suggested implications for planning policies and practices. Although 3D variables have been accepted as normative theories of urban planning and design in

North American cities [21–29], the effects of these variables may vary depending on the urban context. The present study confirmed once again that high job density and facility accessibility (i.e., "density" and "diversity") at the walkshed level were values that planners and policymakers should pursue. However, as explained above, in predicting the increase in pedestrian volume due to the physical environment and the consequent socioeconomic outcomes, it is necessary to sufficiently control for the centrality of the area and the resultant floating population. Furthermore, planners and policymakers must consider the negative externalities of agglomeration such as congestion, air pollution, and even the excessive complexity of landscape. As many studies have shown, compact development has more losses than benefits in high-density Asian cities [81–83].

In terms of "design", more thoughtful policies and plans are required. In contrast to previous studies, our analysis demonstrated that street connectivity (i.e., "design") was not directly associated with street-level pedestrian volume after controlling for regional centrality. This is derived from the distinctive urban context of Seoul, where few cul-de-sacs and loop-like street patterns exist. Thus, the practice of uncritically accepting Western theories centered on North American cities (e.g., planning principles of New Urbanism), as in Korea, is by no means desirable, which is a lesson to be kept in mind in other rapidly growing Asian metropolitan cities.

The city governments should identify the determinants of pedestrian volume and the corresponding urban design principles suitable for their urban context. In this process, to control for regional centrality, they can employ transit ridership data from the highly utilized Smart Card system of the city. Since public transportation in a particular city serves not only the city but also the metropolitan area surrounding the city, transit ridership can be reliable and easily obtainable data to represent regional centrality. Our approach will therefore also be applicable in other cities with similar systems and urban contexts.

Despite these contributions, this study has several limitations. First, as explained in Figure 1, the relationship between the built environment and pedestrian volume may vary depending on the purpose for walking. However, because it is difficult to determine in large-scale PVSs, we could not ascertain how the built environment specifically affects walking activities by purpose. Future studies can address this limitation by securing research data that incorporate a post hoc survey on the counted pedestrians. Alternatively, it would be possible to perform similar analyses using data on the transit ridership and pedestrian volume by time, which narrows the time zone and corresponding travel purpose (e.g., commuting travel during peak hours). Second, we need to consider the endogeneity issue between key variables. That is, pedestrians observed on the street may also be transit riders in the surrounding area. Therefore, in future research, it is necessary to apply a multivariate analysis, such as path analysis, considering the two-way causal relationship. Third, when measuring the built environment variables, the airline distance buffer was used instead of the network distance buffer. Since the streets in Seoul are very dense, the difference between the two buffers is not expected to be relatively large, but this may cause bias in the analysis results. In addition, unlike the transit ridership variables, when measuring the 3D variable, the conventionally used 500 m buffer was uncritically used without testing the appropriate buffer size, which may also be a limitation of our study. The last shortcoming is that the analysis is yet only a single case study. Even though relatively large PVS data from a major metropolitan area were used, their generalizability is still limited due to the distinctive context of the single study area. Thus, key findings in this study would need to be reconfirmed through further research focusing on more diverse cities.

**Author Contributions:** Conceptualization, S.-N.K.; methodology, J.C. and J.L.; software, J.C. and J.L.; validation, S.-N.K.; formal analysis, S.-N.K., J.C. and J.L.; investigation, J.C. and J.L.; resources, J.C. and J.L.; data curation, S.-N.K., J.C. and J.L.; writing—original draft preparation, S.-N.K., J.C. and J.L.; writing—review and editing, S.-N.K.; visualization, J.L.; supervision, S.-N.K.; project administration, S.-N.K.; funding acquisition, S.-N.K. All authors have read and agreed to the published version of the manuscript.

**Funding:** This research was supported by the Chung-Ang University Research Scholarship Grants in 2021. This research was also supported by a grant (21CTAP-C164344-01) from the Technology Advancement Research Program of the Korea Ministry of Land, Infrastructure and Transport.

**Institutional Review Board Statement:** Not applicable.

**Informed Consent Statement:** Not applicable.

**Data Availability Statement:** The data presented in this study are available on request from the corresponding author.

**Acknowledgments:** The authors are grateful to three anonymous reviewers and an academic editor for their excellent suggestions for improving the manuscript.

**Conflicts of Interest:** The authors declare no conflict of interest.

## Appendix A

Regression models of log-transformed street-level pedestrian volume (constrained model).

| Variables | Constrained Model | | | | | | |
|---|---|---|---|---|---|---|---|
| | OLS | | | Spatial Lag | | Spatial Error | |
| | Coef. | t | VIF | Coef. | z | Coef. | z |
| *Rho (ρ)* | | | | 0.678 *** | 30.416 | | |
| *Lambda (λ)* | | | | | | 0.744 *** | 32.652 |
| *Constant* | 8.602 *** | 47.084 | | 2.657 *** | 10.621 | 8.409 *** | 35.374 |
| *Walkshed-level 3D variables* | | | | | | | |
| Density (log_population density) | 0.003 | 0.125 | 1.269 | 0.052 *** | 2.725 | −0.025 | −0.682 |
| Density (log_job density) | 0.216 *** | 10.885 | 1.527 | 0.071 *** | 3.943 | 0.191 *** | 5.623 |
| Diversity (log_facility accessibility index) | 0.563 *** | 9.029 | 1.314 | 0.371 *** | 6.694 | 0.495 *** | 7.292 |
| Design (log_connectivity index) | −0.232 ** | −2.022 | 1.148 | −0.277 *** | −2.756 | −0.343 *** | −2.783 |
| *Street-level variables* | | | | | | | |
| *Land use* | | | | | | | |
| Residential | 0.355 *** | 5.069 | 3.146 | 0.281 *** | 4.572 | 0.376 *** | 4.464 |
| Commercial | 0.763 *** | 9.753 | 3.191 | 0.489 *** | 7.014 | 0.665 *** | 7.208 |
| Other land use (ref.) | | | | | | | |
| *Street type* | | | | | | | |
| Street with a sidewalk | 0.699 *** | 10.745 | 2.627 | 0.575 *** | 10.045 | 0.663 *** | 10.936 |
| Street without a sidewalk | 0.514 *** | 6.073 | 1.666 | 0.435 *** | 5.860 | 0.550 *** | 6.896 |
| Street with a shared sidewalk (ref.) | | | | | | | |
| *Street condition* | | | | | | | |
| Sidewalk width | 0.099 *** | 10.867 | 1.235 | 0.077 *** | 9.589 | 0.080 *** | 9.562 |
| Number of traffic lanes | 0.021 ** | 2.291 | 1.974 | 0.018 ** | 2.237 | 0.022 ** | 2.414 |
| Presence of centerline | −0.342 *** | −5.196 | 2.577 | −0.234 *** | −4.059 | −0.242 *** | −3.986 |
| Presence of sloping road | −0.337 *** | −7.485 | 1.026 | −0.304 *** | −7.662 | −0.369 *** | −8.584 |
| Presence of fence | 0.123 *** | 2.690 | 1.096 | 0.135 *** | 3.390 | 0.127 *** | 3.078 |
| Presence of crosswalk | 0.127 *** | 2.914 | 1.353 | 0.147 *** | 3.871 | 0.185 *** | 4.599 |
| Presence of obstacle | −0.000 | −0.003 | 1.017 | −0.036 | −0.621 | −0.026 | −0.427 |
| *Model Summary* | | | | | | | |
| N | 2889 | | | 2889 | | 2889 | |
| $R^2$ | 0.298 | | | | | | |
| *Adjusted $R^2$* | 0.294 | | | | | | |
| *Pseudo-$R^2$* | | | | 0.460 | | 0.475 | |
| Log likelihood | −4026 | | | −3707 | | −3681 | |
| Akaike Info Criterion (AIC) | 8085 | | | 7448 | | 7394 | |
| Schwarz Criterion (SC) | 8180 | | | 7549 | | 7490 | |
| *Statistics* | | | | | | | |
| Moran's I | 41.182 *** | | | | | | |
| Robust LM (lag) | 10.278 *** | | | | | | |
| Robust LM (error) | 388.289 *** | | | | | | |

** $p < 0.05$, *** $p < 0.01$.

## Appendix B

Regression models of log-transformed street-level pedestrian volume by the transit ridership (unconstrained model).

| Variables | Unconstrained Model | | | | | | |
|---|---|---|---|---|---|---|---|
| | OLS | | | Spatial Lag | | Spatial Error | |
| | Coef. | t | VIF | Coef. | z | Coef. | z |
| *Rho* ($\rho$) | | | | 0.599 *** | 25.151 | | |
| *Lambda* ($\lambda$) | | | | | | 0.706 *** | 28.370 |
| *Constant* | 4.313 *** | 14.441 | | 0.466 | 1.528 | 4.860 *** | 13.465 |
| *Transit ridership (proxy for regional centrality)* | | | | | | | |
| log_bus ridership (400 m buff.) | 0.331 *** | 15.129 | 1.524 | 0.214 *** | 10.492 | 0.267 *** | 10.760 |
| log_subway ridership (300 m buff.) | 0.045 *** | 12.769 | 1.259 | 0.039 *** | 12.085 | 0.043 *** | 11.775 |
| *Walkshed-level 3D variables* | | | | | | | |
| Density (log_population density) | −0.061 *** | −2.955 | 1.306 | 0.003 | 0.151 | −0.069 ** | −2.074 |
| Density (log_job density) | 0.144 *** | 7.643 | 1.597 | 0.039 ** | 2.250 | 0.126 *** | 4.028 |
| Diversity (log_facility accessibility index) | 0.172 *** | 2.829 | 1.451 | 0.097 | 1.760 | 0.200 *** | 3.009 |
| Design (log_connectivity index) | −0.107 | −1.001 | 1.151 | −0.179 | −1.857 | −0.144 | −1.229 |
| *Street-level variables* | | | | | | | |
| *Land use* | | | | | | | |
| Residential | 0.201 *** | 3.039 | 3.244 | 0.193 *** | 3.245 | 0.296 *** | 3.702 |
| Commercial | 0.360 *** | 4.745 | 3.471 | 0.243 *** | 3.527 | 0.433 *** | 4.889 |
| Other land use (ref.) | | | | | | | |
| *Street type* | | | | | | | |
| Street with a sidewalk | 0.690 *** | 11.412 | 2.631 | 0.579 *** | 10.569 | 0.647 *** | 11.199 |
| Street without a sidewalk | 0.603 *** | 7.659 | 1.675 | 0.503 *** | 7.065 | 0.558 *** | 7.342 |
| Street with a shared sidewalk (ref.) | | | | | | | |
| *Street condition* | | | | | | | |
| Sidewalk width | 0.082 *** | 9.634 | 1.246 | 0.067 *** | 8.697 | 0.065 *** | 8.180 |
| Number of traffic lanes | 0.007 | 0.795 | 1.990 | 0.009 | 1.103 | 0.016 | 1.836 |
| Presence of centerline | −0.223 *** | −3.627 | 2.599 | −0.159 *** | −2.865 | −0.169 *** | −2.910 |
| Presence of sloping road | −0.278 *** | −6.626 | 1.032 | −0.267 *** | −7.044 | −0.317 *** | −7.734 |
| Presence of fence | 0.155 *** | 3.650 | 1.098 | 0.159 *** | 4.151 | 0.144 *** | 3.647 |
| Presence of crosswalk | 0.143 *** | 3.525 | 1.356 | 0.161 *** | 4.402 | 0.192 *** | 5.025 |
| Presence of obstacle | 0.070 | 1.128 | 1.020 | 0.019 | 0.333 | 0.024 | 0.414 |
| *Model Summary* | | | | | | | |
| N | 2889 | | | 2889 | | 2889 | |
| $R^2$ | 0.394 | | | | | | |
| *Adjusted* $R^2$ | 0.391 | | | | | | |
| *Pseudo-*$R^2$ | | | | 0.505 | | 0.523 | |
| Log likelihood | −3812 | | | −3565 | | −3532 | |
| Akaike Info Criterion (AIC) | 7661 | | | 7168 | | 7101 | |
| Schwarz Criterion (SC) | 7768 | | | 7282 | | 7208 | |
| *Statistics* | | | | | | | |
| Moran's I | 36.238 *** | | | | | | |
| Robust LM (lag) | 14.415 *** | | | | | | |
| Robust LM (error) | 398.931 *** | | | | | | |

** $p < 0.05$, *** $p < 0.01$.

## Appendix C

The log likelihood of the spatial error models by the transit stop/station's influential area size.

| | | Subway Ridership Within | | | | | |
|---|---|---|---|---|---|---|---|
| | | 100 m | 200 m | 300 m | 400 m | 500 m | 600 m |
| Bus ridership within | 100 m | −3631.534 | −3582.570 | −3554.960 | −3566.286 | −3574.785 | −3619.577 |
| | 200 m | −3624.403 | −3578.292 | −3555.164 | −3567.127 | −3572.268 | −3609.902 |
| | 300 m | −3590.634 | −3552.466 | −3535.724 | −3553.718 | −3557.134 | −3582.928 |
| | 400 m | −3590.328 | −3545.743 | **−3532.321** | −3557.138 | −3566.522 | −3592.585 |
| | 500 m | −3614.468 | −3558.104 | −3542.730 | −3572.256 | −3589.259 | −3619.456 |
| | 600 m | −3665.148 | −3608.897 | −3581.035 | −3604.125 | −3617.418 | −3659.830 |

Note: All models' bus and subway riderships had statistically positive associations with street-level pedestrian volume at $p = 0.01$. Other built environmental factors suggested in Table 2 were applied as control variables in all regression models.

## References

1. Lee, C.; Moudon, A.V. Physical activity and environment research in the health field: Implications for urban and transportation planning practice and research. *J. Plan. Lit.* **2016**, *19*, 147–181. [CrossRef]
2. Kim, H.; Yang, S. Neighborhood walking and social capital: The correlation between walking experience and individual perception of social capital. *Sustainability* **2017**, *9*, 680. [CrossRef]

3. Loukaitou-Sideris, A. Special issue on walking. *Transp. Rev.* **2020**, *40*, 131–134. [CrossRef]
4. Kahn, M.E.; Morris, E.A. Walking the walk: The association between community environmentalism and green travel behavior. *J. Am. Plan. Assoc.* **2009**, *75*, 389–405. [CrossRef]
5. Ogilvie, D.; Bull, F.; Cooper, A.; Rutter, H.; Adams, E.; Brand, C.; Ghali, K.; Jones, T.; Mutrie, N.; Powell, J.; et al. Evaluating the travel, physical activity and carbon impacts of a 'natural experiment' in the provision of new walking and cycling infrastructure: Methods for the core module of the iConnect study. *BMJ Open* **2012**, *2*, e000694. [CrossRef] [PubMed]
6. Sung, H.; Lee, S.; Jung, S. Identifying the relationship between the objectively measured built environment and walking activity in the high-density and transit-oriented city, Seoul, Korea. *Environ. Plan. B Plann. Design* **2014**, *41*, 637–660. [CrossRef]
7. Buehler, R.; Pucher, J.; Gerike, R.; Götschi, T. Reducing car dependence in the heart of Europe: Lessons from Germany, Austria, and Switzerland. *Transp. Rev.* **2016**, *37*, 4–28. [CrossRef]
8. Jacobs, J. *The Death and Life of Great American Cities*; Modern Library Editions & Random House Inc.: New York, NY, USA, 1961.
9. Appleyard, D. *Livable Streets*; University of California Press: Berkley, CA, USA, 1981.
10. Putnam, R. *Bowling Alone, The Collapse and Revival of American Community*; Simon and Schuster: New York, NY, USA, 2000.
11. Handy, S.L.; Boarnet, M.G.; Ewing, R.; Killingsworth, R.E. How the built environment affects physical activity: Views from urban planning. *Am. J. Prev. Med.* **2002**, *23*, 64–73. [CrossRef]
12. Litman, T.A. Economic value of walkability. *Trans. Res. Rec.* **2003**, *1828*, 3–11. [CrossRef]
13. Frumkin, H.; Frank, L.D.; Jackson, R. *Urban Sprawl and Public Health: Designing, Planning, and Building for Healthy Communities*; Island Press: Washington, DC, USA, 2004.
14. Montgomery, C. *Happy City: Transforming Our Lives Through Urban Design*, 1st ed.; Farrar, Straus and Giroux: New York, NY, USA, 2013.
15. Kang, C.D. Spatial access to pedestrians and retail sales in Seoul, Korea. *Habitat Int.* **2016**, *57*, 110–120. [CrossRef]
16. Chung, J.; Kim, S.N.; Kim, H. The impact of PM10 levels on pedestrian volume: Findings from streets in Seoul, South Korea. *Int. J. Environ. Res. Public Health* **2019**, *16*, 4833. [CrossRef] [PubMed]
17. Jacobs, A.B. Great streets. *Access Mag.* **1993**, *1*, 23–27.
18. Ewing, R.; Handy, S. Measuring the unmeasurable: Urban design qualities related to walkability. *J. Urban Des.* **2009**, *14*, 65–84. [CrossRef]
19. Sung, H.G.; Go, D.H.; Choi, C.G. Evidence of Jacobs's street life in the great Seoul city: Identifying the association of physical environment with walking activity on streets. *Cities* **2013**, *35*, 164–173. [CrossRef]
20. Kang, C.D. The effects of spatial accessibility and centrality to land use on walking in Seoul, Korea. *Cities* **2015**, *46*, 94–103. [CrossRef]
21. Cervero, R.; Kockelman, K. Travel demand and the 3Ds: Density, diversity, and design. *Transp. Res. Part D Transp. Environ.* **1997**, *2*, 199–219. [CrossRef]
22. Ewing, R.; Cervero, R. Travel and the built environment: A synthesis. *Transp. Res. Rec.* **2001**, *1780*, 87–114. [CrossRef]
23. Kim, T.; Shin, Y.; Sung, H. The relationship of distance-based TOD planning elements to public transit ridership in Seoul subway station areas. *J. Korea Plan. Assoc.* **2013**, *48*, 51–64.
24. Min, B.; Lee, G.; Kim, S. The effects of land-use characteristics on trip patterns by trip modes and purposes: Focused on Seoul Metropolitan Administrative Division. *JAIK Plan. Des.* **2016**, *32*, 77–87.
25. Ewing, R.; Hajrasouliha, A.; Neckerman, K.M.; Purciel-Hill, M.; Greene, W. Streetscape features related to pedestrian activity. *J. Plan. Educ. Res.* **2015**, *36*, 5–15. [CrossRef]
26. Hajrasouliha, A.; Yin, L. The impact of street network connectivity on pedestrian volume. *Urban Stud.* **2015**, *52*, 2483–2497. [CrossRef]
27. Lee, C.; Moudon, A.V. The 3Ds + R, quantifying land use and urban form correlates of walking. *Transp. Res. Part D Transp. Environ.* **2006**, *11*, 204–215. [CrossRef]
28. Peiravian, F.; Derrible, S.; Ijaz, F. Development and application of the Pedestrian Environment Index (PEI). *J. Transp. Geogr.* **2014**, *39*, 73–84. [CrossRef]
29. Learnihan, V.; Van Niel, K.P.; Giles-Corti, B.; Knuiman, M. Effect of scale on the links between walking and urban design. *Geogr. Res.* **2011**, *49*, 183–191. [CrossRef]
30. Cao, X.; Handy, S.L.; Mokhtarian, P.L. The influences of the built environment and residential self-selection on pedestrian behavior: Evidence from Austin, TX. *Transportation* **2006**, *33*, 1–20. [CrossRef]
31. Vojnovic, I. Building communities to promote physical activity: A multi-scale geographic analysis. *Geogr. Ann. Series B Hum. Geogr.* **2006**, *88*, 67–90. [CrossRef]
32. Salingaros, N.A. Urban space and its information field. *J. Urban Des.* **2007**, *4*, 29–49. [CrossRef]
33. Gehl, J. *Life between Buildings: Using Public Space*, 5th ed.; Arkitektens Forlag: Copenhagen, Denmark, 2001.
34. Lee, H.; Kim, S. Shared space and pedestrian safety: Empirical evidence from pedestrian priority street projects in Seoul, Korea. *Sustainability* **2019**, *11*, 4645. [CrossRef]
35. Lee, H.; Kim, S.N. Perceived safety and pedestrian performance in pedestrian priority streets (PPSs) in Seoul, Korea: A virtual reality experiment and trace mapping. *Int. J. Environ. Res. Public Health* **2021**, *18*, 2501. [CrossRef]
36. Gehl, J.; Svarre, B. *How to Study Public Life*; Island Press: Washington, DC, USA, 2013.
37. Mehta, V. *The Street: A Quintessential Social Public Space*; Routledge: New York, NY, USA, 2013.

38. Talen, E.; Koschinsky, J. The walkable neighborhood: A literature review. *Int. J. Sustain. Land Use Urban Plan.* **2013**, *1*, 42–63. [CrossRef]
39. Zegras, P. Sustainable Urban Mobility: Exploring the Role of the Built Environment. Ph.D. Thesis, Massachusetts Institute of Technology, Cambridge, MA, USA, 2005.
40. Zacharias, J. Pedestrian behavior and perception in urban walking environments. *J. Plan. Lit.* **2001**, *16*, 3–18. [CrossRef]
41. Kim, N.S.; Susilo, Y.O. Comparison of pedestrian trip generation models. *J. Adv. Transp.* **2013**, *47*, 399–412. [CrossRef]
42. Erhardt, G.D.; Mucci, R.A.; Cooper, D.; Sana, B.; Chen, M.; Castiglione, J. Do transportation network companies increase or decrease transit ridership? Empirical evidence from San Francisco. *Transportation* **2021**, *49*, 313–342. [CrossRef]
43. Carmona, M.; Heath, T.; Tiesdell, S.; Oc, T. *Public Places—Urban spaces*; Architectural Press: Oxford, UK, 2003.
44. Rodríguez, D.A.; Brisson, E.M.; Estupiñán, N. The relationship between segment-level built environment attributes and pedestrian activity around Bogota's BRT stations. *Transp. Res. Part D Transp. Environ.* **2009**, *14*, 470–478. [CrossRef]
45. Lee, J.; Koo, J. The effect of physical environment of street on pedestrian volume: Focused on central business district (CBD, GBD, YBD) of Seoul. *J. Korea Plan. Assoc.* **2013**, *48*, 269–286.
46. Lee, H.S.; Kim, J.Y.; Choo, S.H. Analyzing pedestrian characteristics using the Seoul floating population survey: Focusing on 5 urban communities in Seoul. *J. Korean Soc. Transp.* **2014**, *32*, 315–326. [CrossRef]
47. Lee, J.; Lee, H.; Koo, J. The study on factors influencing pedestrian volume based on physical environment of street. *J. Korea Plan. Assoc.* **2014**, *49*, 145–163. [CrossRef]
48. Lee, J.; Kim, H.; Jun, C. Analysis of physical environmental factors that affect pedestrian volumes by street type. *J. Urban Des. Inst. Korea* **2015**, *6*, 123–140.
49. Sung, H.; Go, D.; Choi, C.G.; Cheon, S.; Park, S. Effects of street-level physical environment and zoning on walking activity in Seoul, Korea. *Land Use Policy* **2015**, *49*, 152–160. [CrossRef]
50. Jang, J.Y.; Choi, S.T.; Lee, H.S.; Kim, S.J.; Choo, S.H. A comparison analysis of factors to affect pedestrian volumes by land-use type using Seoul Pedestrian Survey data. *J. Korean Inst. Intell. Transp. Syst.* **2015**, *14*, 39–53. [CrossRef]
51. Miranda-Moreno, L.F.; Morency, P.; El-Geneidy, A.M. The link between built environment, pedestrian activity and pedestrian–vehicle collision occurrence at signalized intersections. *Accid. Anal. Prev.* **2011**, *43*, 1624–1634. [CrossRef] [PubMed]
52. Kim, S.; Park, S.; Lee, J.S. Meso- or micro-scale? Environmental factors influencing pedestrian satisfaction. *Transp. Res. Part D Transp. Environ.* **2014**, *30*, 10–20. [CrossRef]
53. KOSIS (Korean Statistical Information Service). Available online: https://kosis.kr/ (accessed on 10 September 2019).
54. Seoul Metropolitan Government. 2030 Seoul Master Plan. 2014. Available online: https://www.seoulsolution.kr/en/content/2030-seoul-plan (accessed on 10 August 2021).
55. Hansen, W.G. How accessibility shapes land use. *J. Am. Plan. Assoc.* **1959**, *25*, 73–76. [CrossRef]
56. Kim, S.; Mokhtarian, P.; Ahn, K. The Seoul of Alonso: New perspectives on telecommuting and residential location from South Korea. *Urban Geogr.* **2012**, *33*, 1163–1191. [CrossRef]
57. Kim, H. The Effects of Compact City Planning Elements on Travel Behavior of Different Income Levels. Unpublished. Master's Thesis, Department of Civil and Environmental Engineering, Seoul National University, Seoul, Korea, 2009.
58. Seoul Open Data Platform. Stat. City Buses Seoul. Available online: https://data.seoul.go.kr/dataList/248/S/2/datasetView.do (accessed on 19 August 2021).
59. Seoul Open Data Platform. Statistics on the Subway Operation Status of Seoul. Available online: https://data.seoul.go.kr/dataList/247/S/2/datasetView.do (accessed on 19 August 2021).
60. Seoul Open Data Platform. Statistics on the Status of Bus Stops in Seoul. Available online: https://data.seoul.go.kr/dataList/249/S/2/datasetView.do (accessed on 19 August 2021).
61. Seoul Open Data Platform. Commuting Modal Share in Seoul. Available online: https://data.seoul.go.kr/dataList/10283/S/2/datasetView.do (accessed on 19 August 2021).
62. Kim, H.M.; Han, S.S. Seoul. *Cities* **2012**, *29*, 142–154. [CrossRef]
63. Lee, M. Travel pattern analysis using public transportation card data in Seoul metropolitan area. *KRIHS Policy Brief.* **2015**, *536*, 1–6.
64. Seoul Metropolitan Government. *A White Paper on Pedestrian Volume Survey*; Seoul Metropolitan Government: Seoul, Korea, 2009.
65. National Information Society Agency; Seoul Metropolitan Government. *A Report on 2015 Pedestrian Volume Survey*; Seoul Metropolitan Government: Seoul, Korea, 2015.
66. Kim, H. Walking distance, route choice, and activities while walking: A record of following pedestrians from transit stations in the San Francisco Bay area. *Urban Des. Int.* **2015**, *20*, 144–157. [CrossRef]
67. Zhao, J.; Sun, G.; Webster, C. Walkability scoring: Why and how does a three–dimensional pedestrian network matter? *Environ. Plan. B: Urban Anal. City Sci.* **2020**, *48*, 2418–2435. [CrossRef]
68. Crane, R. The influence of urban form on travel: An interpretive review. *J. Plan. Lit.* **2000**, *15*, 3–23. [CrossRef]
69. Kocher, J.; Lerner, M. Walk Score. Available online: https://www.walkscore.com/ (accessed on 5 January 2022).
70. Frank, L.D.; Engelke, P. Multiple impacts of the built environment on public health, walkable places and the exposure to air pollution. *Int. Reg. Sci. Rev.* **2005**, *28*, 193–216. [CrossRef]
71. Saaty, T. *The Analytic Hierarchy Process*; McGraw-Hill: New York, NY, USA, 1980.
72. Steiner, F.; Butler, K. *Planning and Urban Design Standard, Student Edition*; John Wiley & Sons, Inc.: Hoboken, NJ, USA, 2007.

73. Knight, P.L.; Marshall, W.E. The metrics of street network connectivity: Their inconsistencies. *J. Urban. Int. Res. Placemaking Urban Sust.* **2014**, *8*, 241–259. [CrossRef]
74. Kim, H.; Kim, S.N. Shaping suburbia: A comparison of state-led and market-led suburbs in Seoul Metropolitan Area, South Korea. *Urban Des. Int.* **2016**, *21*, 131–150. [CrossRef]
75. O'Sullivan, S.; Morrall, J. Walking distances to and from light–rail transit stations, Transportation research record. *J. Transp. Res. Board.* **1996**, *1538*, 19–26. [CrossRef]
76. Wang, J.; Cao, X. Exploring built environment correlates of walking distance of transit egress in the Twin Cities. *J. Transp. Geogr.* **2017**, *64*, 132–138. [CrossRef]
77. Besser, L.M.; Dannenberg, A.L. Walking to public transit: Steps to help meet physical activity recommendations. *Am. J. Prev. Med.* **2005**, *29*, 273–280. [CrossRef] [PubMed]
78. Park, S. Defining, Measuring, and Evaluating Path Walkability, and Testing its Impacts on Transit Users' Mode Choice and Walking Distance to the Station. Ph.D. Thesis, University of California, Berkeley, CA, USA, 2008.
79. Yun, N.; Choi, C. Relationship between pedestrian volume and pedestrian environmental factors on the commercial streets in Seoul. *J. Korea Plan. Assoc.* **2013**, *48*, 135–150.
80. Tchinda, P.E.; Kim, S.-N. The Paradox of "Eyes on the Street": Pedestrian Density and Fear of Crime in Yaoundé, Cameroon. *Sustainability* **2020**, *12*, 5300. [CrossRef]
81. Zhang, X.Q. High-rise and high-density compact urban form: The development of Hong Kong. In *Compact Cities*; Routledge: London, UK, 2000.
82. Burges, R.; Jenks, M. *Compact Cities Sustainable Urban Forms for Developing Countries*; Spon Press: London, UK, 2000.
83. Kim, S.N.; Lee, K.H.; Ahn, K.H. The effects of compact city characteristics on transportation energy consumption and air quality. *J. Korea Plan. Assoc.* **2009**, *44*, 231–246.

*land*

MDPI

*Article*

# Land Use Pattern Affects Microplastic Concentrations in Stormwater Drains in Urban Catchments in Perth, Western Australia

**Cassandra Bond [1], Hua Li [2] and Andrew W. Rate [1,*]**

[1] School of Agriculture and Environment, The University of Western Australia, 35 Stirling Highway, Perth, WA 6009, Australia

[2] Centre for Microscopy, Characterization, and Analysis, The University of Western Australia, 35 Stirling Highway, Perth, WA 6009, Australia

* Correspondence: andrew.rate@uwa.edu.au

**Abstract:** Stormwater drains act as important vectors for microplastics, enabling the transportation of microplastic polymers from terrestrial systems where they are produced and consumed to aquatic and marine ecosystems. In this study, microplastic concentrations and their size fractions were measured in six stormwater catchments in the Perth and Peel region of Western Australia. Stormwater drains with contrasting land uses and catchment characteristics were selected and two sites along each drain were sampled. Water samples were filtered in situ with a purpose-built fractionation device. Catchment boundaries and contributing drainage areas were derived from a hydrologically enforced digital elevation model. Microplastic concentrations within the sites varied from 8.8 to 25.1 microplastics/L (mean 14.2 microplastics/L). Fibrous microplastics were the most common morphology, followed by fragments. Polymer types identified using Raman spectroscopy included polypropylene (64.6% of samples), polyethylene (64.7%), polytetrafluoroethylene (5.9%) and polyvinylidene fluoride (5.9%). There was no statistically significant variation in microplastic concentrations across or within stormwater catchments. A linear mixed-effect model showed that several components of the land use pattern: catchment area, catchment population, and the proportion of industrial land, natural land and public open space, were positively related to microplastic concentrations. The proportion of residential land was negatively related to microplastic concentrations. The lack of significant variation in microplastic concentration observed both across and within the catchments points to their ubiquitous presence in stormwater systems in the region. This study is the first to examine microplastic contamination in the water of stormwater drainage systems in Perth, Western Australia. These stormwater systems contain considerable concentrations of microplastics, confirming their importance as transport mechanisms for plastics into aquatic and marine ecosystems.

**Keywords:** microplastics; stormwater; drainage

**Citation:** Bond, C.; Li, H.; Rate, A.W. Land Use Pattern Affects Microplastic Concentrations in Stormwater Drains in Urban Catchments in Perth, Western Australia. *Land* **2022**, *11*, 1815. https://doi.org/10.3390/land11101815

Academic Editors: Bindong Sun, Tinglin Zhang, Wan Li, Chun Yin and Honghuan Gu

Received: 14 September 2022
Accepted: 12 October 2022
Published: 17 October 2022



## 1. Introduction

The mass production of plastic products began in the 1940s and over 7800 million tons of plastics have been produced since the year 1950 [1]. Less than 5% of these plastic products have been recovered, resulting in accumulation in various environments including marine, freshwater, urban, remote, agricultural, and industrial systems [2,3].

Microplastics are plastic particles with a diameter of less than 5 mm and have been widely identified as contaminants of concern [3,4]. Microplastics are important pollutants due to their small particle size, resistance to biodegradation and ability to move through various environmental media. Importantly, microplastics have a significant capacity to be readily absorbed and ingested by organisms [5]. As such, microplastic polymers represent a substantial risk to wildlife as the small particles may be mistaken as food and ingested; they are also small enough to be ingested by filter feeders and planktonic organisms giving a high potential for bioaccumulation of plastics themselves and co-contaminants [6,7]. The high surface area and hydrophobicity of microplastics allows the sorption of co-contaminants such as polycyclic aromatic hydrocarbons and heavy metals [8,9].

Microplastics can be classified by shape (fibres, fragments, film, microbeads) [10]. As well as shape, microplastics can also be categorised as either primary (intentionally manufactured as small particles, e.g., cosmetic microbeads) or secondary (derived from degradation and fragmentation of larger plastics).

Microplastic research was initially focused on marine environments, while terrestrial systems received far less attention [11]. Terrestrial microplastic contamination is 4–23 times greater than in marine environments, and land-based inputs are important sources preceding the transport of microplastics to the ocean [12].

Stormwater was first indirectly identified as a source of microplastics to urban lakes, and marine locations near urbanisation, based on mass balance rather than direct measurement [13,14] Stormwater runoff has since been widely identified as an important transport mechanism for microplastic pollution on land to the marine environment [12,15]. Urban stormwater seldom receives treatment which could remove microplastic particles, allowing direct fluxes to riverine and marine systems [16]. Microplastics enter stormwater systems through a combination of atmospheric deposition and overland flow [17,18] following which they may be removed by entanglement with organic materials, biofouling, or sedimentation [17,19,20]. Whether microplastics flow through stormwater systems or settle in sediment is determined by the size, shape and density of the particles [19].

Microplastics found in stormwater are comprised of a wide range of polymers with some of the most common being polyethylene, polypropylene, polyvinyl chloride, polyethylene terephthalate, and polystyrene [17]. The absolute abundance of microplastics, and the relative abundance of polymer types, will to some extent reflect the land use and human activities in the catchment [12]. There is no widespread consensus that land use intensity and population density can adequately account for microplastic concentrations in all areas of the globe [20]. Instead, other factors such as weather, specifically the timing of rainfall before sampling, affect measured microplastic concentrations, but these effects are also inconsistent [21,22].

This study is the first to examine the concentration, polymer type and shape of microplastics found in the water of stormwater drainage in Western Australia. The study aimed to answer the following research questions:

1. What is the concentration of microplastics in selected stormwater drains across the Perth and Peel region?
2. What plastic shapes, sizes, colours and polymer types are identified in the drainage systems?
3. Is there a significant difference in microplastic concentrations within or between any of the drainage systems?
4. Does the land use pattern, as defined by catchment area, catchment population, pre-sampling rainfall, and the proportions of residential, industrial, commercial, agricultural, natural land, and public open spaces, affect stormwater plastic concentrations?

## 2. Materials and Methods

### 2.1. Sample Sites

Six stormwater drainage systems were selected for sampling (Figure 1), based on their accessibility, broad distribution across the Perth area, and pattern of surrounding land uses, based on a shapefile of Water Corporation stormwater drains [23]. Two sites were sampled along each of the Bayswater Main Drain, Claisebrook Main Drain, Kitchener St Drain, Osborne Park Branch Drain, South Belmont Main Drain and South Coolup Main Drain (Figure 1). Details of samples are presented in Table 1. The region sampled has a winter-wet subtropical climate (Köppen Csa), with mean annual rainfall (1993–2021) of 737 mm, with 77% of rainfall occurring May–September.

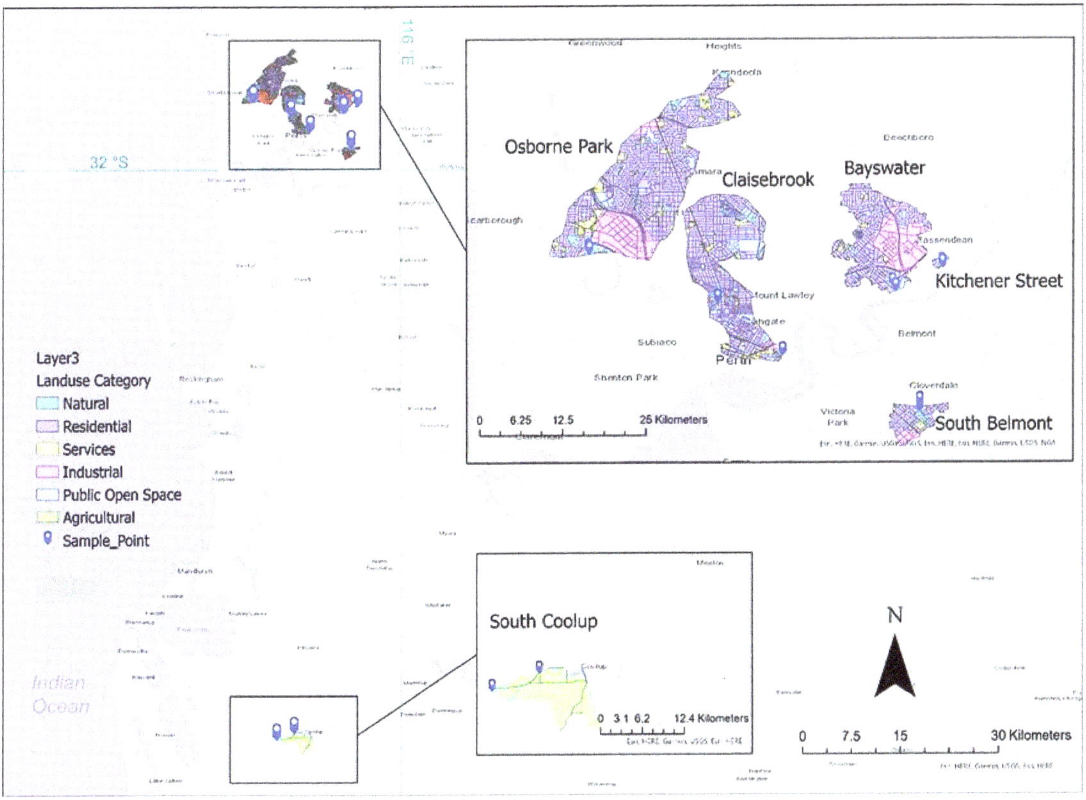


**Figure 1.** The drainage systems sampled, showing sample locations, catchment boundaries and land use categories.

**Table 1.** Sampling location details and the volume of water taken at each site. Two sites were selected at each drain system, denoted Site Number 1 and 2. All samples 3 repeats, 250 mL rinse.

| Sample Catchment | Site Number | Site Code | Coordinates (Decimal Degrees, WGS84) | | Sample Size |
|---|---|---|---|---|---|
| | | | Longitude | Latitude | |
| Bayswater Main Drain | 1 | B1 | 115.92135 | −31.92585 | 10 L |
| Bayswater Main Drain | 2 | B2 | 115.92186 | −31.92730 | 10 L |
| Claisebrook Main Drain | 1 | C1 | 115.85102 | −31.93210 | 10 L |
| Claisebrook Main Drain | 2 | C2 | 115.87702 | −31.95256 | 10 L |
| Osborne Park Branch Drain | 1 | O1 | 115.79987 | −31.91233 | 5 L |
| Osborne Park Branch Drain | 2 | O2 | 115.80029 | −31.91306 | 5 L |
| Kitchener St Drain | 1 | K1 | 115.94133 | −31.91799 | 10 L |
| Kitchener St Drain | 2 | K2 | 115.94171 | −31.91855 | 10 L |
| South Belmont Main Drain | 1 | SB1 | 115.93236 | −31.97443 | 10 L |
| South Belmont Main Drain | 2 | SB2 | 115.93233 | −31.97202 | 5L |
| South Coolup Main Drain | 1 | SC1 | 115.85390 | −32.74431 | 5 L |
| South Coolup Main Drain | 2 | SC2 | 115.83038 | −32.75319 | 5 L |

### *2.2. Sampling Protocol*

The design of the sampler used (Figure 2) was adapted from Ziajahromi et al. [24]. Each sampler used 10 cm diameter stainless steel disks (All-round Mesh, Victoria, Australia), with one of each of the aperture sizes (530 µm, 190 µm, 100 µm, and 25 µm), stacked between five 5 cm long PVC pipe joiners of 10 cm diameter, so each size class had a lower bound corresponding to the aperture size of the filter. The PVC joiners were then fastened together to form one 25 cm pipe with the filtration disks at 5 cm intervals.

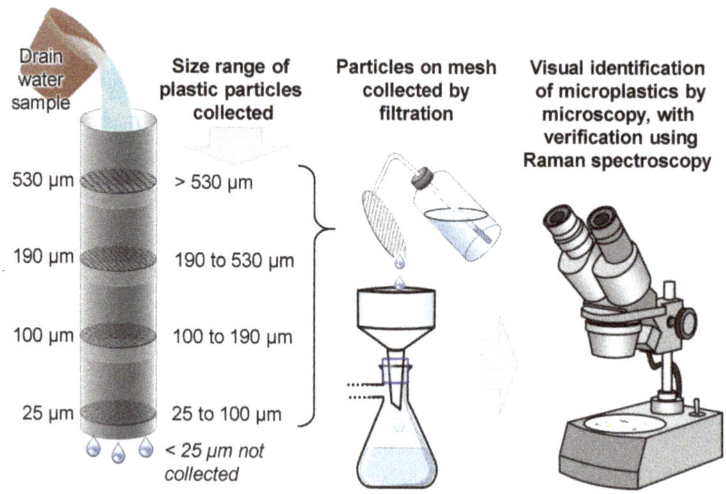


**Figure 2.** Stacked filter design and schematic of microplastic separation and counting procedures. Sampled water is passed through the filter stack so that microplastics are sorted according to size classification. The filtered water is discarded. Particles on each mesh filter are transferred quantitatively onto glass fibre filters for microscopic identification and counting.

Samples of either 5 L or 10 L were taken from the top 10–15 cm of the water body using a bucket. Where suspended solid content was high, and filtering the full 10 L was not feasible, 5 L samples were taken. The samples were then decanted through the stacked filtration device, leaving only suspended particles including any microplastics in the samples on the mesh disks. 250 mL of clean water was then added to rinse any potential microplastics stuck to the sides and poured through the sampling device. Each mesh disk was wrapped in aluminium foil for transport and storage. Sampling was conducted during winter. The timing of sampling was not related to the timing of rainfall.

### 2.3. Sample Processing

The solids on mesh sample filters and Al-foil wrappers were rinsed onto glass fibre filters (Whatman GF/A) for vacuum filtration. A scalpel was used to remove gently any remaining contents onto the filter. The glass filters were then removed carefully from the filter funnel/manifold and placed in clean Petri dishes in prior to microscopic analysis.

### 2.4. Sample Analysis

Microplastics were initially identified using a Nikon TE-PSE30 microscope at 4.5× magnification, according to the criteria identified by Hidalgo-Ruz et al. [25]; in addition, there should be no visible cellular or other organic structures, except for possible biofouling, and a positive reaction to the hot needle test [20,25]. Suspected microplastics on each filter were counted and categorised according to shape and colour. The shape categories included beads (spherical particles), films (thin coatings), fragments (diameter > thickness) or fibres (length > diameter) [10]. Colour categories included clear, white, grey, black, brown, blue, green, yellow, orange, red and pink.

### 2.5. Blanks

Potential background contamination in the laboratory and filtration processes were accounted for by assessing blank samples. Uniformity across the controls and samples themselves was ensured by processing both samples and blanks simultaneously in the laboratory. Five blank samples of 10 mL deionised (DI) water were passed through the glass fibre filters and analysed to determine background contamination from the equipment and DI water [20]. These filters were left uncovered for half an hour to replicate the exposure to atmospheric deposition experienced by the samples during the processing and counting of suspected microplastics on the samples.

### 2.6. Microplastic Identification and Characterisation

A randomly selected subset of visually identified microplastics were analysed using Raman microscopy (WITEC Alpha 300 RA+ Confocal Raman microscope). This representative subsample consisted of 6 fragments and 32 fibres, approximating the proportions of these particle morphologies in the samples. Microplastic particles were adhered to a clean glass slide with a droplet of silicone oil having a known Raman spectrum with minimal interfering bands.

Raman spectroscopy was conducted on a WITec Alpha 300 RA+ system with an Andor iDUS 401 CCD maintained at −60 °C, and a 20× objective. Infrared (785 nm) and red (633 nm) lasers were used with a 600 $mm^{-1}$ grating. Spectra were collected using ProjectFIVE software and cross-referenced manually using reference materials from the Spectral Database Index from the Infrared & Raman Users Group [26]. A false positive frequency was then calculated from the results of this Raman subsample verification and used to correct raw microplastic counts. Raman analysis was selected over ATR-FTIR spectroscopy due to its ability to analyse smaller particle sizes [27].

### 2.7. Data Analysis

Catchment modelling was conducted for each of the drainage systems. Geosciences Australia's Digital Elevation Model (DEM) for the Perth and Peel region was used to delineate areas of low-lying land [28]. Water Corporation and Department of Water and Environmental Regulation drainage channel datasets were compiled, along with the DEM, in ArcGIS 2.6.0 [29]. Watersheds for the entire drainage system were then calculated by mapping the flow direction of overland runoff. These watersheds estimate the total contributing drainage area for the system from the point of discharge. These watershed polygons were then manually altered to match the extent of drainage flow at each specific sampling location. The manual alterations were determined by interpreting the flow accumulation layer and approximating where the flow would funnel into the sample location. This created a more accurate representation of the actual catchment area flowing into each individual sampling location. These site-specific catchments were then converted to polygons, and the internal area of each was calculated, giving the area of land that each of the sample sites serviced. The watersheds calculated for the Claisebrook Main Drain and Osborne Park Branch Drain catchments were originally miscalculated from this catchment modelling process, as the digital elevation model did not account for anthropogenic alterations to the drainage systems in this area. As such, the catchment for these two drainage channels was manually entered into ArcGIS using reference materials on the actual catchment extent from DWER (2014) and Kobryn (2001).

The land use classification layers imported into ArcGIS were adapted from the Australian Land Use Management categories [30] used in the Department of Primary Industries and Regional Development land use dataset [31]. To simplify analysis, land use classifications were regrouped into six categories: residential (urban and rural); industrial; services; agriculture (including horticulture); natural (including water bodies); and public open space. Any misclassifications (e.g., natural land classed as production forestry) were corrected.

The population for each catchment was then determined by multiplying the population density of each suburb within the catchment by the area of the catchment it occupies. The population estimates from the fractions of suburbs within the overall catchment were then summed to give a catchment-wide population. Rainfall amounts were calculated as the cumulative rainfall for the 7 days prior to sampling. Rainfall data were obtained from the Australian Bureau of Meteorology records at the nearest weather station to each location [32].

All data curation and analyses were performed in R [33]. Prior to statistical analysis, microplastic count data were corrected to account for both background contamination from blank measurements, and the false positive rates identified with Raman spectroscopy. Corrections involved random sampling from vectors of experimental blank and false positive count measurements, rather than using mean blank or false positive values.

The differences in mean microplastic concentrations between catchments and sites were assessed using Type-II ANOVA based on linear mixed-effect (LME) models with Tukey pairwise contrasts, cross-checked with non-parametric Kruskal–Wallis tests. LME models were implemented in the R package 'nlme' [34], with contrasts calculated using the R package 'multcomp' [35]. The effects of relevant covariates (catchment population; catchment area; prior rainfall; proportions of the land use categories residential, industrial, services, agricultural, and public open space) on microplastic concentrations were assessed using linear mixed-effect models with sampling site as random effects and each covariate separately as fixed effects. Nested alternatives for each model were (1) using constant variance structure independent of catchment or sampling site; (2) including a variance function with different standard deviations for each catchment, and (3) including a variance function with different standard deviations for each sampling site. The model alternative selected for each covariate was the one giving the lowest Aikake Information Criterion value, if an analysis of variance showed a significant improvement over the next most complex model.

## 3. Results

### 3.1. Catchment Modelling

The catchment modelling output defined the area, estimated population, rainfall total for the week leading up to the date of sampling, and land use proportions expressed as percentages (Table 2).

**Table 2.** Summary of drain and catchment characteristics derived from catchment modelling.

| | B1 | B2 | C1 | C2 | K1 | K2 | O1 | O2 | SB1 | SB2 | SC1 | SC2 |
|---|---|---|---|---|---|---|---|---|---|---|---|---|
| | | | | | Catchment Characteristics | | | | | | | |
| **Area (ha)** | 1089 | 1105 | 1459 | 990 | 20 | 21 | 2227 | 2324 | 223 | 238 | 899 | 631 |
| **Population** | 21039 | 21372 | 34370 | 24548 | 298 | 319 | 45963 | 49291 | 3221 | 3387 | 91 | 64 |
| **Rainfall (mm)** [1] | 31.2 | 31.2 | 56.8 | 11.4 | 31.8 | 31.8 | 44.6 | 6.8 | 31.8 | 5.4 | 14.8 | 14.8 |
| | | | | | Land Use Proportion (%) | | | | | | | |
| **Public open** | 8.5 | 9.2 | 16.6 | 17.9 | 0.00 | 0.00 | 9.6 | 9.3 | 8.7 | 8.1 | 0.00 | 0.00 |
| **Industrial** | 31.4 | 30.9 | 1.4 | 0.00 | 0.00 | 0.00 | 13.0 | 16.0 | 17.4 | 16.3 | 0.47 | 0.50 |
| **Residential** | 58.0 | 58.0 | 76.8 | 77.7 | 84.2 | 84.2 | 68.5 | 65.9 | 64.6 | 67.0 | 1.9 | 2.7 |
| **Services** | 2.2 | 2.1 | 3.9 | 3.0 | 15.8 | 15.8 | 6.7 | 6.4 | 5.1 | 4.8 | 0.00 | 0.00 |
| **Natural** | 0.00 | 0.10 | 1.4 | 1.5 | 0.00 | 0.00 | 2.2 | 2.4 | 4.1 | 3.8 | 2.7 | 2.5 |
| **Agricultural** | 0.00 | 0.00 | 0.00 | 0.00 | 0.00 | 0.00 | 0.00 | 0.00 | 0.00 | 0.00 | 94.9 | 94.3 |

[1] In the 7 days prior to sampling.

### 3.2. Background Contamination and Raman Spectroscopy

Background concentrations (blanks) showed 0, 1, 1, 1 and 2 black suspected microplastic fibres. A value drawn randomly from the vector [0,1,1,1,2] was then subtracted, only from fibre counts, before scaling to concentrations based on sample volumes. Raman spectroscopy indicated that 18 of the 38 suspected microplastics initially counted after visual analysis were correctly identified. The false positive rate of the analysis was therefore 52.6%. This proportion of false positive microplastic identifications is consistent with the findings of other studies, in which between 20 and 70% of polymers have been misidentified [20]. Correctly identified polymers were matched as Polypropylene (PP) (64.7%), Polyethylene (PE) (23.5%), Polytetrafluoroethylene (PTFE) (5.9%), and Polyvinylidene fluoride (PVDF) (5.9%). Natural polymers misidentified as microplastics under visual inspection included cotton and wool fibres and various organic materials derived from plant and animal matter. Raman spectra of the plastics identified, as well as wool and cotton fibres, are presented in Figure 3.

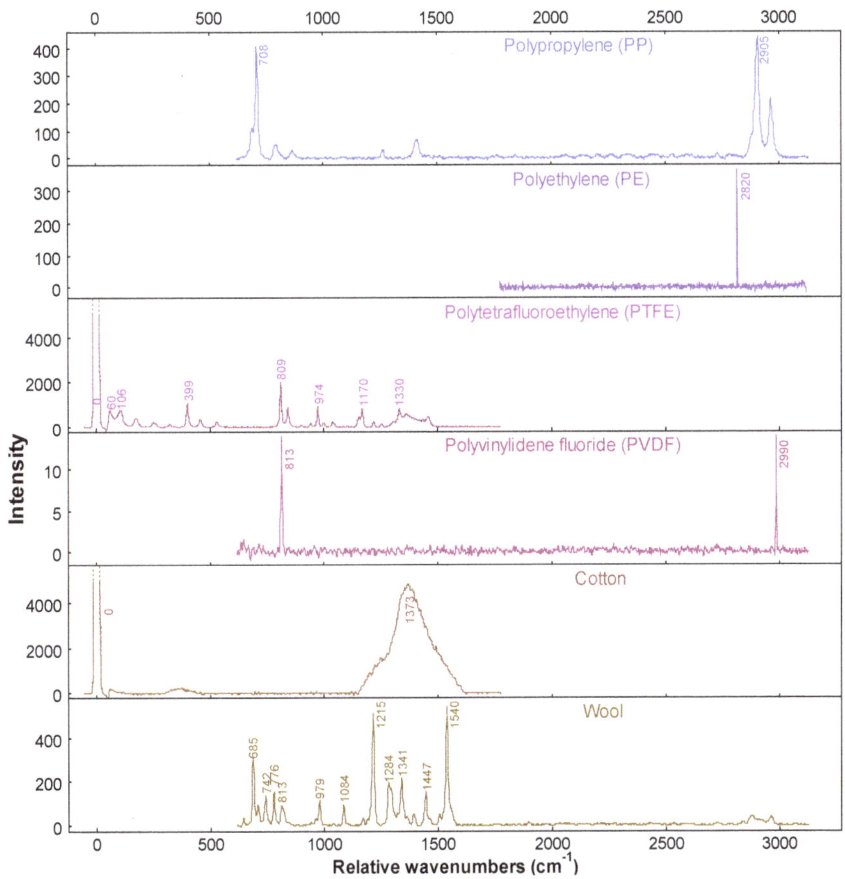


**Figure 3.** Raman spectra (intensity *vs.* relative wavenumbers, cm$^{-1}$) used to classify polymer materials identified from the microplastic subsamples which were identified visually. The main peaks for each spectrum are labelled with their positions in relative wavenumbers.

*3.3. Statistical Analyses of Factors Affecting Microplastic Concentrations*

(a) Microplastic concentrations across and within drainage catchments.

The mean concentration of microplastics across the drainage catchments ranged from 9.22 (Kitchener Road) to 20.1 MP/L (Osborne Park). The catchments with the two highest microplastic concentrations (Osborne Park and South Coolup) also had the greatest intra-catchment variation (relative standard deviations of 58% at Osborne Park and 59% at South Coolup). Table 3 summarises the microplastic concentrations by catchment, and Table 4 summarises the microplastic concentrations by sampling site.

**Table 3.** Mean microplastic concentration (MP/L) and standard deviation at a catchment level.

| | Bayswater | Claisebrook | Kitchener | Osborne | S. Belmont | S. Coolup |
|---|---|---|---|---|---|---|
| Mean (MP/L) | 12.6 | 12.8 | 9.2 | 20 | 12.8 | 18 |
| Standard deviation | 3.6 | 4.7 | 2.1 | 12 | 4.2 | 10 |

**Table 4.** Mean microplastic concentration (MP/L) and standard deviation at a site-specific level.

| | B1 | B2 | C1 | C2 | K1 | K2 | O1 | O2 | SB1 | SB2 | SC1 | SC2 |
|---|---|---|---|---|---|---|---|---|---|---|---|---|
| Mean (MP/L) | 13.9 | 11.2 | 14.7 | 11.0 | 9.7 | 8.8 | 25 | 15.2 | 12.9 | 12.7 | 25 | 10.3 |
| Standard deviation | 3.6 | 3.8 | 5.4 | 4.0 | 2.2 | 2.4 | 16 | 2.6 | 1.5 | 6.5 | 10 | 2.2 |

Within-catchment variation in mean microplastic concentrations was minimal, with no significant difference between the mean microplastic concentrations of site pairs in each catchment (Welch's *t*-test: $0.09 < p < 0.995$). Boxplots of the microplastic concentrations at each drainage catchment and each sampling site are shown in Figure 4.

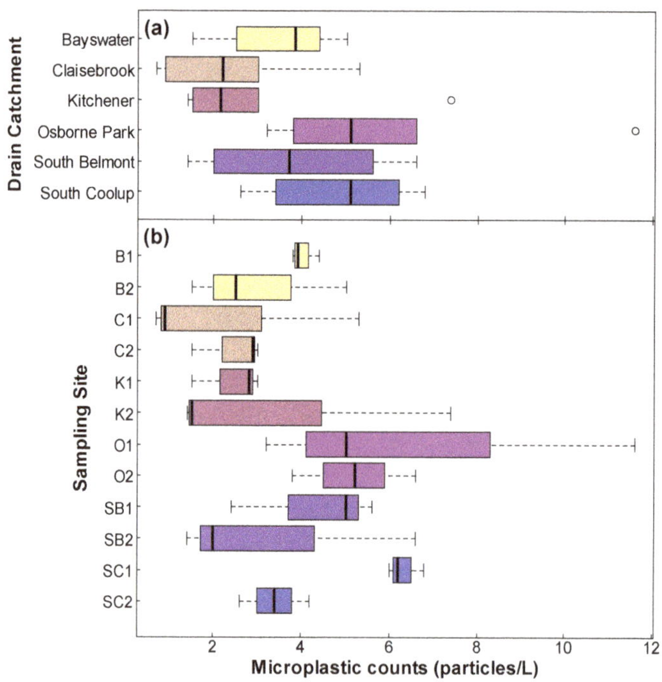


**Figure 4.** Comparison of microplastic concentrations (**a**) in each catchment and (**b**) at each site. The bar colours are simply to provide a visual match between catchments and sites in each catchment.

Neither site nor catchment explain the variation in microplastic concentrations, based on Type-II ANOVA analysis; 3% of the variance was explained by catchment, 23% by site (unrelated to the drainage catchment), and the residual variance was 73%. The Tukey pairwise contrasts, treating catchment as a factor, suggested that none of the catchment pairings had significantly different means from one another ($0.14 < p < 1$).

(b) Analysis of Covariates

Summaries of the LME models for each covariate are listed in Table 5. Catchment area, catchment population, industrial, natural, and public open space land use proportions had significant positive effects on microplastic concentrations. Conversely, the proportion of residential and services land uses had significant negative effects on microplastic concentrations. Preceding rainfall, and the proportion of agricultural land had no significant effect on microplastic concentrations.

**Table 5.** Summaries of linear mixed-effects models predicting microplastic concentrations from each of the covariates reflecting the pattern of land use in each catchment. Coefficients and P-values for covariates with significant effects are in bold type.

| Covariate | Model Specifications | Covariate Coefficient | *p*-Value |
|---|---|---|---|
| Catchment population ÷ 1000 | Heteroskedastic, variation at site level | **0.1041** | **0.005** |
| Catchment area | Heteroskedastic, variation at site level | **0.0025** | **0.0037** |
| Rainfall | Heteroskedastic, variation at site level | −0.0109 | 0.86 |
| Residential proportion | Heteroskedastic, variation at catchment level | **−0.1168** | **0.0169** |
| Agricultural proportion | Heteroskedastic, variation at site level | −0.0120 | 0.60 |
| Industrial proportion | Heteroskedastic, variation at site level | **0.1352** | **0.0107** |
| Natural proportion | Heteroskedastic, variation at site level | **0.7090** | **0.0248** |
| Services proportion | Heteroskedastic, variation at site level | **−0.2761** | **0.0093** |
| Public Open Space proportion | Heteroskedastic, variation at site level | **0.3193** | **0.0051** |

Some features of the data require further exploration. Notably, SC1 had high microplastic concentrations, despite a small catchment population (91 people) and a moderate catchment size (899 ha). Additionally, the negative effect of the proportion of services and residential land uses on microplastic concentrations appears to be influenced by the results at the South Coolup site. Here, the residential and services proportions were low (1.9% and 0%, respectively), yet mean concentrations were high. The trend of increasing microplastic concentration with the increasing proportion of industrial land also had an exception at site SC1, which had the second highest microplastic concentrations but only 0.5% industrial land use. Some covariates were significantly collinear (e.g., catchment area and population, Pearson's r = 0.94; residential-agricultural r = −0.95; etc.), so the individual effects may represent a combination of land use or demographic variables.

### 3.4. Microplastic Characteristics

*Morphology*. The microplastics observed included fibres, fragments, films and beads. Fibres were the most common, accounting for between 86% and 99% of synthetic polymers identified (Figure 5a). Fragments comprised between 1% and 13% of microplastics in the catchments. Films were only identified at the South Belmont Drain catchment, where they contributed 1% of the microplastic count. Claisebrook was the only drainage catchment where microplastic beads were observed but, at 0.01% of the count, microbeads were uncommon.

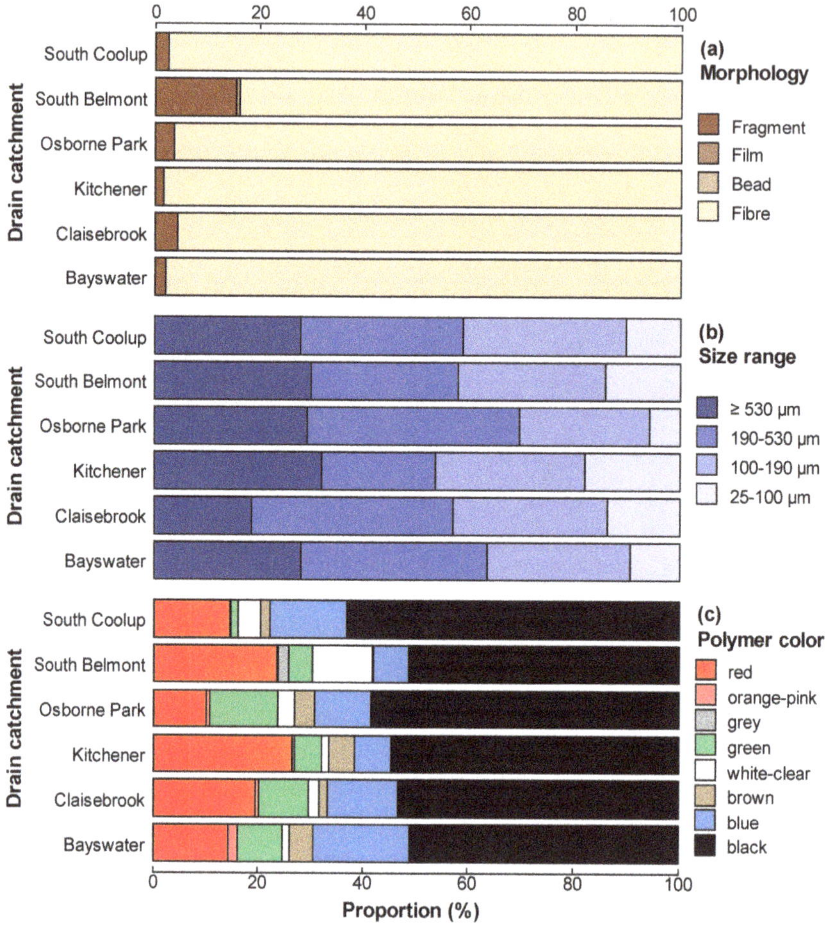


**Figure 5.** Proportions of microplastics in each catchment, classified by (**a**) shape (morphology), (**b**) size, and (**c**) colour.

*Size*. Microplastics in the 190–530 µm size category were the most abundant (33%), followed by 100–190 µm (28%), ≥530 µm (27%) and 25–100 µm (12%). Kitchener and South Belmont were the only drainage catchments having the ≥ 530 µm size category as the most populous size fraction, comprising 32% (K) and 30% (SB) of plastics identified. Kitchener and South Belmont catchments also featured the greatest proportion of 25–100 µm polymers, at 18% (K) and 14% (SB).

*Colour*. The majority of microplastics identified were black (55%), followed by red (18%), blue (12%) and green (7%) (Figure 5c). Pink, orange, brown, clear, white and grey microplastics each represented less than 5% of the total microplastic count. The proportions of microplastics of different morphologies are shown in Figure 5a, different sizes in Figure 5b, and different colours in Figure 5c. Additionally, a subsample of the microplastics identified are presented in Figure 6.

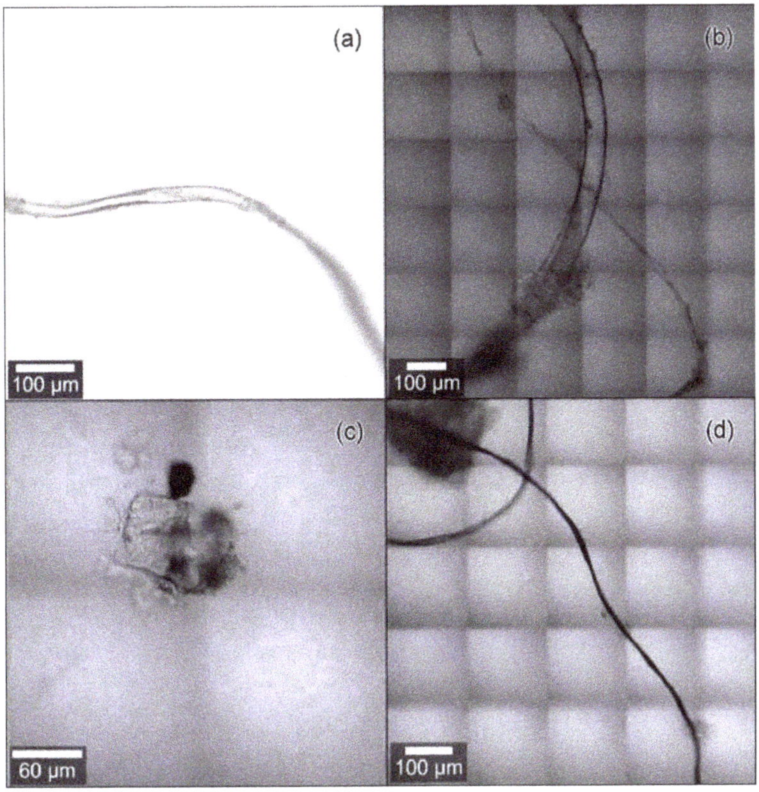


**Figure 6.** Optical micrographs of selected microplastics analysed by Raman spectroscopy: (**a**) clear fibre, (**b**) clear fibre, (**c**) clear fragment, and (**d**) black fibre.

## 4. Discussion

### *4.1. Catchment and Site Variation in Microplastic Concentrations*

The lack of variation in microplastic concentrations across drainage catchments or sites, or within drainage catchment site pairs, reflects the ubiquitous occurrence of microplastics in the Perth region. Similarly, Mora-Teddy and Matthaei [36] found no significant difference between mean microplastic concentrations in drainage catchments in New Zealand. In contrast, Lutz et al. [20] observed substantial variation in microplastic concentrations in stormwater drain sediment in Perth, both within and between catchments. This difference may reflect the more dynamic nature of water, compared with sediments which provide a more stable, longer-term sink for microplastics [20].

### *4.2. Factors Affecting Microplastic Concentrations*

Both catchment area and population were predictive, with positive effects, of stormwater microplastic concentrations in this study. This result is in contrast with microplastic concentrations in stormwater *sediments* in Perth and Melbourne, which were not related to catchment size or population [20,37].

The ability of the proportion of industrial land use to predict microplastic concentrations has been widely reported, consistent with the significant positive effect shown in this study (Table 5). Piñon-Colin et al. [21], Liu et al. [38] and Townsend et al. [37] found that catchments with higher industrial land use proportions had greater microplastic concentrations than catchments dominated by residential land.

The significant positive effect of the proportions of public open space and natural land on microplastic concentrations is somewhat unexpected. In contrast, Townsend et al. [37] found a negative correlation between public open space and microplastics, instead noting that higher concentrations were detected with increasing proportions of urbanisation. Similarly, Lutz et al. [20] measured lower microplastic concentrations in stormwater sediment in areas with greater proportions of public open space and natural land. Some studies conducted in relatively untouched environments, however, have detected high microplastic concentrations in regions with little to no urban development. This indicates that while land use intensity and population can be predictors for microplastic concentrations in some areas, these results cannot be generalised across different studies, regions, or stormwater drainage catchments [20,39,40]. Additionally, the detection of high microplastic concentrations in areas with greater natural land and public open space may be influenced by the inclusion of wetlands and lakes in these land use categories. Wetlands act as sinks for microplastics [38], and areas with greater proportions of natural land and public open spaces included wetland and lake environments and surrounding parks. Therefore, the influence of these wetland sinks may contribute to the significant positive effects of the proportion of natural land and public open space on microplastic concentrations.

Our finding that the agricultural land use fraction is not correlated with microplastic concentrations is unsurprising, given that only one catchment (South Coolup) had any agricultural land. It would usually be expected, however, that agricultural areas that do not use biosolids applications would have lower microplastic concentrations [20].

Rainfall in the preceding 7 days was not predictive of microplastic concentrations in this study. Conversely, Piñon-Colin et al. [21] found the greatest MP concentrations during rainfall events, but their approach differed in that sites were sampled at the beginning of the rainfall event and 10 and 30 min into the event for seven separate weather events. Similarly, Yonkos et al. [40] found that microplastic concentrations were higher after rainfall or other extreme weather conditions such as hurricanes. Different trends may be observed for different microplastic particle morphologies, with fragments increasing during rainfall events but not fibres [22]. This may explain the lack of a rainfall effect in our study, given the predominance of fibres over other microplastic types.

### 4.3. Microplastic Concentrations

The microplastic concentrations measured in the stormwater drains had means varying from 20.1 MP/L at Osborne Park Branch Drain to 9.2 MP/L at Kitchener St Drain. These concentrations match the ranges observed in recent literature, which vary between 15.4 and 30.9 MP/L [22,41,42]. Our microplastic concentrations are lower than those reported by Piñon-Colin et al. [21] for stormwater runoff in semi-arid Tijuana, Mexico with median microplastic concentrations between 66 and 191 MP/L. There may be significant variations in reported concentrations stemming from differing sampling and analytical techniques, and variations in the types of water body sampled. For instance, Liu et al. [38] assessed urban and highway retention ponds receiving stormwater runoff flows from various land uses in Denmark and found microplastic concentrations between 490 and 22,894 MP/L. However, this study included plastics between 10 and 2000 µm, an upper classification twice the size of the approximate 1000 µm upper limit in this study. Similarly, Mora-Teddy and Matthaei [36] assessed stormwater systems in New Zealand and discovered microplastic concentrations in the range of <1000–44,000 MP/L, with an upper size limit of 5,000 µm and without spectroscopic verification of polymer composition or false-positive correction.

As observed in other studies of microplastics in stormwater systems [16,20,43], the dominant microplastic shapes found were fibres. The high proportion of fibres relative to other morphologies relates to the ease of transport by water, since their sedimentation rates are slower than for other particle shapes [20].

### 4.4. Polymer Types

The dominance of polypropylene and polyethylene in this study is consistent with other studies of microplastic in freshwater. Piñon-Colin et al. [21] found that PE and polystyrene (PS) accounted for the greatest proportion of plastics in stormwater, while Liu et al. [38] detected PVE, PS, PP and PE, with PP being the most common polymer. Interestingly, Lutz et al. [20] detected polyethylene terephthalate (PET), nylon polyamide (PA), polyacrylonitrile (PAN) and a synthetic and natural polymer blend, as well as PP and PE in stormwater sediment across Perth. The dominance of PP and PE in our samples can be explained by their densities relative to other plastic polymers. Since PP and PE have lower densities than fresh water, they would float downstream through the drains, rather than settling into the sediments (Lutz et al. 2021). Conversely, more dense polymers such as PA, PVC, polyurethane (PU) and PET, which were not identified in this study, are more likely to become embedded in sediments [44]. Several studies have found that higher-density polymers such as PET, PVC, PA, polyester, and PU were more prevalent in sediment than in water samples [38,42,45]. The dominance of PE and PP in this study indicates some potential sources of microplastics, including single-use plastic bags (PE), and food containers, fabrics, textiles, packaging materials and reusable products (PP) [44,46].

## 5. Conclusions

Our understanding of the fate and transportation of microplastics in terrestrial environments is still limited, and this first study to report stormwater microplastic concentrations in Western Australia provides useful information on the role of stormwater. The study focused on the four research questions stated in the Introduction. First, stormwater drains in Perth, Western Australia, contain concentrations of microplastics similar to other stormwater drainage catchments worldwide. Second, the predominance of fibres and low-density polymers suggests an active role of stormwater drains in microplastic transport. In the case of Perth, stormwater discharge is to the Swan–Canning Estuary, with a direct connection to the Indian Ocean. Third, the consistency of concentrations between different drains, and between sampling sites on the same drain, is consistent with widely occurring microplastic contamination in this region. Finally, the significant effects of catchment population and area and urban land use proportions (residential, industrial, commercial, and public open space) on microplastic concentrations in stormwater are consistent with expectations. The unexpected positive effect of the proportion of natural land on stormwater microplastic concentrations suggests that further research is required to test hypotheses about the role of wetlands or other mechanisms for microplastic transport and retention. Stormwater drains in this urban area, and others worldwide, have a direct connection with estuarine and marine environments. This suggests that connections between stormwater and natural waters need to be interrupted using engineered solutions to limit the transport of microplastic to sensitive ecosystems. This is already widely implemented for macroscopic solids in stormwater, but the small size of plastic particles is an additional threat to urban sustainability.

Future investigations should also sample additional locations, before and after rainfall events and over an extended timeframe, with the timing of sampling designed to understand the effect of weather and seasonality on microplastic concentrations. The possible relationship between sediment microplastic concentrations and stormwater concentrations is also worthy of investigation. The predominance of microplastic fibres in our study also implies that aeolian transport may be important, a further potentially fruitful avenue for research.

**Author Contributions:** Conceptualisation, C.B. and A.W.R.; methodology, C.B., H.L. and A.W.R.; formal analysis, C.B. and A.W.R.; investigation, C.B. and H.L.; data curation, C.B.; writing—original draft preparation, C.B.; writing—review and editing, A.W.R.; visualisation, C.B. and A.W.R.; supervision, H.L. and A.W.R.; project administration, A.W.R. All authors have read and agreed to the published version of the manuscript.

**Funding:** This research received no external funding.

**Data Availability Statement:** The data, with metadata, are available via Mendeley Data: see Rate, Andrew; Bond, Cassandra; Li, Hua (2022), "Data relating microplastics concentrations in stormwater drains to catchment land use and demographics, Perth, Western Australia", *Mendeley Data*, V1, doi:10.17632/vtdwvg34sd.1.

**Acknowledgments:** We acknowledge that The University of Western Australia and the sampling sites are situated on the ancestral lands of the Whadjuk Noongar people. We acknowledge the First Australians as the traditional owners of the lands where these data were collected, and pay respect to their Elders: past, present, and emerging. The authors acknowledge the facilities, and the scientific and technical assistance of the Australian Microscopy & Microanalysis Research Facility at the Centre for Microscopy, Characterisation & Analysis, The University of Western Australia, a facility funded by the University, State and Commonwealth Governments. The background image for the graphical abstract is by Bev Sykes on Flickr and is used under the terms of a CC-BY licence.

**Conflicts of Interest:** The authors declare no conflict of interest.

## References

1. Geyer, R.; Jambeck, J.R.; Law, K.L. Production, use, and fate of all plastics ever made. *Sci. Adv.* **2017**, *3*, e1700782. [CrossRef] [PubMed]
2. Sutherland, W.J.; Clout, M.; Côté, I.M.; Daszak, P.; Depledge, M.H.; Fellman, L.; Fleishman, E.; Garthwaite, R.; Gibbons, D.W.; De Lurio, J.; et al. A horizon scan of global conservation issues for 2010. *Trends Ecol. Evol.* **2010**, *25*, 1–7. [CrossRef] [PubMed]
3. Zhang, K.; Shi, H.; Peng, J.; Wang, Y.; Xiong, X.; Wu, C.; Lam, P.K.S. Microplastic pollution in China's inland water systems: A review of findings, methods, characteristics, effects, and management. *Sci. Total Environ.* **2018**, *630*, 1641–1653. [CrossRef] [PubMed]
4. Arthur, C.; Baker, J.; Bamford, H. (Eds.) *Proceedings of the International Research Workshop on the Occurrence, Effects and Fate of Microplastic Marine Debris*; National Oceanic and Atmospheric Administration (NOAA): Tacoma, WA, USA, 2008; Volume NOS-OR&R-30, p. 528.
5. Qi, R.; Jones, D.L.; Li, Z.; Liu, Q.; Yan, C. Behavior of microplastics and plastic film residues in the soil environment: A critical review. *Sci. Total Environ.* **2020**, *703*, 134722. [CrossRef]
6. Wright, S.L.; Thompson, R.C.; Galloway, T.S. The physical impacts of microplastics on marine organisms: A review. *Environ. Pollut.* **2013**, *178*, 483–492. [CrossRef]
7. Cole, M.; Lindeque, P.; Halsband, C.; Galloway, T.S. Microplastics as contaminants in the marine environment: A review. *Mar. Pollut. Bull.* **2011**, *62*, 2588–2597. [CrossRef]
8. Teuten, E.L.; Rowland, S.J.; Galloway, T.S.; Thompson, R.C. Potential for Plastics to Transport Hydrophobic Contaminants. *Environ. Sci. Technol.* **2007**, *41*, 7759–7764. [CrossRef]
9. Khalid, N.; Aqeel, M.; Noman, A.; Khan, S.M.; Akhter, N. Interactions and effects of microplastics with heavy metals in aquatic and terrestrial environments. *Environ. Pollut.* **2021**, *290*, 118104. [CrossRef]
10. Dris, R.; Gasperi, J.; Tassin, B. Sources and Fate of Microplastics in Urban Areas: A Focus on Paris Megacity. In *Freshwater Microplastics: Emerging Environmental Contaminants?* Wagner, M., Lambert, S., Eds.; Springer International Publishing: Cham, Switzerland, 2018; pp. 69–83.
11. de Souza Machado, A.A.; Kloas, W.; Zarfl, C.; Hempel, S.; Rillig, M.C. Microplastics as an emerging threat to terrestrial ecosystems. *Glob. Change Biol.* **2018**, *24*, 1405–1416. [CrossRef]
12. Horton, A.A.; Walton, A.; Spurgeon, D.J.; Lahive, E.; Svendsen, C. Microplastics in freshwater and terrestrial environments: Evaluating the current understanding to identify the knowledge gaps and future research priorities. *Sci. Total Environ.* **2017**, *586*, 127–141. [CrossRef]
13. Sutton, R.; Mason, S.A.; Stanek, S.K.; Willis-Norton, E.; Wren, I.F.; Box, C. Microplastic contamination in the San Francisco Bay, California, USA. *Mar. Pollut. Bull.* **2016**, *109*, 230–235. [CrossRef]
14. Lasee, S.; Mauricio, J.; Thompson, W.; Karnjanapiboonwong, A.; Kasumba, J.; Subbiah, S.; Morse, A.; Anderson, T. Microplastics in a freshwater environment receiving treated wastewater effluent: Microplastics in Urban Surface Water. *Integr. Environ. Assess. Manag.* **2017**, *13*, 528–532. [CrossRef]
15. Boucher, J.; Friot, D. *Primary Microplastics in the Oceans: A Global Evaluation of Sources*; International Union for Conservation of Nature and Natural Resources (IUCN): Gland, Switzerland, 2017; p. 43.
16. Smyth, K.; Drake, J.; Li, Y.; Rochman, C.; Van Seters, T.; Passeport, E. Bioretention cells remove microplastics from urban stormwater. *Water Res.* **2021**, *191*, 116785. [CrossRef]
17. Horton, A.A.; Dixon, S.J. Microplastics: An introduction to environmental transport processes. *WIREs Water* **2018**, *5*, e1268. [CrossRef]
18. Monira, S.; Bhuiyan, M.A.; Haque, N.; Shah, K.; Roychand, R.; Hai, F.I.; Pramanik, B.K. Understanding the fate and control of road dust-associated microplastics in stormwater. *Process Saf. Environ. Protect.* **2021**, *152*, 47–57. [CrossRef]

19. Kowalski, N.; Reichardt, A.M.; Waniek, J.J. Sinking rates of microplastics and potential implications of their alteration by physical, biological, and chemical factors. *Mar. Pollut. Bull.* **2016**, *109*, 310–319. [CrossRef]
20. Lutz, N.; Fogarty, J.; Rate, A. Accumulation and potential for transport of microplastics in stormwater drains into marine environments, Perth region, Western Australia. *Mar. Pollut. Bull.* **2021**, *168*, 112362. [CrossRef]
21. Piñon-Colin, T.d.J.; Rodriguez-Jimenez, R.; Rogel-Hernandez, E.; Alvarez-Andrade, A.; Wakida, F.T. Microplastics in stormwater runoff in a semiarid region, Tijuana, Mexico. *Sci. Total Environ.* **2020**, *704*, 135411. [CrossRef]
22. Treilles, R.; Gasperi, J.; Gallard, A.; Saad, M.; Dris, R.; Partibane, C.; Breton, J.; Tassin, B. Microplastics and microfibers in urban runoff from a suburban catchment of Greater Paris. *Environ. Pollut.* **2021**, *287*, 117352. [CrossRef]
23. Data WA. Water Corporation: Drainage Open Channel (WCORP-083). Government of Western Australia: Perth, Australia. Available online: https://catalogue.data.wa.gov.au/dataset/drain-open-channel (accessed on 7 October 2022).
24. Ziajahromi, S.; Neale, P.A.; Rintoul, L.; Leusch, F.D.L. Wastewater treatment plants as a pathway for microplastics: Development of a new approach to sample wastewater-based microplastics. *Water Res.* **2017**, *112*, 93–99. [CrossRef]
25. Hidalgo-Ruz, V.; Gutow, L.; Thompson, R.C.; Thiel, M. Microplastics in the Marine Environment: A Review of the Methods Used for Identification and Quantification. *Environ. Sci. Technol.* **2012**, *46*, 3060–3075. [CrossRef]
26. IRUG. Spectral Database Index. Available online: http://www.irug.org/search-spectral-database (accessed on 7 October 2022).
27. Araujo, C.F.; Nolasco, M.M.; Ribeiro, A.M.P.; Ribeiro-Claro, P.J.A. Identification of microplastics using Raman spectroscopy: Latest developments and future prospects. *Water Res.* **2018**, *142*, 426–440. [CrossRef]
28. Geoscience Australia. *Digital Elevation model (DEM) of Australia derived from LiDAR 5 Metre Grid*; Geoscience Australia: Symonston, Australia, 2021.
29. ESRI. *ArcGIS Pro: Release 2.6.0*; Environmental Systems Research Institute: Redlands, CA, USA, 2020.
30. ABARES. *The Australian Land Use and Management Classification Version 8*; Department of Agriculture, Fisheries and Forestry: Canberra, Australia, 2016.
31. DPIRD. *Catchment Scale Land Use Mapping for Western Australia (DPIRD-067)*; Department of Primary Industries and Regional Development: Perth, Autralia, 2018.
32. BOM. Climate Data Online. Available online: http://www.bom.gov.au/climate/data/ (accessed on 15 September 2021).
33. R Core Team. *R: A Language and Environment for Statistical Computing, Version 4.2.0*; R Foundation for Statistical Computing: Vienna, Austria, 2022.
34. Pinheiro, J.; Bates, D. *R Core Team nlme: Linear and Nonlinear Mixed Effects Models*; R Package Version 3.1-157; R Foundation for Statistical Computing: Vienna, Austria, 2022.
35. Hothorn, T.; Bretz, F.; Westfall, P. Simultaneous Inference in General Parametric Models. *Biom. J.* **2008**, *50*, 346–363. [CrossRef]
36. Mora-Teddy, A.K.; Matthaei, C.D. Microplastic pollution in urban streams across New Zealand: Concentrations, composition and implications. *N. Z. J. Mar. FreshWater Res.* **2020**, *54*, 233–250. [CrossRef]
37. Townsend, K.R.; Lu, H.; Sharley, D.J.; Pettigrove, V. Associations between microplastic pollution and land use in urban wetland sediments. *Environ. Sci. Pollut. Res.* **2019**, *26*, 22551–22561. [CrossRef]
38. Liu, F.; Olesen, K.B.; Borregaard, A.R.; Vollertsen, J. Microplastics in urban and highway stormwater retention ponds. *Sci. Total Environ.* **2019**, *671*, 992–1000. [CrossRef]
39. Free, C.M.; Jensen, O.P.; Mason, S.A.; Eriksen, M.; Williamson, N.J.; Boldgiv, B. High-levels of microplastic pollution in a large, remote, mountain lake. *Mar. Pollut. Bull.* **2014**, *85*, 156–163. [CrossRef]
40. Yonkos, L.T.; Friedel, E.A.; Perez-Reyes, A.C.; Ghosal, S.; Arthur, C.D. Microplastics in Four Estuarine Rivers in the Chesapeake Bay, U.S.A. *Environ. Sci. Technol.* **2014**, *48*, 14195–14202. [CrossRef]
41. Grbić, J.; Helm, P.; Athey, S.; Rochman, C.M. Microplastics entering northwestern Lake Ontario are diverse and linked to urban sources. *Water Res.* **2020**, *174*, 115623. [CrossRef]
42. Sang, W.; Chen, Z.; Mei, L.; Hao, S.; Zhan, C.; Zhang, W.b.; Li, M.; Liu, J. The abundance and characteristics of microplastics in rainwater pipelines in Wuhan, China. *Sci. Total Environ.* **2021**, *755*, 142606. [CrossRef]
43. Zhang, K.; Xu, S.; Zhang, Y.; Lo, Y.; Liu, M.; Ma, Y.; Chau, H.S.; Cao, Y.; Xu, X.; Wu, R.; et al. A systematic study of microplastic occurrence in urban water networks of a metropolis. *Water Res.* **2022**, *223*, 118992. [CrossRef] [PubMed]
44. Shruti, V.C.; Pérez-Guevara, F.; Elizalde-Martínez, I.; Kutralam-Muniasamy, G. Current trends and analytical methods for evaluation of microplastics in stormwater. *Trends Environ. Anal. Chem.* **2021**, *30*, e00123. [CrossRef]
45. Ziajahromi, S.; Drapper, D.; Hornbuckle, A.; Rintoul, L.; Leusch, F.D.L. Microplastic pollution in a stormwater floating treatment wetland: Detection of tyre particles in sediment. *Sci. Total Environ.* **2020**, *713*, 136356. [CrossRef] [PubMed]
46. Erni-Cassola, G.; Zadjelovic, V.; Gibson, M.I.; Christie-Oleza, J.A. Distribution of plastic polymer types in the marine environment; A meta-analysis. *J. Hazard. Mater.* **2019**, *369*, 691–698. [CrossRef]

land

Article

# Spatial Patterns Exploration and Impacts Modelling of Carbon Emissions: Evidence from Three Stages of Metropolitan Areas in the YREB, China

Yichen Ding [1,2,3,†], Yaping Huang [1,2,3,†], Lairong Xie [4], Shiwei Lu [1,2,3,5,*], Leizhou Zhu [1,2,3], Chunguang Hu [1,2,3] and Yidan Chen [1,2,3]

1 School of Architecture and Urban Planning, Huazhong University of Science and Technology, Wuhan 430074, China
2 Hubei Engineering and Technology Research Center of Urbanization, Wuhan 430074, China
3 The Key Laboratory of Urban Simulation for Ministry of Natural Resources, Wuhan 430074, China
4 Wuhan Huazhong University of Science and Technology Architectural Planning and Design Institute Co., Ltd., Wuhan 430074, China
5 Key Laboratory of Urban Land Resources Monitoring and Simulation, Ministry of Natural Resources, Shenzhen 518034, China
* Correspondence: lusw@hust.edu.cn; Tel.: +86-27-87543156
† These authors contributed equally to this work.

**Abstract:** Metropolitan areas in China are not only the core spatial carriers of urbanization development but also the main generators of land use carbon emission (LUCE). However, existing research lacks comparative studies on the differential patterns and impact factors of LUCE in different stages of metropolitan areas. Therefore, this paper deeply analyzes the spatial characteristics of LUCE and the coupling coordination degree (CCD) of the economy contributive coefficient (ECC) and ecological support coefficient (ESC) in three different stages of metropolitan areas in the Yangtze River Economic Belt (YREB), China. Moreover, quantitative modelling of the impact factors of LUCE in these different stages of metropolitan areas is furtherly revealed. Results show that: (1) The more mature stage of the metropolitan area, the higher the amount of LUCE, and the more districts or counties with high carbon emissions levels are clustered. (2) At the metropolitan area scale, the more mature the metropolitan area is and the lower the CCD between ECC and ESC is, while at the finer scale, more developed counties have lower CCD. (3) Resident population, per capita GDP, and urbanization rate have good explanatory effects on carbon emissions in these three metropolitan areas; however, except for the urbanization rate, which has a negative effect on LUCE in Nanchang metropolitan area (NMA), the other two factors have positive effects on LUCE in these three metropolitan areas. This study has important implications for different stages of metropolitan areas to formulate targeted LUCE reduction policies.

**Keywords:** land use carbon emissions; metropolitan areas; coupling coordination degree; STIRPAT model; driving factors

**Citation:** Ding, Y.; Huang, Y.; Xie, L.; Lu, S.; Zhu, L.; Hu, C.; Chen, Y. Spatial Patterns Exploration and Impacts Modelling of Carbon Emissions: Evidence from Three Stages of Metropolitan Areas in the YREB, China. *Land* **2022**, *11*, 1835. https://doi.org/10.3390/land11101835

Academic Editors: Bindong Sun, Tinglin Zhang, Wan Li, Chun Yin and Honghuan Gu

Received: 23 September 2022
Accepted: 16 October 2022
Published: 18 October 2022



## 1. Introduction

Climate change brings enormous challenges to the natural environment and human society. Studies have reported that carbon dioxide is one of the dominant contributors to climate change [1], which has become the main area of concern at home and overseas. Since the industrial revolution, land use carbon emissions (LUCE) have contributed around 30% of human carbon emissions (CE) [2,3]. At present, research on LUCE is relatively rich in mainly two parts. Firstly, in terms of research content, the spatio-temporal characteristics of LUCE and the influencing factors [4,5], the relationships of LUCE [6], the network relationship and spillover effects [7], the efficiency [8], the economy contributive coefficient (ECC) and ecological support coefficient (ESC) of LUCE [9,10] are the research concerns of most studies. However, these studies do not focus as much on the CCD (coupling coordination degree) between ECC and ESC of LUCE and are unable to offer specific improvements to this relationship. Secondly, in terms of spatial scales, studies have been carried out on the urban agglomeration [11], provincial or state [12], municipal [13], and watershed scales [14], while few studies have been conducted in metropolitan area scale. However, as the main form of urbanization [15], the land use of metropolitan areas is not only the spatial projection of the main economic activities of human society but also the main generator of carbon emissions [16,17]. Therefore, it is of great urgency to scientifically identify the CCD between ECC and ESC and the impact factors of carbon emissions in metropolitan areas, which is helpful for formulating targeted low-carbon development measures in this kind of important area.

Studies have reported that different regions and areas had different network synergistic capabilities and driving power [18]. Similarly, the LUCE characteristics of metropolitan areas at different development stages may also differ significantly. When Fujii et al. studied the relationship between economic development and $CO_2$ emissions in 276 global metropolitan areas, they assumed that the urban $CO_2$ emissions per capita in the same sector would show differences in different urban economic development stages [19]. However, verifications of the above assumptions have not been conducted. In addition, the current studies on $CO_2$ emissions in urban areas of China are usually focused on a single evolution type of study area [20,21]. The same problem also exists for the Yangtze River Economic Belt (YREB), which is leading China's high-quality economic development. For instance, much attention has been paid to the environmental and economic development of the YREB, which is crucial to both regional ecological security and sustainable development in China [22]. The existing studies mainly concern the patterns [23], the influencing factors [24,25], and the efficiency of LUCE [26] in the YREB. For example, spatial autocorrelation [27], social network analysis (SNA) [7], and information entropy model [28] are often introduced to analyze the spatio-temporal patterns of LUCE in the YREB. The grey relational analysis model [29], regression models regarding spatial lag model, spatial error model [30], or LMDI [31] are used to model the impact factors; DEA, SBM-DEA [32], and SBM-UN model [33] are often used to measure efficiency under the constraint of LUCE. In addition, over 95% population lives in the 34 metropolitan areas in China, but few studies have compared the LUCE characteristics of metropolitan areas in different development stages. Therefore, a systematic study on the spatial differentiation of LUCE in metropolitan areas at different development stages is needed.

In summary, existing studies have made some achievements in regional LUCE in the YREB. However, there is still a lack of comparative investigations in metropolitan areas with different development stages. Simultaneously, the CCD between ECC and ESC of LUCE and impact factors are rarely analyzed. Therefore, the purpose of this study is to conduct a comparative study on the spatial patterns of LUCE in metropolitan areas at various development stages in the YREB and to explore the CCD between ECC and ESC, and the impact factors of LUCE in each metropolitan area. Contributions of this paper are two-fold: Firstly, the spatial differentiation characteristics and the spatial patterns of LUCE in cultivating, developmental and mature metropolitan areas are identified, which is conducive to the determination of sub-regional and differentiated low-carbon sustainable development goals for each metropolitan area. Secondly, differential analysis of the impact factors of LUCE from metropolitan areas is conducive to the targeted formulation of carbon reduction measures for each metropolitan area. The findings of this study can serve implications for the low-carbon development of metropolitan areas at different development stages.

## 2. Materials and Methods

### 2.1. Study Area

The Yangtze River Economic Belt (YREB) is an important strategic area for China's economic development [34,35]. YREB is divided into three parts (namely, the upper, middle, and lower reaches). Chongqing, Sichuan, Guizhou, and Yunnan provinces are in the upper reach. Hubei, Jiangxi, and Hunan provinces are in the middle of reach. Anhui, Zhejiang, Jiangsu, and Shanghai provinces are in the lower reach. The lower reach is the most developed, which has the famous Yangtze River Delta (YRD), while the economic levels of the other two reaches are relatively low [36]. The Chengdu, Nanchang, and Hangzhou metropolitan areas are located in the upper, middle, and lower reaches of the YREB, respectively. According to the China Metropolitan Area Development Report 2021 [37] announced by the China Institute of New Urbanization of Tsinghua University, the Chengdu, Nanchang, and Hangzhou metropolitan areas belong to the developmental type, cultivating type, and mature type respectively. Thus, these three different stages of metropolitan areas are the study cases of this research (as shown in Figure 1).

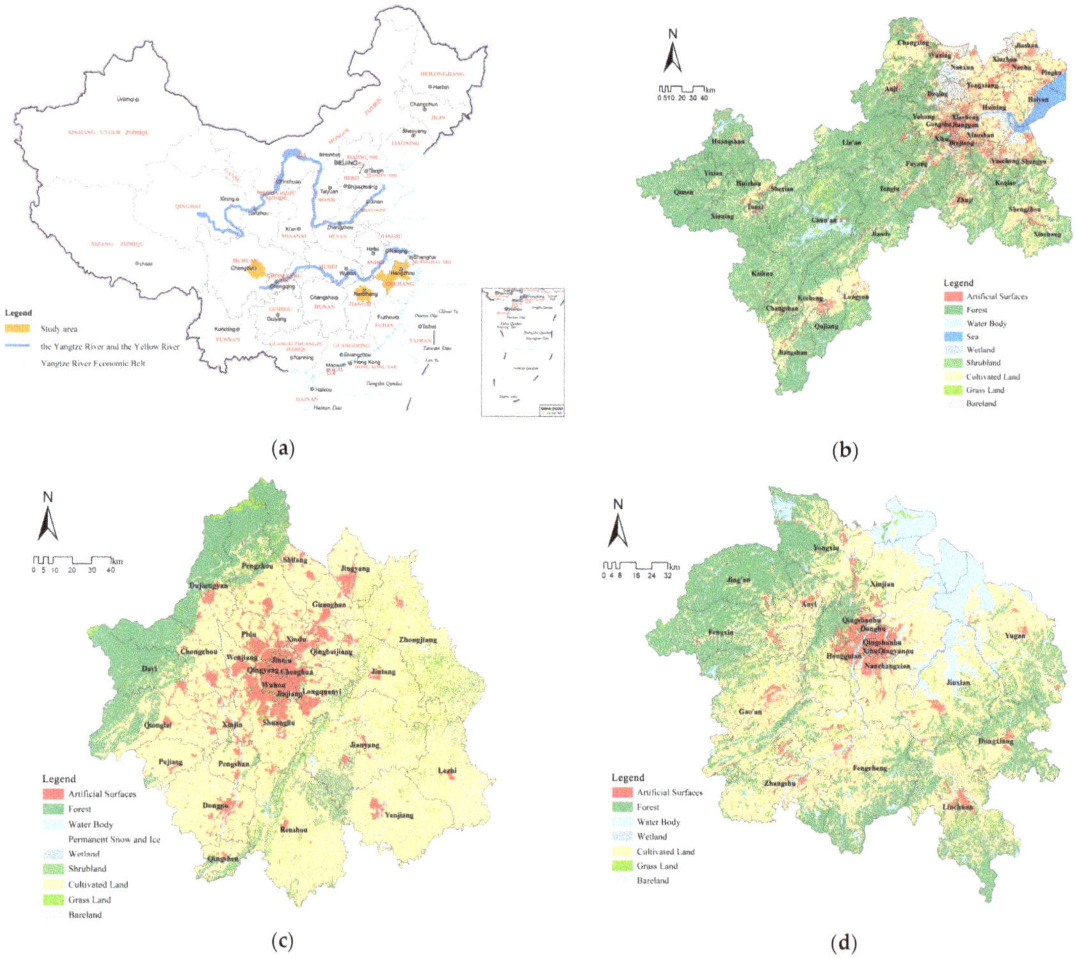


**Figure 1.** Study areas. (**a**) Locations of the study areas in China. (**b**–**d**) land use of the Hangzhou, Chengdu, and Nanchang metropolitan areas, respectively.

Chengdu Metropolitan Area (CMA) is in the upper reach of the YREB and the economic centers of southwestern China. According to the CMA Development Plan, the CMA is centered in Chengdu City and consists of 30 districts or counties with an area of $2.70 \times 10^4$ km$^2$; the resident population of the CMA in 2020 is 27.61 million, and the economic output accounts for 2.11% of China's Gross Domestic Product (GDP).

Nanchang Metropolitan Area (NMA) is in the middle reach of YREB. According to NMA Plan (2015–2030), the NMA consists of 18 districts or counties with a total area of $2.45 \times 10^4$ km$^2$; the resident population of NMA reaches 11.58 million in 2020, and its total GDP accounts for 0.71% of China.

Hangzhou Metropolitan Area (HMA) is located downstream of the YREB. According to the HMA Development Plan (2020–2035), the HMA includes 6 cities, including Hangzhou, Jiaxing, Huzhou, Shaoxing, Quzhou, and Huangshan, with a total of 44 districts or counties and a total area of about $5.48 \times 10^4$ km$^2$; by 2020, the population of HMA was 27.46 million, and its GDP accounted for 3.11% of the country.

### 2.2. Data Sources and Pre-Processing

Land use classification results are 30-meter spatial resolution GlobeLand30 images of year 2020 (http://www.globallandcover.com/) (accessed on 10 May 2022), which has become quite popular for many scholars to conduct related research [38,39]. Social and economic data are respectively derived from the 2020 Statistical Yearbooks of each province and China City Statistical Yearbook involved in the study areas. The resident population data are mainly from the 7th National Census bulletin and the statistical yearbooks and bulletins of the corresponding districts or counties. The energy consumption per unit of GDP was calculated from the total energy consumption and total GDP in the statistical yearbooks of each region.

### 2.3. Methods

In this study, the total LUCE of each metropolitan area is obtained by measuring the number of sources and sinks of LUCE in the three stages of metropolitan areas. The relationship between the ECC and ESC of each district and county in the metropolitan areas is studied by the CCD model, and the STIRPAT model is introduced to investigate the dominant factors affecting LUCE in the three types of metropolitan areas. Figure 2 shows the analysis clue of this study.

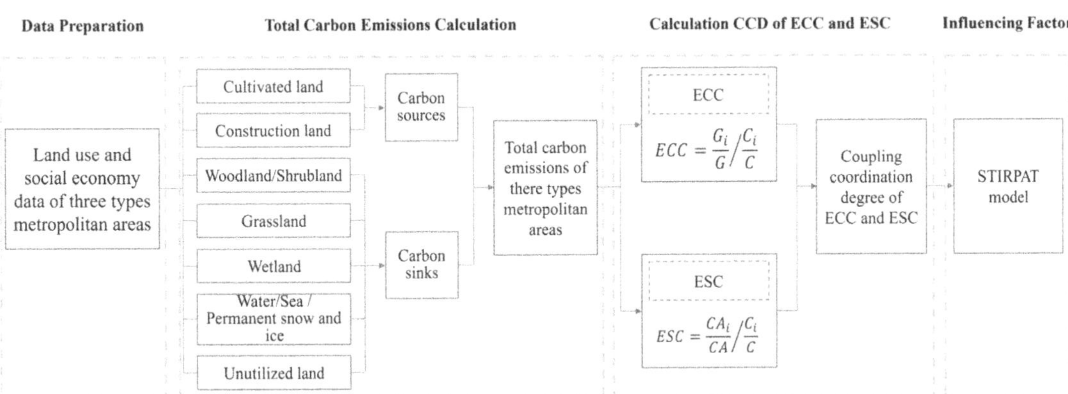


**Figure 2.** Research framework.

#### 2.3.1. Measurement of LUCE

The total amount of LUCE is equal to the sum of carbon sources and sinks [40], as shown in Equation (1):

$$CE = CO_{2\ sources} + CO_{2\ sinks} \quad (1)$$

where $CE$ is the total amount of LUCE; $CO_{2\ sources}$ represents the number of carbon sources of LUCE; and $CO_{2\ sinks}$ represents the number of carbon sinks of LUCE.

(1) Calculation of the number of carbon sources

The number of land use carbon sources for each district and county in the metropolitan areas is calculated from the cultivated and construction land. The number of LUCE from cultivated land use is the area of this kind of land use multiplied by its carbon emission factor, which is taken as 0.0422 according to Sun [41] and Zhang [42]. The number of LUCE from construction land use is usually measured indirectly based on the energy consumption of the city (such as coal, oil, natural gas, electricity, etc.). However, energy consumption data for each district and county in the metropolitan areas is difficult to obtain. According to relevant studies [43,44], since the value of secondary and tertiary industries is mainly contributed by construction land, the number of carbon emissions from construction land in each district and county can be approximated from the total GDP of secondary and tertiary industries. The calculation formula of carbon sources amount is as follows:

$$CO_{2\ sources} = A_c \times \delta_c + P_i \times M_i \times \theta_i \tag{2}$$

where $CO_{2\ sources}$ is the total land use carbon sources; $A_c$ represents the area of cultivated land use; $\delta_c$ represents the coefficient of cultivated land; $P_i$ represents the energy consumption per unit of GDP; $M_i$ represents the total GDP of secondary and tertiary industries in each district and county; and $\theta_i$ is the coefficient of standard coal.

(2) Calculation of the number of carbon sinks

Land with carbon sink function and its corresponding carbon sink coefficient [45] are involved in calculating the carbon sinks of each district and county in the metropolitan areas. The land with carbon sink function includes green lands (such as grassland, woodland, and shrubland), water (such as wetland, water, and sea), unutilized land, permanent snow and ice, and so on. According to relevant research, woodland and shrubland [42] and water and sea were combined [46]. Referring to existing studies, the corresponding carbon sink coefficients $\delta_i$ for different land use types are shown in Table 1. Since the percentage of permanent snow and ice in these study areas is only 0.0017%, this study uses the LUCE coefficient of water to replace its coefficient. The carbon sinks are calculated as follows:

$$CO_{2\ sinks} = \sum e_i = \sum A_i \times \delta_i \tag{3}$$

where $CO_{2\ sinks}$ is the total amount of carbon sink; $e_i$ represents the amount of carbon sink generated by land use type $i$; $A_i$ and $\delta_i$ represent the spatial area; and sink coefficient of land use type, respectively.

**Table 1.** CE coefficient of different land use types (kg·m$^{-2}$·a$^{-1}$).

| Land Use | Coefficient | References |
|---|---|---|
| Woodland/Shrubland | −0.0644 | Zhang et al., [42]; Fang et al., [47] |
| Grassland | −0.0021 | Sun et al., [41]; Zhang et al., [42] |
| Wetland | −0.0001 | Zhang et al., [42] |
| Water/Sea | −0.0253 | Yang et al., [48] |
| Unutilized | −0.0005 | Yang et al., [48] |

2.3.2. Global Moran's *I*

Global Moran's *I* is used to analyze the overall correlation degree of LUCE spatial distribution of each metropolitan area [49], and the calculation formula is as follows:

$$\text{Moran's } I = \frac{n\sum_{i=1}^{n}\sum_{j=1}^{n} W_{ij}(x_i - \overline{x})(x_j - \overline{x})}{\left(\sum_{i=1}^{n}\sum_{j=1}^{n} W_{ij}\right)\sum_{i=1}^{n}(x_i - \overline{x})^2} \tag{4}$$

where $n$ is the number of districts and counties in the metropolitan area; $x_i$ and $x_j$ are the LUCE of district or county $i$ and $j$, respectively; $W_{ij}$ is the spatial weight matrix of district or county $i$ and $j$; and $\overline{x}$ is the average value. The values of Moran's $I$ range from $[-1, 1]$; Moran's $I > 0$, Moran's $I < 0$, and Moran's $I = 0$ represent positive correlation, negative correlation, and no spatial correlation, respectively.

2.3.3. Measurement of CCD

The CCD model is introduced to measure the relationship between the ECC and ESC of CE for each district and county in the metropolitan areas. The calculation formula of CCD [50] is as follows:

$$C = \sqrt{\frac{U_1U_2}{\left(\frac{U_1+U_2}{2}\right)^2}} = \frac{2\sqrt{U_1U_2}}{U_1+U_2} \quad (5)$$

$$T = a_1U_1 + a_2U_2 \quad (6)$$

$$\mathrm{CCD} = \sqrt{C \times T} \quad (7)$$

where CCD is between 0 and 1; $C$ and $T$ are the coupling degree and integrated coordination index between ECC and ESC, respectively; $U_1$ and $U_2$ are the values of ECC and ESC respectively; $a_1$ and $a_2$ are the weights of indicators ECC and ESC, in this study, ECC and ESC are considered equally important, so the weights of both indicators $a_1$ and $a_2$ are taken as 0.5, then $T = 0.5U_1 + 0.5U_2$.

According to the CCD grading method [51,52], the CCD was classified into five classes as shown in Table 2:

**Table 2.** Levels of CCD.

| Development Category | Level | Balanced or Not | Degree |
|---|---|---|---|
| Coordinated | $0.8 < \mathrm{CCD} \leq 1.0$ | Balanced | Highly |
| Transformation | $0.6 < \mathrm{CCD} \leq 0.8$ | Balanced | Moderately |
| | $0.4 < \mathrm{CCD} \leq 0.6$ | Balanced | Basically |
| Uncoordinated | $0.2 < \mathrm{CCD} \leq 0.4$ | Unbalanced | Moderately |
| | $0 < \mathrm{CCD} \leq 0.2$ | Unbalanced | Seriously |

(1) Calculating ECC

ECC is introduced to estimate the equity of economic contribution of CE among districts or counties within a metropolitan area [53] and can reflect the socio-economic benefits that accompany the process of generating carbon emissions. *ECC* is calculated as:

$$ECC = \frac{G_i}{G} / \frac{C_i}{C} \quad (8)$$

where $G_i$ and $G$ are the GDP of each district and county and the whole metropolitan area, respectively; and $C_i$ and $C$ are the carbon emissions of each district and county and the whole metropolitan area, respectively. When the economic contribution of a district or county is greater than its share of carbon emissions ($ECC > 1$), it indicates that the district or county has a high level of economic efficiency and green development. When $ECC$ is less than 1, the economic contribution of the district is smaller than its carbon emissions contribution, and its economic efficiency of carbon emission is relatively low.

(2) Calculating ESC

ESC is introduced to estimate the equity of contribution of carbon ecological capacity among districts or counties in the metropolitan area [54], which could reflect the carbon sink capacity of each district and county [55] as a reflection of ecological benefits. ESC is calculated by the ratio of the carbon sink of each city to the carbon sink of all cities, divided by the ratio of carbon emissions of each city to the carbon emissions of all cities. The calculation formula of *ESC* is:

$$ESC = \frac{CA_i}{CA} / \frac{C_i}{C} \tag{9}$$

where $CA_i$ and $CA$ are the carbon sinks of each district or county and the whole metropolitan area, respectively. Districts with carbon sinks contribution greater than their share of carbon emissions ($ESC > 1$), indicate positive impacts on the absorption of CE in the whole metropolitan area and generate positive externalities that help other districts or counties, while districts with *ESC* less than 1 indicate negative externality to other districts or counties.

(3) Data normalization

Since the distribution range of ECC and ESC values are different, the coupling coordination degree between the two cannot be calculated directly, so they must be normalized. According to existing research [56,57], the formulas of positive and negative standardization are:

$$Y_i = \frac{X_i - \min(X_i)}{\max(X_i) - \min(X_i)} \tag{10}$$

$$Y_i = \frac{\max(X_i) - X_i}{\max(X_i) - \min(X_i)} \tag{11}$$

where $Y_i$ represents the standardized value of $X_i$; $X_i$ represents the actual value of indicator $i$; $\max(X_i)$, $\min(X_i)$ are the maximum and minimum values of $X_i$, respectively.

#### 2.3.4. Impact Factor Measurement Model

STIRPAT is a commonly used model to investigate the impact of population, affluence, and technology on the environment in the field of carbon emissions [58], which is expressed as follows:

$$I = aP^b A^c T^d e \tag{12}$$

where $a$ is a constant variable; $b$, $c$, and $d$ represent the coefficients of population, affluence, and technology, respectively; and $e$ is an error variable.

Equation (11) transforms to Equation (12) by logarithms method:

$$\ln I = a + b \ln P + c \ln A + d \ln T + e \tag{13}$$

where $I$ is the LUCE of each district and county in each metropolitan area; $P$, $A$, and $T$ are the residential population, per capita GDP and urbanization rate of each district or county, respectively; $a$ is a constant variable; $b$, $c$, and $d$ represent the coefficients of $P$, $A$, and $T$, respectively; $e$ is an error variable.

## 3. Results

### 3.1. Spatial Characteristics of LUCE

#### 3.1.1. Structures of LUCE in Metropolitan Areas

Overall, the total amount of LUCE in the metropolitan area is consistent with its development stage. The total LUCE in HMA (the mature metropolitan area) of $7802.7285 \times 10^4$ t is the highest, which is much higher than the total carbon emissions in CMA (the developmental metropolitan area) of $4678.7527 \times 10^4$ t, and in NMA (the cultivating metropolitan area) of $1421.2675 \times 10^4$ t. Construction land use types are the main contributor of LUCE in these three metropolitan areas, accounting for 99.18%, 96.71%, and 98.34%, respectively, while the proportion of LUCE from cultivated land use types is much lower. Regarding the composition structure of carbon sequestration, the carbon sinks of the three metropolitan areas are also very similar, with woodland accounting for the largest proportion of carbon sequestration, 95.29%, 84.93%, and 95.84% of the carbon sinks in HMA, NMA, and CMA, respectively, while wetland and unutilized land are both weaker due to their smaller carbon sink coefficients and weaker carbon sink capacities, as shown in Table 3.

**Table 3.** Land use carbon emissions/sinks of three metropolitan areas ($\times 10^4$ t/a).

| | Cultivated Land | Construction Land | Woodland | Grassland | Wetland | Water | Unutilized Land | Total |
|---|---|---|---|---|---|---|---|---|
| Nanchang Metropolitan Area | 48.4964 | 1424.5591 | −43.9845 | −0.2845 | −0.0006 | −7.5160 | −0.0024 | 1421.2675 |
| Chengdu Metropolitan Area | 77.9751 | 4631.6161 | −29.5562 | −0.2083 | −0.0003 | −1.0737 | 0.0000 | 4678.7527 |
| Hangzhou Metropolitan Area | 65.6841 | 7936.4449 | −190.0182 | −0.2788 | −0.0006 | −9.1028 | −0.0001 | 7802.7285 |

### 3.1.2. Spatial Differentiation of LUCE in Metropolitan Areas

The spatial distribution characteristics of LUCE in NMA, CMA, and HMA metropolitan areas are weak core cluster type, strong core cluster type, and flat extension respectively. Based on the natural breakpoint method, the sources and sinks of LUCE of the three metropolitan areas were reclassified, as shown in Figures 3 and 4. The areas with high carbon emissions (more than $74.62 \times 10^4$ tons) in NMA are mainly concentrated in the southwest, probably due to the higher development level of the economy in its southwest, such as the "Feng-Zhang-Gao" (Fengcheng-Zhangshu-Gaoan) industrial development area. The areas with high carbon sinks (more than $4.49 \times 10^4$ tons) were mainly concentrated in two parts (the northwestern and southeastern parts) of the region. However, the LUCE of districts or counties in the CMA shows an obvious "core-edge" spatial distribution pattern, with carbon emissions decreasing from the middle to the periphery of the metropolitan area. Areas with large carbon absorption are in the northwest of the CMA. The cause of this spatial distribution may be due to the concentration of construction land being mainly distributed in the central CMA, and the fact that the non-agricultural industries are also most developed in the central part of the area and the central city of the metropolitan area is more attractive. The spatial distribution of LUCE in the HMA shows a "flattened" extension. Areas with large carbon emissions (more than $74.62 \times 10^4$ tons) are distributed in the northeast of the metropolitan area. Areas with large carbon absorption (more than $4.49 \times 10^4$ tons) are concentrated in the southwest of this metropolitan area. From the west to the east, the carbon emissions of districts or counties gradually increased.

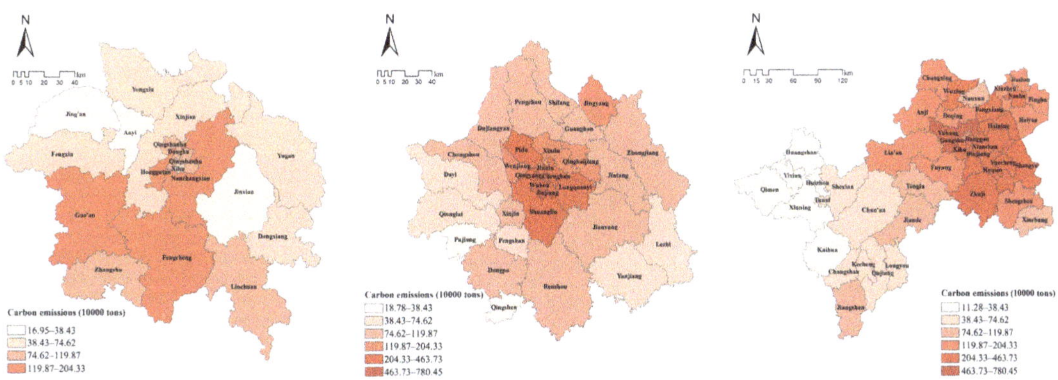

**(a)** Nanchang Metropolitan Area **(b)** Chengdu Metropolitan Area **(c)** Hangzhou Metropolitan Area

**Figure 3.** County-level distribution of LUCE in Nanchang, Chengdu, and Hangzhou metropolitan areas.

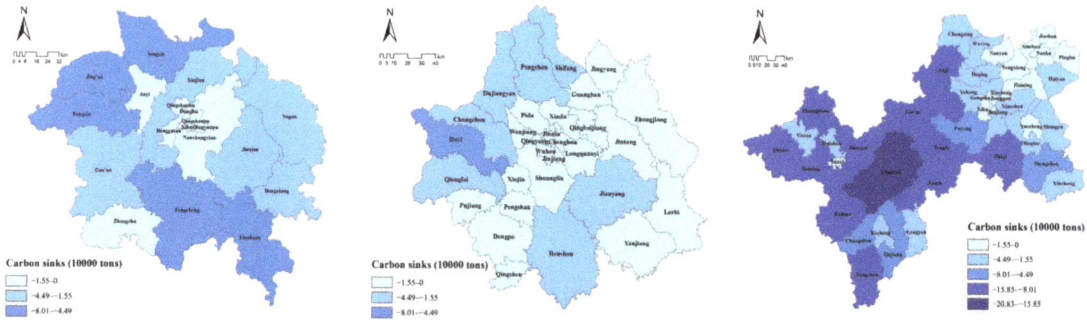


(a) Nanchang Metropolitan Area (b) Chengdu Metropolitan Area (c) Hangzhou Metropolitan Area

**Figure 4.** County-level distribution of LUCE sinks in Nanchang, Chengdu, and Hangzhou metropolitan areas.

3.1.3. Spatial Clustering Characteristics of LUCE in Metropolitan Areas

Except for NMA (the cultivating metropolitan area), which showed no significant positive spatial correlation, the carbon sources in CMA (the developmental metropolitan area) and HMA (the mature metropolitan area) showed significant positive correlation results. The Moran's $I$ value is 0.7957 in the Hangzhou metropolitan area, which is larger than that of 0.7425 in CMA (Table 4). In addition, the carbon sequestration of all three metropolitan areas shows a significant positive correlation result (Table 5). The specific Moran's $I$ value of NMA is 0.8501, which is much higher than 0.4795 of HMA, and 0.4561 of CMA, due to the contiguous distribution of carbon sink spaces, such as woodland and water, within the metropolitan area.

**Table 4.** Global Moran's $I$ of LUCE sources.

| | Moran's $I$ | z-Score | $p$-Value |
|---|---|---|---|
| Nanchang Metropolitan Area | 0.0266 | 0.5065 | 0.6125 |
| Chengdu Metropolitan Area | 0.7425 | 9.0600 | 0.0000 |
| Hangzhou Metropolitan Area | 0.7957 | 7.8238 | 0.0000 |

**Table 5.** Global Moran's $I$ of LUCE sinks.

| | Moran's $I$ | z-Score | $p$-Value |
|---|---|---|---|
| Nanchang Metropolitan Area | 0.8501 | 5.3412 | 0.0000 |
| Chengdu Metropolitan Area | 0.4561 | 5.4008 | 0.0000 |
| Hangzhou Metropolitan Area | 0.4795 | 4.6800 | 0.0000 |

Overall, the more mature the development stage of the metropolitan area, the more clustered districts or counties with high LUCE levels. The districts or counties with high LUCE levels are more distributed in the core circles of the CMA compared to the NMA, while the districts or counties with high carbon emission levels in the HMA are more widely clustered in contiguous areas. This is related to the high concentration of construction land, industry, population, and other elements in and around the core area of the metropolitan area.

*3.2. Coupling Coordination Degree Analysis of LUCE*

3.2.1. ECC of LUCE in Metropolitan Areas

Districts or counties with high economic contribution coefficient (ECC) of LUCE are generally concentrated in the central part or the periphery and edge part of the metropolitan area (Figure 5). The average values of ECC of carbon emissions of districts or counties in NMA, CMA, and HMA are 1.07, 1.04, and 1.13, respectively. The districts or counties with high ECC in Nanchang metropolitan area (cultivating metropolitan area), i.e., districts or counties with ECC higher than 1.23, are mainly located in the periphery of its central city. Nanchang, as the central city, accounts for 22.22% of the number of districts or counties in NMA, indicating that the economic efficiency of LUCE in the periphery of the central city is higher. The counties with higher ECC, i.e., those with ECC higher than 1.11, are mainly located in the periphery of the CMA, accounting for 20.00% of the total number of counties in the CMA, indicating that the contribution of the counties in the periphery to the economy of the entire metropolitan area is greater than the contribution of their LUCE to the LUCE of the entire metropolitan area. The ECC of HMA (mature metropolitan area) gradually decreases from west to east. The districts or counties with high ECC, i.e., districts or counties with ECC higher than 1.23, are concentrated in the western edge of the HMA, accounting for 15.91% of the total number of districts or counties in the HMA. The districts or counties in this area have higher economic efficiency than LUCE.

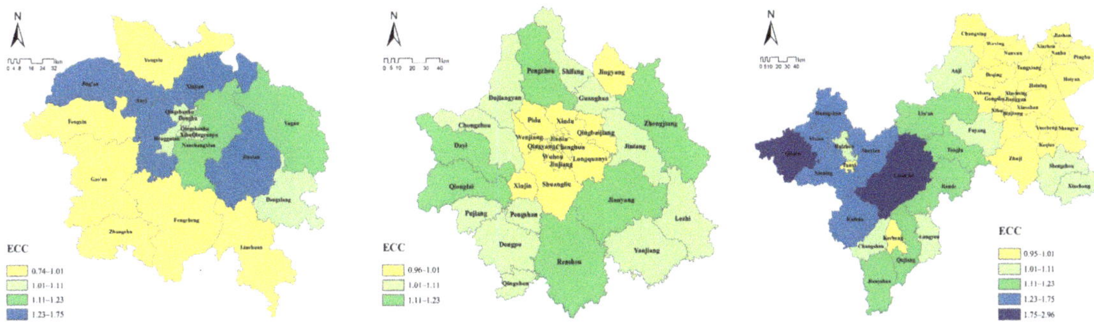


(**a**) Nanchang Metropolitan Area (**b**) Chengdu Metropolitan Area (**c**) Hangzhou Metropolitan Area

**Figure 5.** Economic contribution coefficients of three metropolitan areas.

### 3.2.2. ESC of Carbon Emission in Metropolitan Areas

The districts or counties with high ecological support coefficients (ESC) in the three metropolitan areas are all located in the peripheral regions of each metropolitan area (as shown in Figure 6), indicating that the carbon sinks capacity in the peripheral regions contributes more than their carbon emissions contribute to the carbon emissions of each metropolitan area. The mean values of ESC of NMA, CMA, and HMA are 2.31, 2.18, and 5.61, respectively. In addition, the number of districts or counties with higher ESC value in HMA and CMA, i.e., ESC higher than 3.58 is 29.55% and 26.67% respectively, which are much higher than the corresponding number of districts or counties in NMA, which is 11.11%. The results show that HMA and CMA are not only ahead of NMA in terms of the economic development stage, but also, the proportion of districts or counties with high contribution of carbon sink capacity is higher than the proportion of districts or counties at the corresponding level in NMA.

### 3.2.3. CCD of LUCE in Metropolitan Areas

The CCD (coupling coordination degree) between the ECC and ESC of each metropolitan area is closely bound up to the development stage of the metropolitan area. On the one hand, the more developed the metropolitan area, the lower the CCD. On the other hand, the more developed the economic districts or counties within the metropolitan area, the lower the CCD. The average CCD of NMA, CMA, and HMA is 0.27, 0.34, and 0.18 respectively. The more mature the development stage of the metropolitan area, the more unbalanced the economy contributive coefficient of LUCE and the ESC. Regarding the spatial distribution, there is an extreme imbalance between the ECC and ESC of carbon emissions of districts or counties in the core circles of CMA and NMA. The number of districts or counties with extreme imbalance accounts for 40.00% and 44.44% in each metropolitan area, respectively. The spatial distribution characteristics of the CCD of ECC and ESC of the developmental metropolitan area (CMA) and cultivating metropolitan area (NMA) show similarity, except that the number of seriously unbalanced districts or counties in the developmental metropolitan area accounts for a larger proportion. The eastern part of HMA is seriously unbalanced and the western part is more balanced overall. The amount of seriously unbalanced districts or counties account for 65.91% of the total amount of districts or counties in HMA (as shown in Figure 7).

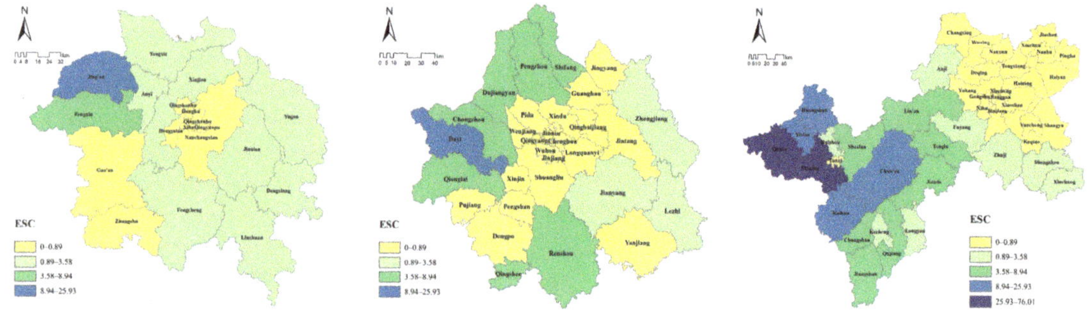


(**a**) Nanchang Metropolitan Area (**b**) Chengdu Metropolitan Area (**c**) Hangzhou Metropolitan Area

**Figure 6.** Ecological support coefficients of three metropolitan areas.

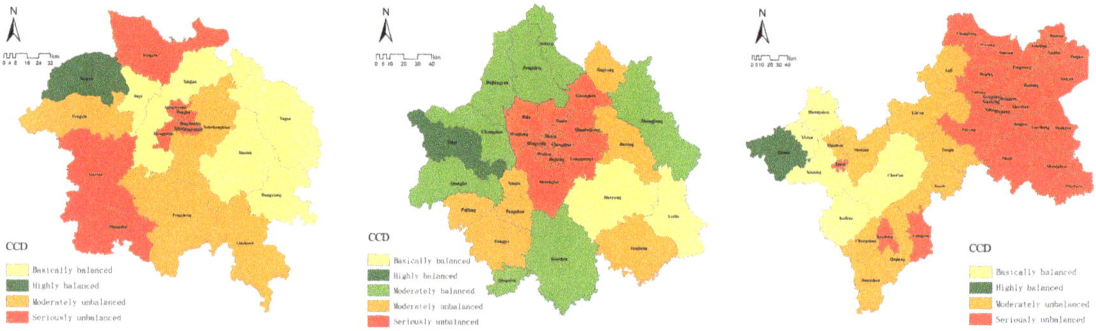


(**a**) Nanchang Metropolitan Area (**b**) Chengdu Metropolitan Area (**c**) Hangzhou Metropolitan Area

**Figure 7.** CCD between ECC and ESC in three metropolitan areas.

*3.3. Driving Factors of LUCE in Different Stages of Metropolitan Areas*

The LUCE of each district and county in the metropolitan areas are taken as the dependent variables, while the resident population, per capita GDP, and urbanization rate are the independent variables, and the carbon emissions of the three metropolitan areas are analyzed by the STIRPAT model. Regression results are shown in Table 6. The R-Square of all three models is close to 1, and the influencing factors of the model have good explanatory power. The lowest value of R-Square is 0.926 in NMA, indicating that the influencing factors in the model can explain carbon emissions to a degree of 92.60%, and the remaining 7.40% is not explained by the influencing factors selected by the model. The highest value of R-square is 0.998 in the CMA, and the influencing factors of its model can explain the carbon emissions of its districts or counties to 99.80%.

**Table 6.** Models Summary Table of Three Metropolitan Areas.

| Model | R | R-Square | Adjusted R-Square | Standard Error in Estimation | Durbin-Watson |
|---|---|---|---|---|---|
| Nanchang Metropolitan Area (NMA) | 0.962 | 0.926 | 0.910 | 0.230 | 1.604 |
| Chengdu Metropolitan Area (CMA) | 0.999 | 0.998 | 0.998 | 0.034 | 1.865 |
| Hangzhou Metropolitan Area (HMA) | 0.991 | 0.982 | 0.981 | 0.155 | 2.619 |

The results of NMA are shown in Table 7. The order of influence of carbon emission influencing factors is per capita GDP (A) > resident population (P) > urbanization rate (T). Meanwhile, the influence of resident population and per capita GDP on the LUCE of the NMA is positive, and the significance test value is 0.000 < 0.050, which both pass the significance test. However, the influence of the urbanization rate on the LUCE of districts or counties is negative, and the significance test value is 0.183 > 0.050, which does not pass the significance test. Therefore, for the NMA (the cultivating metropolitan area), increasing the urbanization rate of its districts or counties will be a good strategy to reduce LUCE in the region.

**Table 7.** Model test results of Nanchang metropolitan area.

| Variables | Non-Standardized Coefficient | | Standardized Coefficient | | | Colinearity Statistics | |
|---|---|---|---|---|---|---|---|
| | B | Std. Error | Beta | t | Significance | Tolerance | VIF |
| CONSTANT | −2.847 | 0.655 | | −4.343 | 0.001 | | |
| Resident population | 1.084 | 0.092 | 0.916 | 11.795 | 0.000 | 0.876 | 1.141 |
| Per capita GDP | 1.376 | 0.186 | 0.758 | 7.393 | 0.000 | 0.503 | 1.988 |
| Urbanization rate | −0.390 | 0.279 | −0.138 | −1.401 | 0.183 | 0.541 | 1.847 |

The regression results for CMA are shown in Table 8. The order of influence of carbon emission factors is resident population (P) > per capita GDP (A) > urbanization rate (T). The influence of all three factors on carbon emission in the CMA is positive. The significance test values of influencing factors are all 0.000, and all pass the significance test. Therefore, there exists a significant positive correlation between resident population, per capita GDP, urbanization rate, and carbon emissions for the CMA. In the process of continuous concentration of population, industry, and other factors, the carbon emissions of this area should be reduced by other strategies such as improving the carbon emission efficiency and optimizing the energy structure of districts or counties.

**Table 8.** Model test results of Chengdu metropolitan area.

| Variables | Non-Standardized Coefficient | | Standardized Coefficient | | | Colinearity Statistics | |
|---|---|---|---|---|---|---|---|
| | B | Std. Error | Beta | t | Significance | Tolerance | VIF |
| CONSTANT | −1.308 | 0.134 | | −9.760 | 0.000 | | |
| Resident population | 0.984 | 0.015 | 0.748 | 66.841 | 0.000 | 0.501 | 1.996 |
| Per capita GDP | 0.966 | 0.033 | 0.455 | 28.921 | 0.000 | 0.254 | 3.936 |
| Urbanization rate | 0.242 | 0.050 | 0.088 | 4.812 | 0.000 | 0.187 | 5.345 |

The results for the HMA are shown in Table 9. Similar to CMA, the order of influence of the factors affecting carbon emissions in the districts or counties of the HMA is resident population (P) > per capita GDP (A) > urbanization rate (T); besides, the impact of all three factors on LUCE in HMA is positive, except for the significance test value of urbanization rate on carbon emissions which is 0.054 > 0.050, which does not pass the significance test. The significance of resident population and per capita GDP all passed the test. Therefore, for the HMA (the mature metropolitan area), the influence degree of the resident population on LUCE of districts or counties is the strongest among the three metropolitan areas, and its urbanization rate has already reached a relatively high level. While in the process of continuous concentration of population and industries, the LUCE efficiency of districts or counties should be improved by other ways to reduce carbon emissions in the region.

**Table 9.** Model test results of Hangzhou metropolitan area.

| Variables | Non-Standardized Coefficient | | Standardized Coefficient | | | Colinearity Statistics | |
|---|---|---|---|---|---|---|---|
| | B | Std. Error | Beta | t | Significance | Tolerance | VIF |
| CONSTANT | −2.117 | 0.256 | | −8.266 | 0.000 | | |
| Resident population | 1.142 | 0.041 | 0.700 | 28.129 | 0.000 | 0.717 | 1.394 |
| Per capita GDP | 1.078 | 0.066 | 0.438 | 16.314 | 0.000 | 0.614 | 1.627 |
| Urbanization rate | 0.265 | 0.134 | 0.059 | 1.984 | 0.054 | 0.509 | 1.966 |

## 4. Discussion

### *4.1. Factors Influencing the LUCE of Different Stages Metropolitan Area*

Previous studies on the spatial patterns and impact modeling of LUCE have mainly concentrated on a single aspect of the research object [59–61], such as analyzing the spatial evolution characteristics [62] or the influencing factors of carbon emissions [63–65]. Few studies have focused on research objects at different development stages at the same time, thus failing to identify the spatial differentiation patterns and influencing factors of LUCE among different types of research objects. Different from previous studies, this study quantifies the LUCE of three different development stages of metropolitan areas in the lower, middle, and upper reaches of the YREB and measures the CCD and influencing factors of LUCE in each metropolitan area. The findings of this study show that the urbanization rate only had a negative impact on the LUCE of the cultivating metropolitan area (such as NMA), while the resident population and per capita GDP had a positive impact on the LUCE of the three stages of metropolitan areas. Similarly, Chel et al. found that population and per capita GDP are also positively related to LUCE after studying 103 metropolitan statistical areas (MSAs) in the United States [66]. The possible reason is that the level of social and economic development and urbanization of the cultivating metropolitan area (such as NMA) are lower than those of developmental and mature metropolitan areas (such as CMA and HMA respectively). Besides, the distribution of urban and rural populations, industry, land use, and other factors are not intensive and efficient enough. Therefore, it is necessary to formulate corresponding strategies to reduce LUCE for different types of metropolitan areas.

### *4.2. Policy Implications*

For cultivating metropolitan areas, large numbers of the rural population will be continuously promoted to gather into central cities in the future to increase the urbanization rate. Meanwhile, changing spatial planning from incremental to stock will be accompanied by the development trend of more intensive urban and rural land use and more concentrated industries, etc. The above changes may make the carbon emissions from construction land in cultivating metropolitan area decrease to a certain extent.

For mature and developmental metropolitan areas, it may be possible to learn from the research of Yang et al. on metropolitan areas in the United States, that is, to increase the number of high-density areas (urban centers) in the metropolitan area to reduce commuting time and distance [67], thereby reducing traffic carbon emissions in the metropolitan area.

### 4.3. Limitations

However, there are some limitations and potential uncertainties in this study:

(1) Due to the tiny percentage of permanent snow and ice in the research area, we substituted the permanent snow and ice emission coefficient with the water emission coefficient. However, if it occurs in other research regions where there is a sizable amount of permanent snow and ice or sea, it might lead to unreasonable results. Therefore, future studies should further explore the coefficient of permanent snow and ice and sea.
(2) This study is only based on the STIRPAT model, which examines the effects on LUCE in metropolitan areas at each development stage regarding population (P), affluence (A), and technology (T), without selecting control variables, which may result in an incomplete analysis of the influencing factors.
(3) Related studies show that regional climate and carbon cycle changes affect $CO_2$ emission pathways [68]. In addition, some studies have reported that the $CO_2$ emission level of metropolitan areas with a high level of sprawl is generally high [69]. However, climate change, urban form, and other aspects of the metropolitan area are not considered in this research. Therefore, further relevant studies are needed in the future.

## 5. Conclusions

Based on the GlobeLand30 land use type data of 2020 with 30-meter spatial resolution, this study calculated the total LUCE, analyzed the spatial characteristics, and revealed the relationship between the ECC and ESC of LUCE in each district and county of three metropolitan areas at different development stages, namely NMA, CMA and HMA in YREB. The STIRPAT model was further introduced to explore the impacts of various socio-economic factors on land use carbon emissions in these three metropolitan areas. The main conclusions of this work are drawn as follows.

(1) The more mature the stage of the metropolitan area, the higher the amount of LUCE is. Meanwhile, the spatial distribution patterns of LUCE of Nanchang, Chengdu, and Hangzhou metropolitan areas show weak core grouping, strong core clustering, and flattening extension patterns respectively. In general, the more mature the development stage of the metropolitan area, the more concentrated the districts or counties with high carbon emission levels.
(2) The districts or counties with a higher economy contributive coefficient (ECC) are generally concentrated in the central cities or on the periphery or edge of the metropolitan areas. The districts or counties with higher ecological support coefficient (ESC) in the three metropolitan areas are in the peripheral areas of each metropolitan area. Meanwhile, the more developed the metropolitan area, the lower the CCD between ECC and ESC. The more economically developed districts or counties are within the metropolitan area, the lower CCD is.
(3) Based on the STIRPAT model, the resident population, per capita GDP, and urbanization rate have good explanatory effects on the carbon emissions of the three metropolitan areas. All these three factors have positive effects on carbon emissions, except for the urbanization rate, which contributes to a negative effect on the LUCE of NMA.

Our empirical study revealed the spatial patterns and CCD between the ECC and ESC of LUCE, and verified the effects of the resident population, per capita GDP, and urbanization rate on LUCE in each metropolitan area by applying the STIRPAT model. Both the characteristics of LUCE in metropolitan areas and the influencing factors show specific correlations with the development stage of metropolitan areas. Findings can help to identify sustainable development strategies [70] and formulate corresponding carbon reduction measures for metropolitan areas at different stages of development and different regions within the metropolitan areas. In future studies, we will conduct further comparative analysis on the spatio-temporal evolution patterns of LUCE and other drivers in metropolitan areas at different development stages.

**Author Contributions:** Conceptualization, Y.D., Y.H. and S.L.; methodology, Y.D. and S.L.; software, Y.D., S.L. and L.Z.; validation, Y.D., S.L. and L.Z.; formal analysis, S.L. and L.X.; investigation, Y.D., L.X., C.H. and Y.C.; resources, Y.H.; data curation, Y.D. and S.L.; writing—original draft preparation, Y.D. and Y.H.; writing—review and editing, S.L.; visualization, Y.D. and S.L.; supervision, S.L.; project administration, Y.H.; funding acquisition, Y.H. All authors have read and agreed to the published version of the manuscript.

**Funding:** This research was funded by the National Natural Science Foundation of China (No. 51978299 and No. 41901390), the Natural Science Foundation of Hubei Province (No. 2021CFB012), Key Program of the National Social Science Foundation of China (21AZD048), and Open Fund of Key Laboratory of Urban Land Resources Monitoring and Simulation, Ministry of Natural Resources (KF-2020-05-005).

**Institutional Review Board Statement:** Not applicable.

**Informed Consent Statement:** Not applicable.

**Data Availability Statement:** The data used in this study can be requested from the authors.

**Conflicts of Interest:** The authors declare no conflict of interest.

## References

1. Akimoto, K.; Sano, F.; Oda, J.; Kanaboshi, H.; Nakano, Y. Climate Change Mitigation Measures for Global Net-Zero Emissions and the Roles of $CO_2$ Capture and Utilization and Direct Air Capture. *Energy Clim. Chang.* **2021**, *2*, 100057. [CrossRef]
2. Lai, L.; Huang, X.; Yang, H.; Chuai, X.; Zhang, M.; Zhong, T.; Chen, Z.; Chen, Y.; Wang, X.; Thompson, J.R. Carbon Emissions from Land-Use Change and Management in China between 1990 and 2010. *Sci. Adv.* **2016**, *2*, e1601063. [CrossRef] [PubMed]
3. Houghton, R.A.; Hackler, J.L. Emissions of Carbon from Forestry and Land-Use Change in Tropical Asia. *Glob. Chang. Biol.* **1999**, *5*, 481–492. [CrossRef]
4. Garofalo, D.F.T.; Novaes, R.M.L.; Pazianotto, R.A.A.; Maciel, V.G.; Brandão, M.; Shimbo, J.Z.; Folegatti-Matsuura, M.I.S. Land-Use Change CO2 Emissions Associated with Agricultural Products at Municipal Level in Brazil. *J. Clean. Prod.* **2022**, *364*, 132549. [CrossRef]
5. Cameron, C.; Hutley, L.B.; Friess, D.A.; Munksgaard, N.C. Hydroperiod, Soil Moisture and Bioturbation Are Critical Drivers of Greenhouse Gas Fluxes and Vary as a Function of Landuse Change in Mangroves of Sulawesi, Indonesia. *Sci. Total Environ.* **2019**, *654*, 365–377. [CrossRef]
6. Ahmad, M.; Rehman, A.; Shah, S.A.A.; Solangi, Y.A.; Chandio, A.A.; Jabeen, G. Stylized Heterogeneous Dynamic Links among Healthcare Expenditures, Land Urbanization, and CO2 Emissions across Economic Development Levels. *Sci. Total Environ.* **2021**, *753*, 142228. [CrossRef]
7. Yu, Z.; Chen, L.; Tong, H.; Chen, L.; Zhang, T.; Li, L. Spatial Correlations of Land-Use Carbon Emissions in the Yangtze River Delta Region: A Perspective from Social Network Analysis. *Ecol. Indic.* **2022**, *142*, 109147. [CrossRef]
8. Miura, T.; Tamaki, T.; Kii, M.; Kajitani, Y. Efficiency by Sectors in Areas Considering CO2 Emissions: The Case of Japan. *Econ. Anal. Policy* **2021**, *70*, 514–528. [CrossRef]
9. Ghosh, S.; Dinda, S.; Das Chatterjee, N.; Dutta, S.; Bera, D. Spatial-Explicit Carbon Emission-Sequestration Balance Estimation and Evaluation of Emission Susceptible Zones in an Eastern Himalayan City Using Pressure-Sensitivity-Resilience Framework: An Approach towards Achieving Low Carbon Cities. *J. Clean. Prod.* **2022**, *336*, 130417. [CrossRef]
10. Liu, G.; Zhang, F. Land Zoning Management to Achieve Carbon Neutrality: A Case Study of the Beijing–Tianjin–Hebei Urban Agglomeration, China. *Land* **2022**, *11*, 551. [CrossRef]
11. Liu, J.; Peng, K.; Zuo, C.; Li, Q. Spatiotemporal Variation of Land-Use Carbon Emissions and Its Implications for Low Carbon and Ecological Civilization Strategies: Evidence from Xiamen-Zhangzhou-Quanzhou Metropolitan Circle, China. *Sustain. Cities Soc.* **2022**, *86*, 104083. [CrossRef]
12. Breetz, H.L. Regulating Carbon Emissions from Indirect Land Use Change (ILUC): U.S. and California Case Studies. *Environ. Sci. Policy* **2017**, *77*, 25–31. [CrossRef]
13. Gim, T.H.T. Analyzing the City-Level Effects of Land Use on Travel Time and CO2 Emissions: A Global Mediation Study of Travel Time. *Int. J. Sustain. Transp.* **2022**, *16*, 496–513. [CrossRef]
14. Marescaux, A.; Thieu, V.; Garnier, J. Carbon Dioxide, Methane and Nitrous Oxide Emissions from the Human-Impacted Seine Watershed in France. *Sci. Total Environ.* **2018**, *643*, 247–259. [CrossRef]
15. Ali, G.; Pumijumnong, N.; Cui, S. Valuation and Validation of Carbon Sources and Sinks through Land Cover/Use Change Analysis: The Case of Bangkok Metropolitan Area. *Land Use Policy* **2018**, *70*, 471–478. [CrossRef]

16. Ali, G.; Nitivattananon, V. Exercising Multidisciplinary Approach to Assess Interrelationship between Energy Use, Carbon Emission and Land Use Change in a Metropolitan City of Pakistan. *Renew. Sustain. Energy Rev.* **2012**, *16*, 775–786. [CrossRef]
17. Aryapratama, R.; Pauliuk, S. Life Cycle Carbon Emissions of Different Land Conversion and Woody Biomass Utilization Scenarios in Indonesia. *Sci. Total Environ.* **2022**, *805*, 150226. [CrossRef]
18. Shen, W.; Liang, H.; Dong, L.; Ren, J.; Wang, G. Synergistic $CO_2$ Reduction Effects in Chinese Urban Agglomerations: Perspectives from Social Network Analysis. *Sci. Total Environ.* **2021**, *798*, 149352. [CrossRef]
19. Fujii, H.; Iwata, K.; Chapman, A.; Kagawa, S.; Managi, S. An Analysis of Urban Environmental Kuznets Curve of $CO_2$ Emissions: Empirical Analysis of 276 Global Metropolitan Areas. *Appl. Energy* **2018**, *228*, 1561–1568. [CrossRef]
20. Shi, Y.; Wang, H.; Shi, S. Relationship between Social Civilization Forms and Carbon Emission Intensity: A Study of the Shanghai Metropolitan Area. *J. Clean. Prod.* **2019**, *228*, 1552–1563. [CrossRef]
21. Luo, Z.; Wu, Y.; Zhou, L.; Sun, Q.; Yu, X.; Zhu, L.; Zhang, X.; Fang, Q.; Yang, X.; Yang, J.; et al. Trade-off between Vegetation $CO_2$ Sequestration and Fossil Fuel-Related CO2 Emissions: A Case Study of the Guangdong–Hong Kong–Macao Greater Bay Area of China. *Sustain. Cities Soc.* **2021**, *74*, 103195. [CrossRef]
22. China National Development and Reform Commission National Development and Reform Commission on the Issuance of the "14th Five-Year Plan" Implementation Plan for the Development of the Urban Agglomeration in the Middle Reaches of the Yangtze River. Available online: http://www.gov.cn/zhengce/zhengceku/2022-03/16/content_5679303.htm (accessed on 15 February 2022).
23. Chuai, X.; Feng, J. High Resolution Carbon Emissions Simulation and Spatial Heterogeneity Analysis Based on Big Data in Nanjing City, China. *Sci. Total Environ.* **2019**, *686*, 828–837. [CrossRef]
24. Xia, C.; Li, Y.; Xu, T.; Chen, Q.; Ye, Y.; Shi, Z.; Liu, J.; Ding, Q.; Li, X. Analyzing Spatial Patterns of Urban Carbon Metabolism and Its Response to Change of Urban Size: A Case of the Yangtze River Delta, China. *Ecol. Indic.* **2019**, *104*, 615–625. [CrossRef]
25. Guo, R.; Leng, H.; Yuan, Q.; Song, S. Impact of Urban Form on CO2 Emissions under Different Socioeconomic Factors: Evidence from 132 Small and Medium-Sized Cities in China. *Land* **2022**, *11*, 713. [CrossRef]
26. Danling, C.; Xinhai, L.U.; Bing, K.; Ulue, U. Dynamic Evolution and Spatial Convergence of Urban Land Use Efficiency in the Middle Reaches of the Yangtze River. *China Popul. Environ.* **2018**, *28*, 106–114.
27. Cui, Y.; Li, L.; Chen, L.; Zhang, Y.; Cheng, L.; Zhou, X.; Yang, X. Land-Use Carbon Emissions Estimation for the Yangtze River Delta Urban Agglomeration Using 1994–2016 Landsat Image Data. *Remote Sens.* **2018**, *10*, 1334. [CrossRef]
28. Yang, S.; Fu, W.; Hu, S.; Ran, P. Watershed Carbon Compensation Based on Land Use Change: Evidence from the Yangtze River Economic Belt. *Habitat Int.* **2022**, *126*, 102613. [CrossRef]
29. Liu, C.; Sun, W.; Li, P. Characteristics of Spatiotemporal Variations in Coupling Coordination between Integrated Carbon Emission and Sequestration Index: A Case Study of the Yangtze River Delta, China. *Ecol. Indic.* **2022**, *135*, 108520. [CrossRef]
30. Yang, B.; Wang, Z.; Zou, L.; Zou, L.; Zhang, H. Exploring the Eco-Efficiency of Cultivated Land Utilization and Its Influencing Factors in China's Yangtze River Economic Belt, 2001–2018. *J. Environ. Manag.* **2021**, *294*, 112939. [CrossRef]
31. Shen, X.; Zheng, H.; Jiang, M.; Yu, X.; Xu, H.; Zhong, G. Multidimensional Impact of Urbanization Process on Regional Net $CO_2$ Emissions: Taking the Yangtze River Economic Belt as an Example. *Land* **2022**, *11*, 1079. [CrossRef]
32. Tan, S.; Hu, B.; Kuang, B.; Zhou, M. Regional Differences and Dynamic Evolution of Urban Land Green Use Efficiency within the Yangtze River Delta, China. *Land Use Policy* **2021**, *106*, 105449. [CrossRef]
33. Wu, H.; Fang, S.; Zhang, C.; Hu, S.; Nan, D.; Yang, Y. Exploring the Impact of Urban Form on Urban Land Use Efficiency under Low-Carbon Emission Constraints: A Case Study in China's Yellow River Basin. *J. Environ. Manag.* **2022**, *311*, 114866. [CrossRef] [PubMed]
34. Portal, C.C.G. The Outline of the Development Plan of the Yangtze River Economic Belt Was Officially Printed and Distributed. Available online: http://www.gov.cn/xinwen/2016-09/12/content_5107501.htm (accessed on 12 September 2016).
35. China, N.D. and R.C of The National Development and Reform Commission Issued the Development Plan of Urban Agglomeration in the Middle Reaches of the Yangtze River. Available online: http://www.gov.cn/xinwen/2015-04/16/content_2848120.htm (accessed on 16 April 2015).
36. Wang, M.; Wang, Y.; Wu, Y.; Yue, X.; Wang, M.; Hu, P. Identifying the Spatial Heterogeneity in the Effects of the Construction Land Scale on Carbon Emissions: Case Study of the Yangtze River Economic Belt, China. *Environ. Res.* **2022**, *212*, 113397. [CrossRef] [PubMed]
37. Yin, Z.; Lu, Q.; Lv, X.; Wang, Q. *China Metropolitan Area Development Report 2021*, 1st ed.; Tsinghua University: Beijing, China, 2021.
38. Jokar Arsanjani, J. Characterizing, Monitoring, and Simulating Land Cover Dynamics Using GlobeLand30: A Case Study from 2000 to 2030. *J. Environ. Manag.* **2018**, *214*, 66–75. [CrossRef] [PubMed]
39. Kim, D.; Lim, C.H.; Song, C.; Lee, W.K.; Piao, D.; Heo, S.; Jeon, S. Estimation of Future Carbon Budget with Climate Change and Reforestation Scenario in North Korea. *Adv. Sp. Res.* **2016**, *58*, 1002–1016. [CrossRef]
40. Fattah, M.A.; Morshed, S.R.; Morshed, S.Y. Impacts of Land Use-Based Carbon Emission Pattern on Surface Temperature Dynamics: Experience from the Urban and Suburban Areas of Khulna, Bangladesh. *Remote Sens. Appl. Soc. Environ.* **2021**, *22*, 100508. [CrossRef]
41. Sun, H.; Liang, H.M.; Chang, X.L.; Cui, Q.C.; Tao, Y. Land Use Patterns on Carbon Emission and Spatial Association in China. *Econ. Geogr.* **2015**, *35*, 154–162.

42. Zhang, C.-Y.; Zhao, L.; Zhang, H.; Chen, M.-N.; Fang, R.-Y.; Yao, Y.; Zhang, Q.-P.; Wang, Q. Spatial-Temporal Characteristics of Carbon Emissions from Land Use Change in Yellow River Delta Region, China. *Ecol. Indic.* **2022**, *136*, 108623. [CrossRef]
43. Hussain, M.; Wang, W.; Wang, Y. Natural Resources, Consumer Prices and Financial Development in China: Measures to Control Carbon Emissions and Ecological Footprints. *Resour. Policy* **2022**, *78*, 102880. [CrossRef]
44. Zhang, A.; Wen, L.; Chatalova, L.; Gao, X. Reduction of Carbon Emissions through Resource-Saving and Environment-Friendly Regional Economic Integration–Evidence from Wuhan Metropolitan Area, China. *Technol. Forecast. Soc. Chang.* **2021**, *166*, 120590. [CrossRef]
45. Carpio, A.; Ponce-Lopez, R.; Lozano-García, D.F. Urban Form, Land Use, and Cover Change and Their Impact on Carbon Emissions in the Monterrey Metropolitan Area, Mexico. *Urban Clim.* **2021**, *39*, 100947. [CrossRef]
46. Xu, J.; Pan, H.; Huang, P. Carbon Emission and Ecological Compensation of Main Functional Areas in Sichuan Province Based on LUCC. *Chin. J. Eco-Agric.* **2019**, *27*, 142–152. [CrossRef]
47. Fang, J.Y.; Guo, Z.D.; Piao, S.L.; Chen, A.P. Terrestrial Vegetation Carbon Sinks in China, 1981–2000. *Sci. China Ser. D Earth Sci.* **2007**, *50*, 1341–1350. [CrossRef]
48. Yang, B.; Chen, X.; Wang, Z.; Li, W.; Zhang, C.; Yao, X. Analyzing Land Use Structure Efficiency with Carbon Emissions: A Case Study in the Middle Reaches of the Yangtze River, China. *J. Clean. Prod.* **2020**, *274*, 123076. [CrossRef]
49. Yao, L.; Yu, Z.; Wu, M.; Ning, J.; Lv, T. The Spatiotemporal Evolution and Trend Prediction of Ecological Wellbeing Performance in China. *Land* **2021**, *10*, 12. [CrossRef]
50. Tomal, M. Evaluation of Coupling Coordination Degree and Convergence Behaviour of Local Development: A Spatiotemporal Analysis of All Polish Municipalities over the Period 2003–2019. *Sustain. Cities Soc.* **2021**, *71*, 102992. [CrossRef]
51. Ariken, M.; Zhang, F.; Liu, K.; Fang, C.; Kung, H. Te Coupling Coordination Analysis of Urbanization and Eco-Environment in Yanqi Basin Based on Multi-Source Remote Sensing Data. *Ecol. Indic.* **2020**, *114*, 106331. [CrossRef]
52. Yang, Z.; Zhan, J.; Wang, C.; Twumasi-Ankrah, M.J. Coupling Coordination Analysis and Spatiotemporal Heterogeneity between Sustainable Development and Ecosystem Services in Shanxi Province, China. *Sci. Total Environ.* **2022**, *836*, 155625. [CrossRef]
53. Yang, Y.; Fan, M. Analysis of the Spatial-Temporal Differences and Fairness of the Regional Energy Ecological Footprint of the Silk Road Economic Belt (China Section). *J. Clean. Prod.* **2019**, *215*, 1246–1261. [CrossRef]
54. Rong, T.; Zhang, P.; Jing, W.; Zhang, Y.; Li, Y.; Yang, D.; Yang, J.; Chang, H.; Ge, L. Carbon Dioxide Emissions and Their Driving Forces of Land Use Change Based on Economic Contributive Coefficient (ECC) and Ecological Support Coefficient (ESC) in the Lower Yellow River Region (1995–2018). *Energies* **2020**, *13*, 2600. [CrossRef]
55. Yuan, S.; Tang, Y. Spatial Differentiation of Land Use Carbon Emission in the Yangtze River Economic Belt Based on Low Carbon Perspective. *Econ. Geogr.* **2019**, *39*, 190–198.
56. Martyka, A.; Jopek, D.; Skrzypczak, I. Analysis of the Sustainable Development Index in the Communes of the Podkarpackie Voivodeship: A Polish Case Study. *Sustainability* **2022**, *14*, 10237. [CrossRef]
57. Liu, Y.; Yang, R.; Sun, M.; Zhang, L.; Li, X.; Meng, L.; Wang, Y.; Liu, Q. Regional Sustainable Development Strategy Based on the Coordination between Ecology and Economy: A Case Study of Sichuan Province, China. *Ecol. Indic.* **2022**, *134*, 108445. [CrossRef]
58. Vélez-Henao, J.A.; Font Vivanco, D.; Hernández-Riveros, J.A. Technological Change and the Rebound Effect in the STIRPAT Model: A Critical View. *Energy Policy* **2019**, *129*, 1372–1381. [CrossRef]
59. Hutyra, L.R.; Yoon, B.; Hepinstall-Cymerman, J.; Alberti, M. Carbon Consequences of Land Cover Change and Expansion of Urban Lands: A Case Study in the Seattle Metropolitan Region. *Landsc. Urban Plan.* **2011**, *103*, 83–93. [CrossRef]
60. Raihan, A.; Tuspekova, A. The Nexus between Economic Growth, Renewable Energy Use, Agricultural Land Expansion, and Carbon Emissions: New Insights from Peru. *Energy Nexus* **2022**, *6*, 100067. [CrossRef]
61. Sowby, R.B.; Capener, A. Reducing Carbon Emissions through Water Conservation: An Analysis of 10 Major U.S. Cities. *Energy Nexus* **2022**, *7*, 100094. [CrossRef]
62. Baldassini, P.; Bagnato, C.E.; Paruelo, J.M. How May Deforestation Rates and Political Instruments Affect Land Use Patterns and Carbon Emissions in the Semi-Arid Chaco, Argentina? *Land Use Policy* **2020**, *99*, 104985. [CrossRef]
63. Zhao, Y.; Chen, R.; Sun, T.; Yang, Y.; Ma, S.; Xie, D.; Zhang, X. Urbanization Influences CO 2 Emissions in the Pearl River Delta: A Perspective of the "Space of Flows". *Land* **2022**, *11*, 1373. [CrossRef]
64. Domon, S.; Hirota, M.; Kono, T.; Managi, S.; Matsuki, Y. The Long-Run Effects of Congestion Tolls, Carbon Tax, and Land Use Regulations on Urban $CO_2$ Emissions. *Reg. Sci. Urban Econ.* **2022**, *92*, 103750. [CrossRef]
65. Dumortier, J.; Elobeid, A. Effects of a Carbon Tax in the United States on Agricultural Markets and Carbon Emissions from Land-Use Change. *Land Use Policy* **2021**, *103*, 105320. [CrossRef]
66. Chel, M.; Kang, M.; Kim, S. Urban Climate Does Polycentric Development Produce Less Transportation Carbon Emissions? Evidence from Urban Form Identified by Night-Time Lights across US Metropolitan Areas. *Urban Clim.* **2022**, *44*, 101223. [CrossRef]
67. Yang, J.; French, S.; Holt, J.; Zhang, X. Measuring the Structure of U.S. Metropolitan Areas, 1970–2000 Measuring the Structure of U.S. Metropolitan Areas, 1970–2000. *J. Am. Plan. Assoc.* **2012**, *78*, 197–209. [CrossRef]
68. Nohara, D.; Yoshida, Y.; Misumi, K.; Ohba, M. Dependency of Climate Change and Carbon Cycle on $CO_2$ Emission Pathways. *Environ. Res. Lett.* **2013**, *8*, 014047. [CrossRef]

69. Bereitschaft, B.; Debbage, K. Urban Form, Air Pollution, and $CO_2$ Emissions in Large U.S. Metropolitan Areas. *Prof. Geogr.* **2013**, *65*, 612–635. [CrossRef]
70. Li, M.; Li, X.; Liu, S.; Lyu, X.; Dang, D.; Dou, H.; Wang, K. Analysis of the Spatiotemporal Variation of Landscape Patterns and Their Driving Factors in Inner Mongolia from 2000 to 2015. *Land* **2022**, *11*, 1410. [CrossRef]

*land*

Article

# Assessment and Decomposition of Regional Land Use Efficiency of the Service Sector in China

Mingzhi Zhang [1], Hongyu Liu [1], Yangyue Su [2,*], Xiangyu Zhou [1], Zhaocheng Li [1] and Chao Chen [1]

[1] School of Economics, Institute of Population and Economic Development, Shandong University of Finance and Economics, Jinan 250014, China
[2] School of Management Engineering, Shandong Jianzhu University, Jinan 250101, China
* Correspondence: 1710260@tongji.edu.cn

**Abstract:** High land use efficiency is the key to improving total factor productivity, and also an important force behind achieving sustained economic growth. Existing studies have mainly focused on the land use efficiency of the industry sector. Yet, the issue of land use efficiency of the service sector (SLUE) has been largely overlooked. This study examines regional differences and efficiency decomposition by using a slack based model (SBM) of undesirable output, and the Malmquist productivity index (MPI) under a data envelopment analysis framework. The results reveal that: (1) In China, the land use efficiency of the service sector is unbalanced, showing an inverted growth law of "low in developed areas and high in backward areas". (2) The land use efficiency of the service sector can be decomposed into technical progress, pure technical efficiency, and scale efficiency. From the decomposition results, the growth rate of pure technical efficiency presents a trend of "low in the east and high in the west"; the scale efficiency also falls into the situation of weak group growth. Technological progress has maintained steady improvement. (3) The coordinated improvement of land use efficiency of the service sector needs to focus on resolving the "beggar-thy-neighbor" issue caused by existing large regional differences. In this article, the puzzle of land use efficiency differences in the service industry is well solved, and thus provides valuable enlightenment for the benign growth of service industries in countries and regions around the world.

**Keywords:** service industry; land use efficiency; regional differences; efficiency decomposition

**Citation:** Zhang, M.; Liu, H.; Su, Y.; Zhou, X.; Li, Z.; Chen, C. Assessment and Decomposition of Regional Land Use Efficiency of the Service Sector in China. *Land* **2022**, *11*, 1911. https://doi.org/10.3390/land11111911

Academic Editors: Bindong Sun, Tinglin Zhang, Wan Li, Chun Yin and Honghuan Gu

Received: 29 September 2022
Accepted: 24 October 2022
Published: 27 October 2022



## 1. Introduction

Land use efficiency (LUE) refers to economic benefits of each unit of land, which is an important indicator to determine the efficiency of land allocation. With the acceleration of urbanization, LUE plays a crucial role in urban operation efficiency and sustainable development [1–3]. Sustainable development, first proposed in 1972, is an economic growth model that focuses on long-term development. It refers to the development that not only meets the needs of contemporary people, but also does not harm future generations to meet their needs. It is one of the basic requirements of the Scientific Outlook on Development. As an important type of urban land, land for the service section plays an important role in the process of sustainable and high-quality urban development, carrying important functions, such as improving urban management level, enhancing the overall function of the city, improving the living environment, and promoting the optimization, integration, and rational allocation of land resources [4–6]. As an indicator reflecting the degree of scientific utilization of land for the service section, land use efficiency of the service sector (SLUE) should be paid attention to and researched. SLUE, namely the land use efficiency of the service industry, is an indicator that should be used effectively in the current resource utilization. SLUE is an important issue that the service sector pays close attention to.

With the rapid development of urbanization, the high frequency exchange and drastic transformation of urban land function attributes become a land use law of urbanization [7–9]. Economic development oriented towards the service industry plays a unique role in the adjustment of urban land use structure and the transformation of urban landscape in China [10,11]. This leads to the large-scale expansion of land for the service section. According to the China Urban Construction Statistical Yearbook, the area of service industry construction land increased by 36% from 19,902.4 km$^2$ in 2011 to 27,092.5 km$^2$ in 2017. Two intuitive facts are: (1) Urban sprawl is intensifying, urban boundaries are extending, and a large amount of agricultural land is being converted into urban land [12], especially land for the service section. (2) Industrial land in cities is gradually giving way to land for the service section; more and more factories are moving out of cities. These facts have much to do with the government's regulation and control of land. On the one hand, facing the incentive of economic performance assessment, local governments have a high dependence on land finance. The government tends to seek more land transfer benefits by arranging urban planning space and land planning indicators, which are important means for the government to obtain financial revenue [11]. On the other hand, compared with the manufacturing industry, the service industry is environmentally friendly and has a high degree of land intensification. The supply of land for the service industry is more in line with the demand under the dual pressure of land resource scarcity and environmental pollution [13]. However, the transformation of land structure caused by the acceleration of urbanization may lead to inefficient land use status, such as disordered land use and fragmented land use [14–17]. In this context, it is urgent to study the spatial cooperative improvement of SLUE.

At present, the research on SLUE is relatively scarce, but the research on LUE of other land types can provide certain reference for us to measure SLUE. In terms of efficiency evaluation methods, the Data Envelopment Analysis (DEA) and the Stochastic Frontier Analysis are commonly used [12,18]. Among them, the DEA is a widely used method in LUE evaluation because it does not need to set the form of production function and can consider multiple inputs and outputs simultaneously [19]. With people's increasing attention to environmental issues, it has become a trend for LUE evaluation to include undesirable output factors in the efficiency measurement [19,20]. Since the development of China's service industry still has the extensive characteristics of high emissions [21], the measurement of SLUE should also consider environmental pollution and other undesirable outputs. Compared with the defects of the traditional DEA model, the slack based model (SBM) proposed by Tone [22] can not only handle the undesirable outputs, but also solve the input-output slack issue in efficiency evaluation [19]. This model can measure efficiency more accurately [23]. On this basis, the Malmquist productivity index (MPI) can describe the change trend and evolution characteristics of efficiency [24]. Thus, the combination of the SBM model of undesirable output and the Malmquist productivity index can accurately measure SLUE and its dynamic trend under environmental constraints.

In terms of the measurement of regional differences, quantitative analysis methods for the degree of regional differences mostly involve the Theil index [25], the Gini coefficient [26], the kernel density estimation [27] and spatial autocorrelation [28]. The Theil index, proposed by Theil [29], was first used to describe income differences among countries. Since it could measure the contribution of different types of differences to total differences, it was later widely used to describe individual differences. From the perspective of research, many scholars discussed the regional differences of land use from the perspectives of urban agglomeration [30], basin [31], and specific provinces [32,33]. In reality, there is a significant regional difference in the economic output of land for the service section. For example, the unit output value of land for the service section in Beijing was nearly four times that of Xinjiang in 2017. Therefore, it is necessary to measure the regional difference of SLUE. The division of four traditional regions in China provides us with a research perspective.

In addition, in the selection of research objectives, existing research mainly focuses on the measurement of the overall urban construction land efficiency [34], industrial LUE [35], and agricultural LUE [36]. However, few studies have measured LUE from the perspective of the service industry. As the service industry has the highest proportion of added value in GDP in China, it is necessary to measure SLUE and its regional differences. This raises two challenges. First, can the distribution pattern and regional differences of LUE in different provinces and four major regions of China be analyzed from the perspective of the service industry? Second, what effective strategies can be used to improve both SLUE and synergistic development in these regions? If these two problems cannot be solved in time, the benign growth in terms of services will be hard to achieve.

The main objectives of this study are as follows: First, this article measures the SLUE of 30 provinces in China and compares the differences between the provinces. Second, the Malmquist productivity index method is used to decompose the SLUE and find the source of the driving force that affects the spatial differences of the SLUE. Third, the Theil index method is used to quantitatively analyze regional differences in SLUE and investigate the size and source of regional differences in SLUE. Finally, this article proposes a differentiated regional governance strategy to narrow the spatial differences of SLUE and reduce the deadweight efficiency loss. The research of this article is expected to form and identify the reasons behind the spatial differences in SLUE and the path to achieve spatial coordinated promotion. Furthermore, theoretical support is provided for countries and regions at different development stages that will help these areas to improve their SLUE. Therefore, our efforts are beneficial in terms of filling the gaps in academic differences in related fields.

The remaining sections of this article are arranged as follows: Section 2 introduces the model and data used in this paper. Section 3 analyzes the regional differences and the dynamic trend and efficiency decomposition of SLUE. Section 4 provides a discussion of the findings. Section 5 presents the research contributions and implications, and Section 6 presents the study's conclusions and policy suggestions.

## 2. Methods and Data

### 2.1. Efficiency Assessment Model

#### 2.1.1. The Global SBM-Undesirable Model

The DEA method, first proposed by Charnes et al. [37] in the United States, is a commonly used model for measuring efficiency. The advantage of DEA is that there is no requirement to set the specific form of function, and it can objectively evaluate the actual production process of multiple inputs and multiple outputs [38]. Traditional DEA models, such as the CCR (Charnes, Cooper, and Rhodes) [37] and the BCC (Banker, Charnes, and Cooper) [39], do not consider the input-output slack improvements in angle and radial selection, and they cannot accurately measure the efficiency with undesirable outputs. However, within the background of resource and environment constraints, SLUE evaluation should emphasize the coordination among the economy, resources, and the environment. In addition to the single economic benefit indicators, the DEA's evaluation indicator system should also include ecological benefit indicators, such as the waste discharged in the land use process of a service industry, and other undesirable outputs. With regard to considering undesirable outputs, we hope to obtain as many desirable outputs as possible while minimizing the number of undesirable outputs.

Compared with the defects of the traditional DEA model, the undesirable output SBM model proposed by Tone [22] can not only solve the slack issue of variable in angle and radial selection, but also accurately measure the efficiency with undesirable output. This provides a new idea for the measurement of efficiency under environmental constraints [23]. In addition, due to the differences in production frontiers in different periods, the efficiency measured based on the best production frontiers in different periods does not have intertemporal comparability and circularity. The global DEA method proposed by Pastor and Lovell [40] effectively solves this problem by constructing frontiers of production technology based on global reference.

Taking provinces as decision making units (DMUs), we assume that there are $K$ DMUs. Each DMU uses $M$ inputs $x_m$ to produce $N$ desirable outputs $y_n$ and $L$ undesirable outputs $b_l$. The global SBM-undesirable model is constructed as follows:

$$\begin{aligned}
\rho = & \min_{s^x,s^y,s^b,\lambda} \frac{1-\frac{1}{M}\sum_{m=1}^{M}\frac{s_m^x}{x_{km}}}{1+\frac{1}{N+L}\left(\sum_{n=1}^{N}\frac{s_n^y}{y_{kn}}+\sum_{l=1}^{L}\frac{s_l^b}{b_{kl}}\right)} \\
x_{km} = & \sum_{k=1}^{K}\sum_{p=1}^{P}\lambda_k^p x_{km}^p + s_m^x, m = 1,2,\cdots,M \\
y_{kn} = & \sum_{k=1}^{K}\sum_{p=1}^{P}\lambda_k^p y_{kn}^p - s_n^y, n = 1,2,\ldots,N \\
b_{kl} = & \sum_{k=1}^{K}\sum_{p=1}^{P}\lambda_k^p b_{kl}^p + s_l^b, l = 1,2,\cdots,L
\end{aligned} \quad (1)$$

where $s_m^x \geq 0, s_n^y \geq 0, s_l^b \geq 0, \lambda^p \geq 0$. $\rho$ is the efficiency of the unit being evaluated and $0 < \rho \leq 1$; $(x_{km}, y_{kn}, b_{kl})$ represents the mth input, nth desirable output and lth undesirable output in the kth DMU, respectively. Also, $\left(s_m^x, s_n^y, s_l^b\right)$ represents the slack variables of input, desirable output and undesirable output, respectively, and $\lambda_k$ represents the weight.

2.1.2. Malmquist Productivity Index (MPI)

This research further constructs the DEA-Malmquist productivity model to investigate the efficiency evolution of land use in China's service industry, based on a static analysis of the global SBM-undesirable model. The Malmquist index between two periods $t$ and $t+1$ can be formulated as follows:

$$M(x_{t+1}, y_{t+1}, x_t, y_t) = \sqrt{\left(\frac{D^t(x_{t+1}, y_{t+1})}{D^t(x_t, y_t)}\right) \cdot \left(\frac{D^{t+1}(x_{t+1}, y_{t+1})}{D^{t+1}(x_t, y_t)}\right)} \quad (2)$$

where $y$ represents the output vector, and $x$ is the input vector; $D$ is the distance function, $M$ refers to the Malmquist productivity index. The $MPI$ is calculated as the relationships between the distances of DMUs from technological frontiers. Also, $MPI > 1$ represents improvements in SLUE, and vice versa. In order to identify the different components that contribute to improvements in productivity, $MPI$ can be decomposed into efficiency change (EC) and technical change (TC) [41]. That is:

$$M(x_{t+1}, y_{t+1}, x_t, y_t) = \frac{D^{t+1}(x_{t+1}, y_{t+1})}{D^t(x_t, y_t)} \sqrt{\frac{D^t(x_{t+1}, y_{t+1})}{D^{t+1}(x_{t+1}, y_{t+1})} \cdot \frac{D^t(x_t, y_t)}{D^{t+1}(x_t, y_t)}} \quad (3)$$

where the first term denotes the efficiency changes between the period $t$ and the period $t+1$. This factor is called $EC$. Here, $EC > 1$ means that technical efficiency improves, and vice versa. The second term denotes the technology changes between two periods. This factor is called $TC$. In this case, $TC > 1$ means the technology has advanced, or otherwise degenerated. According to the CRS and VRS decomposition of the traditional Malmquist productivity index, we can further decompose efficiency change ($EC$) into scale efficiency change ($SE$) and pure technical efficiency change ($PTE$) [39]. That is: $MPI = SE \times PTE \times TC$. Details are as follows:

$$SE = \left[\frac{D_{CRS}^t(x_{t+1}, y_{t+1})/D_{VRS}^t(x_{t+1}, y_{t+1})}{D_{CRS}^t(x_t, y_t)/D_{VRS}^t(x_t, y_t)} \times \frac{D_{CRS}^{t+1}(x_{t+1}, y_{t+1})/D_{VRS}^{t+1}(x_{t+1}, y_{t+1})}{D_{CRS}^{t+1}(x_t, y_t)/D_{VRS}^{t+1}(x_t, y_t)}\right]^{\frac{1}{2}} \quad (4)$$

$$PTE = \frac{D_{VRS}^{t+1}(x_{t+1}, y_{t+1})}{D_{VRS}^t(x_t, y_t)} \quad (5)$$

$$TC = \left[ \frac{D^{t}_{VRS}(x_t, y_t)}{D^{t+1}_{VRS}(x_t, y_t)} \times \frac{D^{t}_{VRS}(x_{t+1}, y_{t+1})}{D^{t+1}_{VRS}(x_{t+1}, y_{t+1})} \right]^{\frac{1}{2}} \tag{6}$$

The $SE$ component evaluates the effect of the change in scale of DMUs on their productivity. The $PTE$ component measures whether the DMU under evaluation is closer to (or further away from) the frontiers of production technology. The $TC$ component indicates whether the frontiers of production technology have shifted over time. If the value of any of the components is less than 1, it means regress; a value greater than 1 denotes progress, while a value of 1 implies a constant situation.

### 2.2. Regional Difference Analysis

This paper uses the Theil index to measure regional differences in SLUE. Then, the additivity of the Theil index is used to decompose the overall differences into intra-regional differences and inter-regional differences. The smaller the Theil index is, the smaller are the regional differences, and vice versa. Here, China is divided into east, central, west, and northeast regions. According to Theil [29], Bourguignon [42], Cowell [43] and Shorrocks [44], the Chinese SLUE Theil index construction and decomposition formulas are defined as follows:

$$\begin{aligned} T = T_w + T_b &= \frac{1}{n}\sum_{i=1}^{n} \frac{y_i}{y} \log\left(\frac{y_i}{y}\right) \\ T_w &= \sum_{k=1}^{m} \left(\frac{n_k}{n}\frac{\overline{y_k}}{\overline{y}}\right) \times T_k \\ T_b &= \sum_{k=1}^{m} \frac{n_k}{n}\left(\frac{\overline{y_k}}{\overline{y}}\right) \log\left(\frac{\overline{y_k}}{\overline{y}}\right) \end{aligned} \tag{7}$$

where $T$ represents the overall Theil index, $T_w$ is the intra-regional Theil index, $T_b$ is the inter-regional Theil index, and $T = T_b + T_w$. Suppose $n$ provinces are divided into $m$ groups; $n_k$ represents the number of provinces in the $k$ region $(k = 1, 2, \ldots\ldots, m)$; $y_i$ represents the SLUE of $i$ province, and $\overline{y_k}$ and $\overline{y}$ represent the average value of SLUE in the $k$ region and the whole region, respectively. Next, $T_k$ is the overall Theil index in the $k$ region. In addition, the intra-regional contribution is the rate of the intra-regional and the overall Theil index. That is: $T_w/T$; the inter-regional contribution is the rate of inter-regional and overall Theil index. That is: $T_b/T$.

### 2.3. Dataset and Variables

#### 2.3.1. Dataset

The SLUE of 30 provinces in China, from 2011 to 2017, was evaluated. Tibet, Hong Kong, Macao, and Taiwan were excluded from the study sample due to a lack of data. According to the traditional area partition method, China can be divided into the eastern area, the central area, the western area, and the northeastern area [45]. Table 1 shows the division according to the classification method of the National Bureau of Statistics of China. Then, Figure 1 shows the visual distribution of the four regions in the research area. The data in this paper are from the China Urban Construction Statistical Yearbook, provincial statistical yearbooks, the National Bureau of Statistics, economic databases, and the Carbon Emission Accounts & Datasets (CEADs). Among them, $CO_2$ data are directly from CEADs.

**Table 1.** Regional division in China.

| Region | Provinces, Municipalities, and Autonomous Regions |
|---|---|
| Eastern | Beijing, Tianjin, Shanghai, Hebei, Shandong, Jiangsu, Zhejiang, Fujian, Guangdong, Hainan |
| Central | Henan, Shanxi, Anhui, Hubei, Hunan, Jiangxi |
| Western | Gansu, Guizhou, Ningxia, Qinghai, Shaanxi, Yunnan, Xinjiang, Sichuan, Chongqing, Inner Mongolia, Guangxi, Tibet |
| Northeastern | Liaoning, Heilongjiang, Jilin |

**Figure 1.** Visual distribution map of four regions in China.

### 2.3.2. Input and Output Variables

Indicators should be selected to fully reflect the economic, social, and environmental aspects of urbanization [46]. Therefore, according to land use characteristics and existing research results [47], this article constructs a SLUE evaluation index system with three inputs and three outputs. Specifically, the input index must fully reflect the three factors of land, capital, and labor, while the output index starts from the three aspects of economic benefit, social benefit, and negative environmental effect.

The inputs include the area of built districts in the service industry, fixed capital stock in the service industry, and the number of people in employment in the service industry. Areas of built districts in a service industry include the following six types of land use: (1) commercial and business facilities; (2) logistics and warehouse; (3) road, street and transportation; (4) administration and public services; (5) municipal utilities; (6) green space and square. The fixed capital stock of a service industry is calculated by using the perpetual inventory method.

The outputs include value-added of tertiary industry, the average wage of employed persons, and the amount of $CO_2$ emissions in the service industry. The value-added of tertiary industry reflects the direct economic performance of land use in the service sector. The average wage of employed persons reflects the social benefits obtained by the service industry in the process of urbanization. Considering the availability of data, the undesirable outputs mainly involve the $CO_2$ emissions from the land use processes of the service industry. Table 2 shows the indicator system that has been constructed for this article to measure SLUE.

**Table 2.** Indicator system for evaluating SLUE.

| Category | Indicator | Specific Indicator | Unit |
|---|---|---|---|
| Inputs | Land input | Area of built districts | Square kilometers |
| | Capital input | Fixed capital stock | 100 million yuan |
| | Workforce input | Employment | 10 thousand persons |
| Desirable outputs | Economic output | Value-added of tertiary industry | 100 million yuan |
| | Social output | Average wage of employed persons | Yuan |
| Undesirable outputs | Environmental output | $CO_2$ emissions | Million tons |

## 3. Results

### 3.1. Temporal and Spatial Pattern of Regional Differences in SLUE

According to the results shown in Table 3, from 2011 to 2017, the SLUE showed an upward trend in China, with a national average increase of 27.21%. Also, the SLUE of the four major regions showed different degrees of growth. However, the overall SLUE in China is still at a low level, at an average of 0.4258. The average SLUE values of the eastern, western, central, and northeastern regions are 0.5554, 0.4556, 0.2500, and 0.2357, respectively. Regardless of any year, the efficiency ratings in the eastern and western regions are much higher than those of the central and northeastern regions. This finding indicates that there is still a long way to go in terms of improving the SLUE in the central and northeastern regions. In terms of regional differences, obvious regional differences do exist, and only a few provinces have achieved high levels of efficiency. Among the 30 provinces, only seven had SLUE values above the national average level. Of those seven provinces, five are located in the eastern region, namely Shanghai, Beijing, Tianjin, Jiangsu, and Hainan. The other two provinces are located in the western region, namely Qinghai and Ningxia. The remaining 23 regions that do not exceed the national average level are mainly concentrated in nine western provinces, six central provinces, and five eastern provinces. In particular, the average efficiency ratings of provinces in the central and northeastern regions are lower than the national average. These results intuitively show that large differences exist in SLUE among China's different regions. Moreover, one can clearly see that, whether looking at the four major regions or just within the eastern and western regions in Figure 2, the SLUE has obvious unbalanced distribution characteristics.

**Table 3.** SLUE in 30 Chinese provinces from 2011 to 2017.

| Region | Province | 2011 | 2012 | 2013 | 2014 | 2015 | 2016 | 2017 | Average |
|---|---|---|---|---|---|---|---|---|---|
| Eastern | Shanghai | 1.0000 | 1.0000 | 1.0000 | 0.9441 | 1.0000 | 1.0000 | 1.0000 | 0.9920 |
| | Beijing | 0.7977 | 0.8166 | 0.8633 | 0.8823 | 0.9216 | 0.9428 | 1.0000 | 0.8892 |
| | Tianjin | 0.6708 | 0.7204 | 0.8884 | 0.9019 | 0.9176 | 0.9413 | 1.0000 | 0.8629 |
| | Jiangsu | 0.5164 | 0.5213 | 0.5616 | 0.5788 | 0.6548 | 0.8484 | 1.0000 | 0.6688 |
| | Hainan | 0.5722 | 0.6034 | 0.6049 | 0.6586 | 0.5871 | 0.6829 | 0.7244 | 0.6334 |
| | Zhejiang | 0.3381 | 0.3524 | 0.3768 | 0.4013 | 0.4328 | 0.4892 | 0.5666 | 0.4224 |
| | Fujian | 0.3142 | 0.3397 | 0.3947 | 0.4081 | 0.4229 | 0.4481 | 0.4717 | 0.3999 |
| | Guangdong | 0.2476 | 0.2592 | 0.2759 | 0.2653 | 0.2512 | 0.2456 | 0.2589 | 0.2577 |
| | Hebei | 0.1907 | 0.2071 | 0.2221 | 0.2520 | 0.2731 | 0.2722 | 0.3168 | 0.2477 |
| | Shandong | 0.0941 | 0.0964 | 0.1843 | 0.1987 | 0.2164 | 0.2368 | 0.2320 | 0.1798 |
| | Average | 0.4742 | 0.4916 | 0.5372 | 0.5491 | 0.5678 | 0.6107 | 0.6570 | 0.5554 |
| Central | Jiangxi | 0.3546 | 0.3602 | 0.3378 | 0.3492 | 0.3354 | 0.3565 | 0.3617 | 0.3508 |
| | Shanxi | 0.2765 | 0.2811 | 0.2925 | 0.2992 | 0.2926 | 0.2878 | 0.3060 | 0.2908 |
| | Anhui | 0.2704 | 0.2355 | 0.2344 | 0.2366 | 0.2455 | 0.2617 | 0.2709 | 0.2507 |
| | Hunan | 0.2126 | 0.2494 | 0.2244 | 0.2295 | 0.2338 | 0.2451 | 0.2609 | 0.2365 |
| | Henan | 0.1905 | 0.1907 | 0.1789 | 0.1836 | 0.1836 | 0.1978 | 0.2153 | 0.1915 |
| | Hubei | 0.1471 | 0.1570 | 0.1812 | 0.1795 | 0.1996 | 0.1940 | 0.1983 | 0.1795 |
| | Average | 0.2419 | 0.2457 | 0.2415 | 0.2463 | 0.2484 | 0.2572 | 0.2688 | 0.2500 |
| Western | Qinghai | 0.8159 | 1.0000 | 1.0000 | 1.0000 | 1.0000 | 1.0000 | 1.0000 | 0.9737 |
| | Ningxia | 0.8185 | 0.8933 | 1.0000 | 1.0000 | 0.9929 | 0.9795 | 1.0000 | 0.9549 |
| | Gansu | 0.4457 | 0.4560 | 0.4119 | 0.4087 | 0.4154 | 0.4167 | 0.4191 | 0.4248 |
| | Yunnan | 0.3239 | 0.3370 | 0.3999 | 0.4051 | 0.4514 | 0.4812 | 0.5343 | 0.4190 |
| | Xinjiang | 0.4196 | 0.4246 | 0.3918 | 0.3877 | 0.3658 | 0.3555 | 0.3473 | 0.3846 |
| | Chongqing | 0.3324 | 0.3426 | 0.3540 | 0.3970 | 0.3893 | 0.4047 | 0.4218 | 0.3774 |
| | Guizhou | 0.3557 | 0.3628 | 0.3582 | 0.3644 | 0.3639 | 0.3636 | 0.3643 | 0.3619 |
| | Shaanxi | 0.2556 | 0.3156 | 0.3641 | 0.3502 | 0.3482 | 0.3881 | 0.3974 | 0.3456 |
| | Inner Mongolia | 0.2722 | 0.2866 | 0.2866 | 0.2716 | 0.3035 | 0.3541 | 0.4128 | 0.3125 |
| | Guangxi | 0.2341 | 0.2327 | 0.2971 | 0.2789 | 0.2840 | 0.2897 | 0.2966 | 0.2733 |
| | Sichuan | 0.1704 | 0.1790 | 0.1782 | 0.1904 | 0.2025 | 0.1826 | 0.1856 | 0.1841 |
| | Average | 0.4040 | 0.4391 | 0.4584 | 0.4594 | 0.4652 | 0.4741 | 0.4890 | 0.4556 |
| Northeastern | Jilin | 0.2874 | 0.2974 | 0.3220 | 0.3523 | 0.3262 | 0.3398 | 0.3699 | 0.3279 |
| | Heilongjiang | 0.2131 | 0.1995 | 0.1987 | 0.1998 | 0.1980 | 0.2003 | 0.2184 | 0.2040 |
| | Liaoning | 0.1496 | 0.1544 | 0.1712 | 0.1682 | 0.1825 | 0.1917 | 0.2104 | 0.1754 |
| | Average | 0.2167 | 0.2171 | 0.2306 | 0.2401 | 0.2356 | 0.2439 | 0.2662 | 0.2357 |
| Nationwide | Average | 0.3763 | 0.3957 | 0.4185 | 0.4248 | 0.4331 | 0.4533 | 0.4787 | 0.4258 |

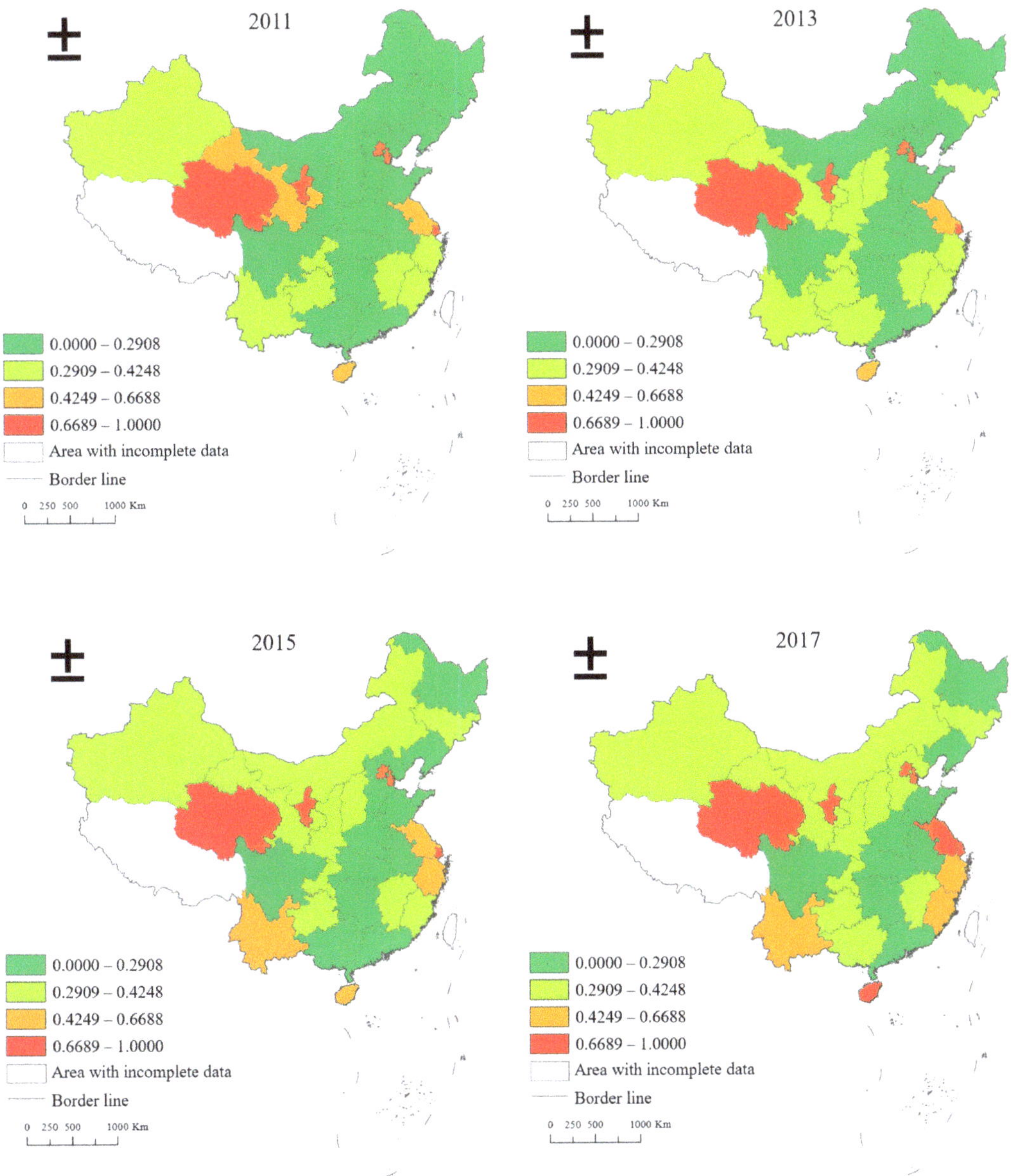


**Figure 2.** Spatial and temporal distribution and evolution trend of SLUE in China.

Furthermore, Figure 3 plots the dynamic density distribution variations of the SLUE index for the whole country and the four regions in 2011, 2014, and 2017. This is used to track the changes in SLUE over time. Nationally, SLUE values were mainly concentrated between 0.2 and 0.4. In addition, the density function curve has the characteristics of moving to the right, with the left peak falling and the right peak moving upward. This indicates that the average value of national SLUE trended upward from 2011 to 2017. At the same time, SLUE in the four regions showed a trend of improvement. There are also great development differences, both between regions and within regions. The analysis results are consistent with the above.

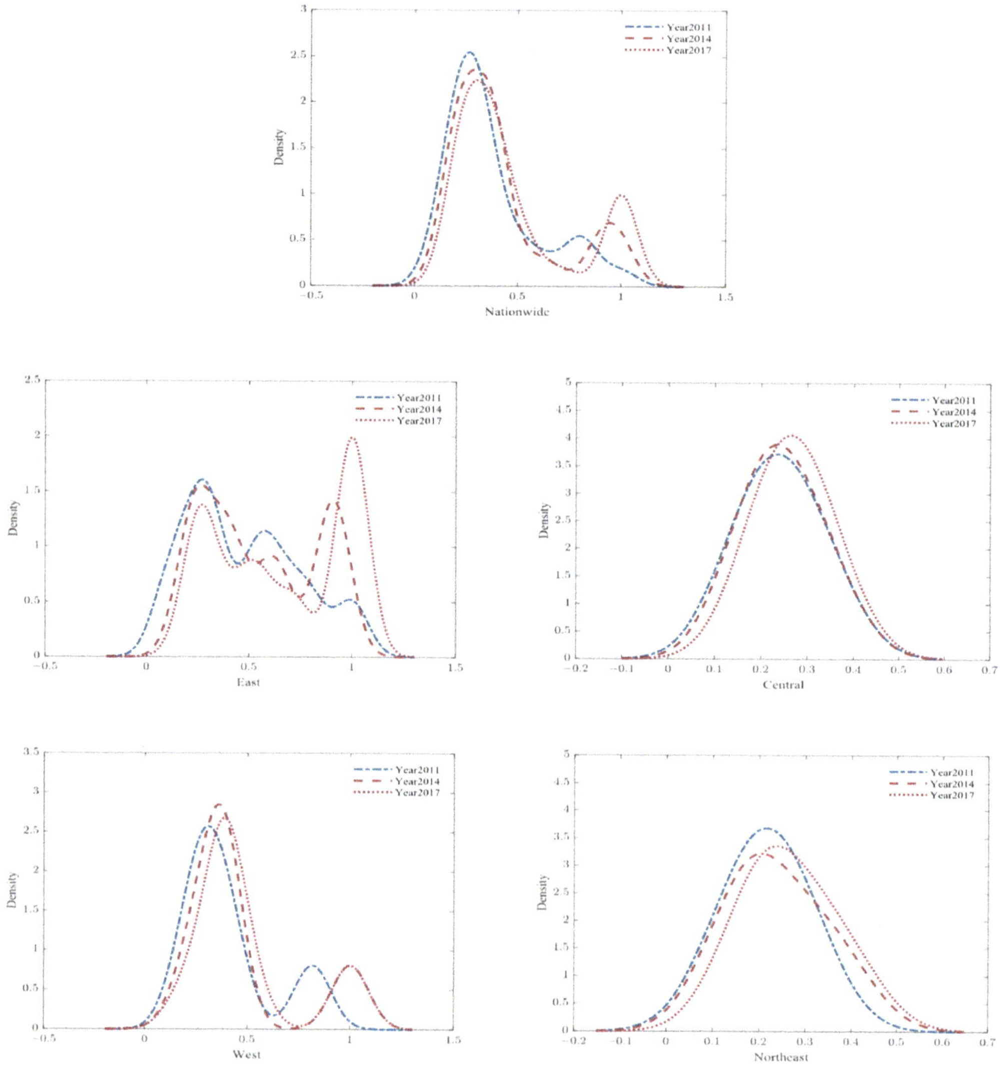


**Figure 3.** Distribution of SLUE density by region.

### 3.2. The "Beggar-Thy-Neighbor" Situation in the SLUE Regional Differences

According to the above analysis, SLUE in China has obvious characteristics of unbalanced regional development. In order to further explore the sources of this unbalance, the Theil index is used to measure the regional differences of SLUE in China, as shown in Figure 4. The results show that the overall regional differences in Chinese SLUE have narrowed; the intra-regional differences are the main source of the overall regional differences, and the intra-regional differences in the eastern and western regions have made an important contribution to the intra-regional differences.

First, as shown in Figure 4a, the SLUE's Thiel index fell from 0.1627 in 2011 to 0.1601 in 2017, with regional differences rising in 2012 and then slowly declining. Second, with regard to the sources of regional differences, as shown in Figure 4b, the overall regional differences of SLUE in China mainly come from intra-regional differences, with the contribution rate remaining at between 65.00% and 78.00%. Both the overall regional differences and the intra-regional differences of SLUE show the same decreasing trend of fluctuation, while the inter-regional differences show an increasing trend. The contribution rate of intra-regional differences to the overall regional differences of SLUE is always greater than that of inter-regional differences. This result indicates that it is difficult for inter-regional differences to become the decisive force affecting the overall regional differences of SLUE in a short time. Therefore, narrowing the intra-regional differences is the main path to further realizing the coordinated development of regional SLUE. However, attention must still be paid to changes in inter-regional differences. Finally, the regional contribution of intra-regional differences is further analyzed through the decomposition of the sources of intra-regional differences. As shown in Figure 4c, the intra-regional differences are relatively large. The intra-regional differences in the eastern and central regions show an unstable state over time, while the intra-regional differences in the central and northeastern regions are relatively stable. In general, the eastern and western regions contribute the most to the intra-regional differences, while the central and northeastern regions contribute less to the intra-regional differences.

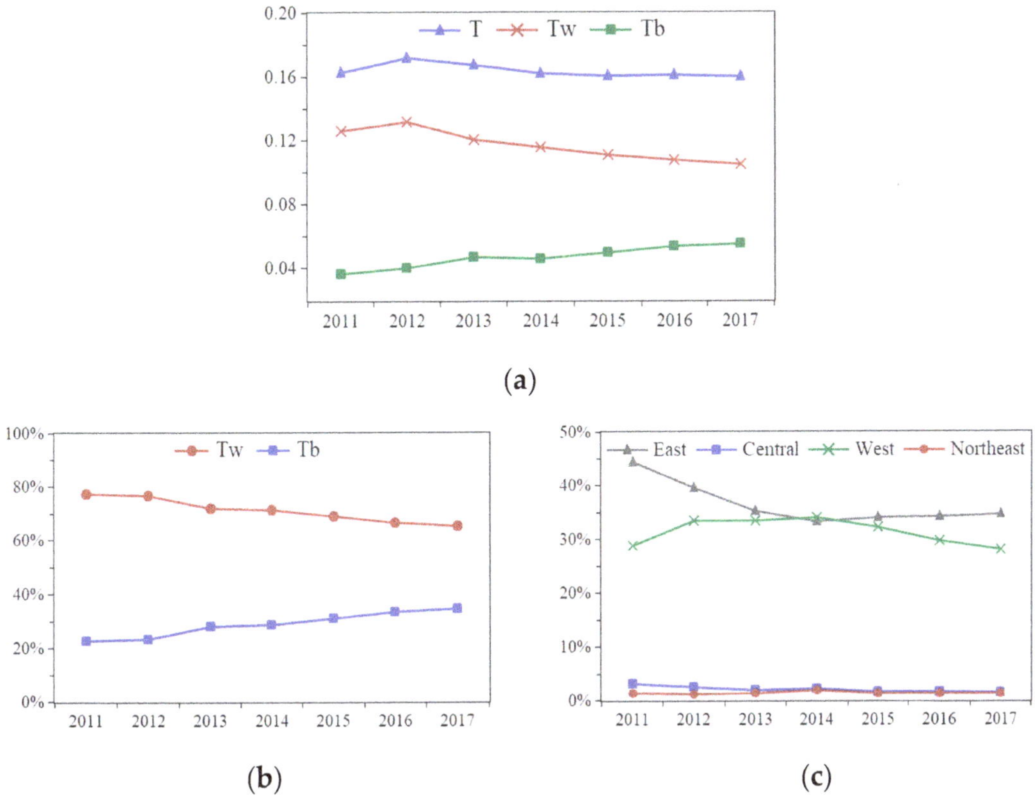


**Figure 4.** Spatial differences and contribution rates in SLUE in China. (**a**) Variation trend in the Theil index in China; (**b**) Contribution rates of $T_w$ and $T_b$; (**c**) Contribution rates of regional differences.

### 3.3. *Dynamic Trend and Efficiency Decomposition of SLUE in China*

The above analysis is only a static comparative study of SLUE. In order to further explore the trends and sources of SLUE changes in different periods in China, we calculated the MPI of land use in the service sector and its decomposition index to observe whether a positive catch-up effect of land use in the service sector exists in different provinces concerning technology and scale. The results are as shown in Table 4 and Figure 5.

**Table 4.** Annual growth rate of MPI, TC, PTE, SE in 30 provinces of China.

| Province | MPI | TC | PTE | SE | Province | MPI | TC | PTE | SE |
|---|---|---|---|---|---|---|---|---|---|
| Guangxi | 14.3 | 3.8 | 9.9 | 0.2 | Sichuan | 1.5 | 2.8 | 0.5 | −1.7 |
| Jilin | 12.6 | 3.3 | 10.1 | −0.9 | Chongqing | 0.6 | 4.2 | −2.1 | −1.4 |
| Hunan | 11.7 | 3.7 | 7.4 | 0.2 | Henan | 0.5 | 2.7 | 2.0 | −4.0 |
| Shanxi | 10.6 | 4.4 | 3.8 | 2.1 | Shandong | 0.1 | 3.8 | −4.8 | 1.3 |
| Gansu | 9.3 | 5.1 | 6.0 | −1.8 | Anhui | 0.0 | 2.3 | −3.5 | 1.2 |
| Liaoning | 8.7 | 3.4 | 4.0 | 1.0 | Inner Mongolia | 0.0 | 3.1 | 0.9 | −3.9 |
| Jiangxi | 8.2 | 3.1 | 4.8 | 0.2 | Hainan | −0.4 | 3.4 | −5.4 | 1.8 |
| Zhejiang | 6.4 | 4.7 | 1.5 | 0.1 | Heilongjiang | −0.5 | 4.0 | −0.2 | −4.1 |
| Ningxia | 4.0 | 4.0 | 0.0 | 0.0 | Xinjiang | −0.6 | 3.9 | −2.4 | −1.9 |
| Guizhou | 3.8 | 2.6 | 0.1 | 1.1 | Shaanxi | −0.7 | 3.9 | −3.8 | −0.7 |
| Tianjin | 3.1 | 3.1 | 0.0 | 0.0 | Jiangsu | −1.4 | 1.5 | −1.6 | −1.2 |
| Qinghai | 3.1 | 3.1 | 0.0 | 0.0 | Shanghai | −2.1 | 2.9 | −3.5 | −1.4 |
| Hubei | 2.6 | 3.3 | 5.2 | −5.6 | Hebei | −2.1 | 2.2 | −4.1 | −0.2 |
| Yunnan | 2.0 | 3.1 | 0.2 | −1.3 | Beijing | −3.5 | 5.0 | −8.0 | −0.1 |
| Fujian | 1.9 | 3.2 | −1.9 | 0.6 | Guangdong | −8.5 | 2.7 | −9.5 | −1.5 |

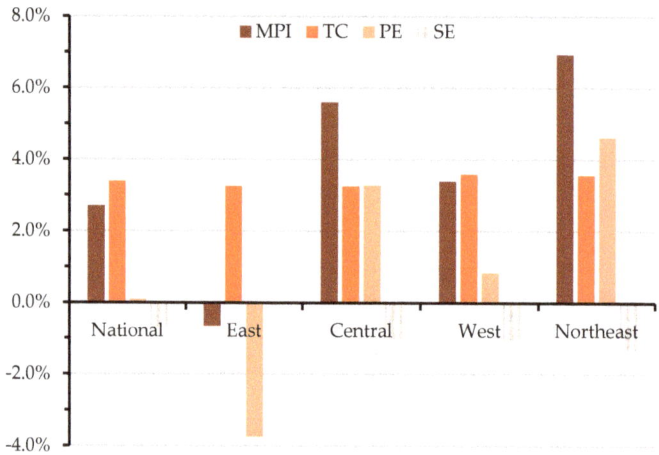


**Figure 5.** Drivers of SLUE growth by region.

Table 4 shows that the overall allocation of land use related resources in the service sector is reasonable, but it is unbalanced in terms of regional growth. Low-efficiency regions are shown to have a positive catch-up effect, while some high-efficiency regions fall into negative growth. From 2011 to 2017, the national SLUE growth rate trended upward, with an average annual growth rate of 2.70%. Except for the negative growth in the eastern region, the other three regions showed positive growth. Among them, northeastern China ranked first in growth, with an annual growth rate of 6.93%, followed by the central, western, and eastern regions, with annual growth rates of 5.60%, 3.39%, and −0.65%, respectively. Specifically, the SLUE of most provinces in China showed positive growth. Provinces with low efficiency, such as Hunan, Liaoning, and Hubei, are shown to have a positive catch-up effect. In addition, nine provinces also fell into negative growth; the average efficiency of four of the nine provinces, namely Beijing, Shanghai, Jiangsu, and Hainan, was greater than the national average. From the perspective of spatial distribution, six of the nine provinces are from the eastern region, showing that SLUE in the eastern region is in a state of weak growth. In short, China's SLUE trended upward while remaining at a low level. At the same time, SLUE in China has experienced unbalanced regional development, showing an inverted growth law of "low in developed areas and high in backward areas".

The structural composition of SLUE shows that, during the study period, all provinces had positive growth in technological progress, while most regions are in the double dilemma of pure technical efficiency and scale efficiency deterioration. With regard to technological progress, the average annual growth rate of China and the four main regions all maintained steady improvement. The growth rate of pure technical efficiency presents a trend of "low in the east and high in the west". With the exception of the negative growth experienced in the eastern region, the other three regions had a positive growth trend. Every region faces the severe situation of weak growth of scale efficiency, and is deteriorating to different degrees. Among them, the northeastern region deteriorated the most. As can be observed in Beijing, Shanghai, Guangdong, Chongqing, and other developed regions where the service industry is widely concentrated, scale efficiency showed a downward trend. In addition, one can find that the fluctuation of scale efficiency is weak. One possible reason for this finding is policy stability.

## 4. Discussion

### *4.1. Spatio-Temporal Distribution of SLUE*

In this study, the undesirable output SBM model was used to measure the SLUE of each province and the four major regions in China. As shown in Table 3 and Figure 2, SLUE in China presents unbalanced distribution characteristics. Most of the efficient provinces are located in the developed eastern regions and a few western regions. Among them, Shanghai, Beijing, Tianjin, Qinghai, and Ningxia are relatively stable in high-value regions. The allocation of capital, labor, and land in the service industry in these provinces is relatively reasonable. Therefore, the SLUE in these provinces has always been in the forefront of the region. Among them, due to the developed service industry, high level of industrial agglomeration, and dense population, the intensive degree of land for the service section in the eastern region is relatively high. It is worth noting that, although Qinghai and Ningxia are located in the western regions with relatively backward economic development, the SLUE of these regions is relatively high. The possible reasons for this finding are that although the speed of economic development in these regions is relatively slow, the land for service section is better able to meet the needs of economic development. Therefore, the development speed of the urban service industry is relatively synchronized with the region's economic development speed. Most of the inefficient provinces are located in the central and northeastern regions, such as Liaoning, Hubei, and Henan, and the potential for the intensive use of land resources in the service sector needs to be further explored. The land supply of the service industry in these areas is sufficient, but the vitality of the stock land is not enough, and the incremental land use is decentralized, leading to the low SLUE. However, as one of the four major municipalities, Chongqing is relatively poor in terms of land use in the service sector. One reason is that Chongqing has a high proportion of secondary industry and lacks innovation power [48].

### *4.2. Regional Differences of SLUE*

This paper uses the Theil index to further explore the regional differences in SLUE and their sources. As is shown in Figure 4, intra-regional differences are the main source of regional differences. The large intra-regional differences in SLUE lead to significant regional differences, which can be further explained from the "beggar-thy-neighbor" perspective of land use in the service sector. Similar to Gao et al. [49], we believe that advanced service regions or developed regions may siphon off superior resources in adjacent regions. Such actions intensify the vicious competition between adjoining regions, and lead to a decline of LUE in these regions. However, the vicious competition for service resources in adjacent areas will aggravate the differences between regions. Therefore, resolving the "beggar-thy-neighbor" problem of SLUE is the key to realizing the coordinated improvement of spatial utility. The "beggar-thy-neighbor" issue of SLUE refers to the fact that the performance of regional SLUE is often subject to the region's adjacent central cities; the crowding out effect is also formed to transfer inferior resources to adjacent regions. For example, the

IT service sector in India is concentrated in first-tier cities, such as Bangalore and New Delhi. The industry attracts superior resources to the metropolises and further aggravates resource differences [50]. In fact, the reverse situation can be explained by the siphon effect. The siphon effect refers to the ability of a city to attract superior resources from adjacent small cities [51], thereby crowding out inferior resources. The siphon effect results in the centralization of land for service section supply due to the regional aggregation capacity of urban resources. Meanwhile, the land for service section in the adjacent area tends to be relatively scattered and less intensive. That combination of factors leads to low LUE, thus exacerbating the imbalance of SLUE in the region.

### 4.3. Dynamic Trend of SLUE

The results of MPI and its decomposition index show the variation trend and source of SLUE. As is shown in Table 4 and Figure 5, it was found that the growth rate of pure technical efficiency shows a trend of "low in the east and high in the west", and scale efficiency also belongs to the situation of weak group growth. As is well known, the eastern region is most developed in China, while the western region is relatively backward [52]. However, the pure technical efficiency and scale efficiency of land use in the service sector in the eastern region trended downward during the study period, while the pure technical efficiency in the western region increased slowly. This finding seems to be contrary to the level of economic development. This phenomenon is worth pondering. One reason for this finding may be that, as a large number of service factors are concentrated in the eastern region, the negative utility generated by the unreasonable management and utilization of land for the service section offsets the positive effect brought by high quality technical bases and factors. In recent years, with the transfer of many industries from eastern China to western China, the western region can promote economic growth and attract the inflow of high-end technologies by taking in these industries [53]. This same approach could also promote the improvement of SLUE.

In addition, the scale efficiency of land use in China's service sector generally shows a downward trend. This article further analyzes this phenomenon from the perspective of the excessive agglomeration of the service industry. Some scholars [54] believe that the factor behind the decline of scale efficiency in the service industry is the insufficient agglomeration of factors. However, this view does not apply to developed regions, where the service industry is highly developed and agglomerated. For example, the high degree of agglomeration of the service industries in Beijing, Tokyo [55], and Silicon Valley [56] has produced adverse reactions. The high agglomeration of the service industry in developed regions is currently an issue that needs to be addressed. Excessive agglomeration of a service industry means that, with the acceleration of urbanization, the service industry is highly concentrated in several regions, and presents an efficiency loss mode of a high attrition rate, fast rotation rate, and short periodicity. Those conditions will make it difficult for the service industry to play the role of scale economy in the process of land use. The strong economic foundation of developed regions attracts a large-scale inflow of population and resources, and industries are more likely to show the characteristics of agglomeration development. However, the people flowing to these regions need the support of the service industry. Yet, due to rapid population growth, the service industry's supply speed can easily lead to a situation of excessive growth under the pressure of enormous demand. This, in turn, results in a series of negative situations, such as the excessive agglomeration of the service industry. At present, Beijing, Shanghai, Guangdong, and other developed regions in China have fallen into the dilemma of scale efficiency deterioration. This conclusion can provide a warning for Silicon Valley in the United States, Tokyo in Japan, and other high-agglomeration regions of service industry in the world. For backward regions, the scale of service industry agglomeration is still not high. Owing to the resource endowment, rapid urbanization expansion, and the orientation of government policies, it is difficult to form a reasonable layout of factors and achieve the orderly development of the service industry in these regions. For example, in order to pursue urban expansion, Ethiopia transferred some farmland to urban commercial, residential, and other land use in the service sector. However, this type of land for the service section is not fully utilized [57,58]. Therefore, these regions have large growth space in scale efficiency.

In fact, the continuous reduction of scale efficiency may be related to the regional differentiated land supply policy. Land supply policies favoring the central and western regions lead to spatial misallocation of land resources [59], which hinders the improvement of scale efficiency. The same is true of SLUE. From 2011 to 2017, compared with the average value of the central and western regions, the average annual growth rate of land area of the service industry in the eastern region was nearly 0.5 times that of the central and western regions (China Urban Construction Statistics Yearbook). For backward areas, the rapid expansion of land for the service industry aggravates the situation of disordered and fragmented land use. Because this decentralized land use mode difficultly exerts the agglomeration effect on the service industry, the loss of scale efficiency is inevitable. Especially in areas with population outflow, such as northeast China, the redundancy of land for the service industry makes it difficult to form economies of scale in land use, leading to inefficient use of land for service industry. Moreover, the disordered land use will also result in an invalid matching between the supply of land resources and the production efficiency [59]. The aggravation of land resource mismatch will further lead to excess land supply or overcapacity [5], which will inhibit the improvement of SLUE. In fact, the large supply of land will also lead to the repeated construction of homogeneous industries [60], which will form a dependent path for inefficient land use in the service industry. For the eastern region, although there is a high degree of intensive land use for the service industry, the excessive agglomeration of the service industry hinders the scale effect due to insufficient land supply.

## 5. Research Implications and Research Contributions

### *5.1. Research Implications*

The evaluation of SLUE in this paper has rich practical significance for the optimization of land management systems and related policies, the layout planning of the service industry, cross-regional coordination and interaction, and the equalization of services.

We thus examine the promotion of the optimization and improvement of land management systems and related policies. Through accurate evaluation of the SLUE in 30 provinces, this paper captures the real situation of the SLUE of each region, and discovers the unreasonable situation of the land resource distribution of the service industry in China. The above analysis provides valuable data reference for further improvement of the land management system and related policies. The improvement of land systems and related policies is conducive to reducing idle land waste, revitalizing the stock of land resources, and thus improving SLUE.

We also examine the promotion of the scientific layout of the service industry development plan. Based on the analysis of SLUE in 30 provinces, this paper further discusses the dynamic evolution trend and efficiency decomposition of SLUE in each region, and fully explores the regional characteristics of the service industry. The above analysis provides scientific reference for the government in the inter-regional and the intra-regional development norms of the service industry, and then puts forward targeted development planning suggestions for the service industry. Urban development in developed areas should formulate a set of long-term service industry layout plans. A pattern of multiple agglomeration regions should be formed instead of multiple scattered points. In this way, the development of the service industry can avoid falling into a situation of having too high a local agglomeration level. The inefficiency of service industry land caused by too many scattered points could also be avoided. The promotion of SLUE in backward areas should focus on solving the medium- and long-term spatial layout problem so as to form an intensive and efficient mode of service industry agglomeration. Moreover, the development quality of the service industry cannot be measured only by superficial indicators, such as the increase of the proportion of the service industry and the expansion of service industry scale. Finally, an effective plan will avoid the occurrence of low-level circular lock and the excessive agglomeration of the service industry.

We also examine the promotion of the establishment of services across the regional coordination and interaction mechanism. In this paper, the source of regional disparity in SLUE is deeply discussed, and the results show that the intra-regional disparity in SLUE is quite serious. The above analysis provides directional guidance and policy reference for the establishment of an institutional coordination mechanism for the service industry, and helps to ensure the effective implementation of a scientific development plan for the service industry. At the same time, this paper puts forward some practical suggestions for the establishment of cross-regional coordination and interaction mechanism within the service industry. In the process of service industry development, the layout of the regional industrial structure should be optimized, and the integrated development between regions should be strengthened according to regional resource endowments and geographical location. Through the industrial division of labor and cooperation, the complementary and dislocation linkage development of interregional industries can be promoted. Also, inferior repeated construction and homogeneous competition of service industries must be avoided so as to weaken the negative impact of the siphon effect and to improve SLUE.

We also examine the promotion of the service industry development policy of "balance". To realize the sustainable improvement of SLUE, it is necessary to actively cultivate labor force with advanced technology and management experience. Taking into account the differences between developed and backward regions, a "balanced" development policy needs to be fully considered. For example, a relative equalization of urban capital and technology support policies will further encourage the transfer and agglomeration of superior production factor resources, such as knowledge, technology, and innovation, to relatively backward regions [61]. This would be conducive to the intensive development of land use in the service industry, and would thus narrow the regional differences in SLUE.

We also examine the promotion of the service industry growth pole to nurture and develop competitive advantage. This paper focuses on the low-efficiency areas of land use in the service sector, compares and analyzes the high-efficiency areas, and seeks the catch-up path for the low-efficiency areas according to their own factors. Low-efficiency regions need to focus on building their own service industry levels and characteristics, attracting and retaining more and superior production factors with their own advantages, and accelerating the agglomeration of production factors. By constantly absorbing the superior resources of adjacent high-efficiency regions, they can catch up with those high-efficiency regions and form the advantage of backwardness.

### 5.2. Research Contributions

This study is an early literature on the land use efficiency gap in the service industry. The imbalance of regional development in China provides a good model for the study of the spatial land use efficiency gap in the service industry. We innovatively evaluate the SLUE of 30 provinces and four regions in China, which makes for a very rich sample. In addition, we have adopted a good measurement method, combining the SBM model of undesirable output with the Malmquist productivity index. This combined method can better overcome the incomplete problem of index construction and the research of spatial-temporal dimensions.

This article provides some valuable implications for the SLUE improvement problem, particularly given the increasing shortage of industrial land all around the world. First, developed countries and regions should pay more attention to the low efficiency of land use caused by the excessive agglomeration of service industries. Meanwhile, developing countries and regions should focus more on the transformation of technology application and the spatial optimization of service industry layout so as to avoid some unnecessary efficiency loss. Second, for the coordinated improvement of SLUE, countries need to pay attention to the "beggar-thy-neighbor" problem caused by the large differences within a region. These practical contributions will provide important enlightenments for a new pattern of service industry development in the future.

## 6. Conclusions and Limitations

This article evaluates the SLUE of 30 provinces from 2011 to 2017 in China using the SBM model of undesirable output and the Malmquist productivity index. Then, the regional differences are further investigated, and suggestions are made for the regional collaborative improvement of SLUE. The main conclusions are as follows: (1) The SLUE in China presents obvious unbalanced distribution characteristics; the SLUE in the eastern and western region is also much higher than that in the central and northeastern regions. Further analysis shows that obvious regional differences exist in Chinese SLUE, and the intra-regional differences are the main source of overall regional differences. (2) From 2011 to 2017, China's SLUE generally achieved positive growth, with an average annual growth rate of 2.70%. The SLUE of the four regions shows an inverted growth law of "low in developed areas and high in backward areas". (3) The SLUE is decomposed into technological progress, pure technical efficiency, and scale efficiency. From the decomposition results, technological progress is shown to have maintained steady improvement. The growth rate of pure technical efficiency presents a trend of "low in the east and high in the west", while the scale efficiency shows a downward trend in the four regions. (4) Finally, the study discusses the path of SLUE, with a view to achieving collaborative improvement. The coordinated improvement of SLUE needs to focus on resolving the issues in large differences within regions and also on reduced scale efficiency.

However, some research limitations remain. First, evaluating SLUE at the provincial level is relatively macro. Future studies could add some micro-evidence at the enterprise level. In particular, what kind of internal structure exists in the loss of land use efficiency within service enterprises? This question should be thoroughly investigated and answered. Second, this study only calculates the LUE of the overall service industry during the period from 2011 to 2017. Future studies could be expanded to sub-sectors of the service industry, as well as a longer time horizon. In addition, as the trend of industrial integration intensifies, the interaction between service and manufacturing industries will become increasingly stronger. There will be some linkage laws between SLUE and the efficiency of manufacturing land, which is another research direction for the future.

**Author Contributions:** Conceptualization, M.Z. and Y.S.; methodology, M.Z. and Z.L.; software, M.Z. and H.L.; validation, M.Z. and H.L.; investigation, M.Z.; resources, M.Z.; writing—original draft preparation, M.Z.; funding acquisition, M.Z.; formal analysis, H.L. and C.C.; writing—review and editing, Y.S. and Z.L.; supervision, Y.S. and Z.L.; data curation, X.Z.; visualization, Z.L. All authors have read and agreed to the published version of the manuscript.

**Funding:** This research was funded by the "Study on the evolution mechanism and optimization strategy of population distribution structure under the "flow space" effect of high-speed rail network" (Grant No. ZR2022QG005); "Effect of high-speed rail network on urban population distribution pattern in Shandong province" (Grant No. 20DRKJ02); "Theoretical and economic research-oriented innovation team" of the youth innovation talent introduction and education plan of colleges and universities in Shandong Province (Grant No. 201901); Taishan Scholars Program (Grant No. tsqn201909135); "Employment polarization effect of low-carbon development constraint from the perspective of unequal opportunities" (Grant No. ZR2020QG040).

**Data Availability Statement:** The land input data were obtained from the China Urban Construction Statistical Yearbook. The data of capital input, workforce input, and desirable output were from the National Bureau of Statistics, provincial statistical yearbooks, and economic databases. Undesirable output data came from the China Carbon Emission Accounts & Datasets (CEADs).

**Acknowledgments:** The author thanks the support of the "Theoretical Economics Research Innovation Team" of Shandong Provincial University Youth Innovation Talent Introduction and Education Plan, the Shandong Digital Yellow River Industrial Chain System Construction Innovation Team of Youth Innovation Team of Shandong Colleges and Universities.

**Conflicts of Interest:** The authors declare no conflict of interest.

## References


1. Jiang, X.; Lu, X.H.; Liu, Q.; Chang, C.; Qu, L.L. The effects of land transfer marketization on the urban land use efficiency: An empirical study based on 285 cities in China. *Ecol. Indic.* **2021**, *132*, 108296. [CrossRef]
2. Li, H.J.; Qu, J.S.; Wang, D.; Meng, P.; Lu, C.Y.; Zeng, J.J. Spatial-temporal integrated measurement of the efficiency of urban land use in Yellow River Basin. *Sustainability* **2021**, *13*, 8902. [CrossRef]
3. Ge, X.J.; Liu, X.X. Urban land use efficiency under resource-based economic transformation—A case study of Shanxi Province. *Land* **2021**, *10*, 850. [CrossRef]
4. Muhammad, S.; Pan, Y.C.; Agha, M.H.; Umar, M.; Chen, S.Y. Industrial structure, energy intensity and environmental efficiency across developed and developing economies: The intermediary role of primary, secondary and tertiary industry. *Energy* **2022**, *247*, 123576. [CrossRef]
5. Zhou, D.; Huang, Q.; Chong, Z.H. Analysis on the effect and mechanism of land misallocation on carbon emissions efficiency: Evidence from China. *Land Use Policy* **2022**, *121*, 106336. [CrossRef]
6. Song, X.Q.; Feng, Q.; Xia, F.Z.; Li, X.Y.; Scheffran, J. Impacts of changing urban land-use structure on sustainable city growth in China: A population-density dynamics perspective. *Habitat Int.* **2021**, *107*, 102296. [CrossRef]
7. Huang, Y.C.; Yang, B.G.; Wang, M.; Liu, B.W.; Yang, X.D. Analysis of the future land cover change in Beijing using CA-Markov chain model. *Environ. Earth Sci.* **2020**, *79*, 60. [CrossRef]
8. Shi, L.; Halik, U.; Mamat, Z.; Aishan, T.; Abliz, A.; Welp, M. Spatiotemporal investigation of the interactive coercing relationship between urbanization and ecosystem services in arid northwestern China. *Land Degrad. Dev.* **2021**, *32*, 4105–4120. [CrossRef]
9. Li, Y.; Ye, H.P.; Gao, X.; Sun, D.Q.; Li, Z.H.; Zhang, N.H.; Leng, X.J.; Meng, D.; Zheng, J. Spatiotemporal patterns of urbanization in the three most developed urban agglomerations in China based on continuous nighttime light data (2000–2018). *Remote Sens.* **2021**, *13*, 2245. [CrossRef]
10. Lin, G.C.S. Toward a post-Socialist city? Economic tertiarization and urban reformation in the Guangzhou Metropolis, China. *Eurasian Geogr. Econ.* **2004**, *45*, 18–44. [CrossRef]
11. Chen, J.L.; Gao, J.L.; Chen, W. Urban land expansion and the transitional mechanisms in Nanjing, China. *Habitat Int.* **2016**, *53*, 274–283. [CrossRef]
12. Yang, L.; Han, B.Y.; Ma, Z.L.; Wang, T.; Lin, Y.C. Analysis of the urban land use efficiency in the new-type urbanization process of China's Yangtze River Economic Belt. *Int. J. Environ. Res. Public Health* **2022**, *19*, 8183. [CrossRef] [PubMed]
13. Kang, R.; Feng, C. Developing service industry to solve environment conflict: A case of Yuyang district, China. *Procedia–Soc. Behav. Sci.* **2016**, *220*, 159–165. [CrossRef]
14. Bai, Y.; Zhou, W.; Guan, Y.J.; Li, X.; Huang, B.H.; Lei, F.C.; Yang, H.; Huo, W.M. Evolution of policy concerning the readjustment of inefficient urban land use in China based on a content analysis method. *Sustainability* **2020**, *12*, 797. [CrossRef]
15. Dong, G.L.; Ge, Y.B.; Jia, H.W.; Sun, C.Z.; Pan, S.Y. Land use multi-suitability, land resource scarcity and diversity of human needs: A new framework for land use conflict identification. *Land* **2021**, *10*, 1003. [CrossRef]
16. Xu, W.Y.; Jin, X.B.; Liu, J.; Zhou, Y.K. Analysis of influencing factors of cultivated land fragmentation based on hierarchical linear model: A case study of Jiangsu Province, China. *Land Use Policy* **2021**, *101*, 105119. [CrossRef]
17. Cui, X.F.; Huang, S.S.; Liu, C.C.; Zhou, T.T.; Shan, L.; Zhang, F.Y.; Chen, M.; Li, F.; De Vries, W.T. Applying SBM-GPA model to explore urban land use efficiency considering ecological development in China. *Land* **2021**, *10*, 912. [CrossRef]
18. Kottas, A.T.; Bozoudis, M.N.; Madas, M.A. Turbofan aero-engine efficiency evaluation: An integrated approach using VSBM two-stage network DEA. *Omega* **2020**, *92*, 102167. [CrossRef]
19. Wang, A.P.; Lin, W.F.; Liu, B.; Wang, H.; Xu, H. Does smart city construction improve the green utilization efficiency of urban land? *Land* **2021**, *10*, 657. [CrossRef]
20. Fu, Y.H.; Zhou, T.T.; Yao, Y.Y.; Qiu, A.G.; Wei, F.Q.; Liu, J.Q.; Liu, T. Evaluating efficiency and order of urban land use structure: An empirical study of cities in Jiangsu, China. *J. Clean. Prod.* **2021**, *283*, 124638. [CrossRef]
21. Wang, R.Z.; Hao, J.; Wang, C.N.; Tang, X.; Yuan, X.Z. Embodied $CO_2$ Emissions and Efficiency of the Service Sector: Evidence from China. *J. Clean. Prod.* **2020**, *247*, 119116. [CrossRef]
22. Tone, K. Dealing with undesirable outputs in DEA: A slacks-based measure (SBM) approach. In Proceedings of the North American Productivity Workshop, Toronto, ON, Canada, 23–25 June 2004; pp. 44–45.
23. Tang, Y.K.; Wang, K.; Ji, X.M.; Xu, H.; Xiao, Y.Q. Assessment and spatial-temporal evolution analysis of urban land use efficiency under green development orientation: Case of the Yangtze River Delta urban agglomerations. *Land* **2021**, *10*, 715. [CrossRef]
24. Färe, R.; Grosskopf, S.; Norris, M.; Zhang, Z. Productivity growth, technical progress, and efficiency change in industrialized countries. *Am. Econ. Rev.* **1994**, *84*, 66–83.
25. Xue, D.; Yue, L.; Ahmad, F.; Draz, M.U.; Chandio, A.A.; Ahmad, M.; Amin, W. Empirical investigation of urban land use efficiency and influencing factors of the Yellow River basin Chinese cities. *Land Use Policy* **2022**, *117*, 106117. [CrossRef]
26. Duro, J.A.; Lauk, C.; Kastner, T.; Erb, K.; Haberl, H. Global inequalities in food consumption, cropland demand and land-use efficiency: A decomposition analysis. *Glob. Environ. Change* **2020**, *64*, 102124. [CrossRef]
27. Lu, X.H.; Kuang, B.; Li, J. Regional difference decomposition and policy implications of China's urban land use efficiency under the environmental restriction. *Habitat Int.* **2018**, *77*, 32–39. [CrossRef]

28. Cao, X.; Liu, Y.; Li, T.; Liao, W. Analysis of spatial pattern evolution and influencing factors of regional land use efficiency in China based on ESDA-GWR. *Sci. Rep.* **2019**, *9*, 520. [CrossRef]
29. Theil, H. *Economics and Information Theory*; North-Holland Publishing Company: Amsterdam, The Netherlands, 1967.
30. Ge, K.; Zou, S.; Chen, D.L.; Lu, X.H.; Ke, S.A. Research on the spatial differences and convergence mechanism of urban land use efficiency under the background of regional integration: A case study of the Yangtze River Economic Zone, China. *Land* **2021**, *10*, 1100. [CrossRef]
31. Chen, Z.H.; Zhang, Q.X.; Li, F.; Shi, J.L. Comprehensive evaluation of land use benefit in the Yellow River Basin from 1995 to 2018. *Land* **2021**, *10*, 643. [CrossRef]
32. Pang, Y.Y.; Wang, X.J. Land-use efficiency in Shandong (China): Empirical analysis based on a super-SBM model. *Sustainability* **2020**, *12*, 10618. [CrossRef]
33. Yao, M.C.; Zhang, Y.H. Evaluation and optimization of urban land-use efficiency: A case study in Sichuan Province of China. *Sustainability* **2021**, *13*, 1771. [CrossRef]
34. Gao, J.X.; Song, J.B.; Wu, L.F. A new methodology to measure the urban construction land-use efficiency based on the two-stage DEA model. *Land Use Policy* **2022**, *112*, 105799. [CrossRef]
35. Jiang, H.L. Spatial-temporal differences of industrial land use efficiency and its influencing factors for China's central region: Analyzed by SBM model. *Environ. Technol. Innov.* **2021**, *22*, 101489. [CrossRef]
36. Souza, G.D.E.; Gomes, E.G.; Alves, E.R.D.; Gasques, J.G. Technological progress in the Brazilian agriculture. *Socio-Econ. Plan. Sci.* **2020**, *72*, 100879. [CrossRef]
37. Charnes, A.; Cooper, W.W.; Rhodes, E. Measuring the efficiency of decision making units. *Eur. J. Oper. Res.* **1978**, *2*, 429–444. [CrossRef]
38. Cooper, W.W.; Seiford, L.M.; Tone, K. *Data Envelopment Analysis: A Comprehensive Text with Models, Applications, References and DEA-Solver Software*; Kluwer Academic Publishers: New York, NY, USA, 2007.
39. Banker, R.D.; Charnes, A.; Cooper, W.W. Some models for estimating technical and scale inefficiencies in data envelopment analysis. *Manag. Sci.* **1984**, *30*, 1078–1092. [CrossRef]
40. Pastor, J.T.; Lovell, C.A.K. A global Malmquist productivity index. *Econ. Lett.* **2005**, *88*, 266–271. [CrossRef]
41. Fare, R.; Grosskopf, S. Malmquist productivity indexes and fisher ideal indexes. *Econ. J.* **1992**, *102*, 158–160. [CrossRef]
42. Bourguignon, F. Decomposable income inequality measures. *Econometrica* **1979**, *47*, 901–920. [CrossRef]
43. Cowell, F.A. On the structure of additive inequality measures. *Rev. Econ. Stud.* **1980**, *47*, 521–531. [CrossRef]
44. Shorrocks, A.F. The class of additively decomposable inequality measures. *Econometrica* **1980**, *48*, 613–625. [CrossRef]
45. Lin, S.F.; Lin, R.Y.; Sun, J.; Wang, F.; Wu, W.X. Dynamically evaluating technological innovation efficiency of high-tech industry in China: Provincial, regional and industrial perspective. *Socio-Econ. Plan. Sci.* **2021**, *74*, 100939. [CrossRef]
46. Wey, W.; Hsu, J. New urbanism and smart growth: Toward achieving a smart National Taipei University District. *Habitat Int.* **2014**, *42*, 164–174. [CrossRef]
47. Zhou, Y.; Chen, Y.; Hu, Y. Assessing efficiency of urban land utilisation under environmental constraints in Yangtze River Delta, China. *Int. J. Environ. Res. Public Health* **2021**, *18*, 12634. [CrossRef] [PubMed]
48. Yu, J.Q.; Zhou, K.L.; Yang, S.L. Land use efficiency and influencing factors of urban agglomerations in China. *Land Use Policy* **2019**, *88*, 104143. [CrossRef]
49. Gao, X.; Zhang, A.L.; Sun, Z.L. How regional economic integration influence on urban land use efficiency? A case study of Wuhan metropolitan area, China. *Land Use Policy* **2020**, *90*, 104329. [CrossRef]
50. Rao, P.M.; Balasubrahmanya, M.H. The rise of IT services clusters in India: A case of growth by replication. *Telecommun. Policy* **2017**, *41*, 90–105. [CrossRef]
51. Zhao, Q.Y.; Xu, H.; Wall, R.S.; Stavropoulos, S. Building a bridge between port and city: Improving the urban competitiveness of port cities. *J. Transp. Geogr.* **2017**, *59*, 120–133. [CrossRef]
52. Gai, M.; Wang, X.Q.; Qi, C.L. Spatiotemporal evolution and influencing factors of ecological civilization construction in China. *Complexity* **2020**, *14*, 8829144. [CrossRef]
53. Chen, W.; Shen, Y.; Wang, Y.A.; Wu, Q. The effect of industrial relocation on industrial land use efficiency in China: A spatial econometrics approach. *J. Clean. Prod.* **2018**, *205*, 525–535. [CrossRef]
54. Hu, Y.H.; Wang, M.M. Impact of resource reconfiguration on total factor productivity of service industry. *Int. J. Electr. Eng. Educ.* **2020**. [CrossRef]
55. Zheng, X.P. Determinants of agglomeration economies and diseconomies: Empirical evidence from Tokyo. *Socio-Econ. Plan. Sci.* **2001**, *35*, 131–144. [CrossRef]
56. Kemeny, T.; Osman, T. The wider impacts of high-technology employment: Evidence from US cities. *Res. Policy* **2018**, *47*, 1729–1740. [CrossRef]
57. Dadi, D.; Azadi, H.; Senbeta, F.; Abebe, K.; Taheri, F.; Stellmacher, T. Urban sprawl and its impacts on land use change in Central Ethiopia. *Urban For. Urban Green.* **2016**, *16*, 132–141. [CrossRef]
58. Koroso, N.H.; Zevenbergen, J.A.; Lengoiboni, M. Urban land use efficiency in Ethiopia: An assessment of urban land use sustainability in Addis Ababa. *Land Use Policy* **2020**, *99*, 105081. [CrossRef]

59. Yang, L.L.; Wang, J.; Feng, Y.H.; Wu, Q. The impact of the regional differentiation of land supply on Total Factor Productivity in China: From the perspective of Total Factor Productivity decomposition. *Land* **2022**, *11*, 1859. [CrossRef]
60. Peng, S.; Wang, J.; Sun, H.; Guo, Z. How does the spatial misallocation of land resources affect urban industrial transformation and upgrading? evidence from China. *Land* **2022**, *11*, 1630. [CrossRef]
61. Zhang, Y.H.; Chen, J.L.; Gao, J.L.; Jiang, W.X. The impact mechanism of urban land use efficiency in the Yangtze River Delta from the perspective of economic transition. *J. Nat. Resour.* **2019**, *34*, 1157–1170. [CrossRef]

*land*

Article

# Evaluation, Recognition and Implications of Urban–Rural Integration Development: A Township-Level Analysis of Hanchuan City in Wuhan Metropolitan Area

**Shiwei Lu [1,2,3], Yaping Huang [1,2,3], Xiaoqing Wu [1,4] and Yichen Ding [1,2,3,*]**

1 School of Architecture and Urban Planning, Huazhong University of Science and Technology, Wuhan 430074, China
2 Hubei Engineering and Technology Research Center of Urbanization, Wuhan 430074, China
3 The Key Laboratory of Urban Simulation for Ministry of Natural Resources, Wuhan 430074, China
4 Real Estate Management Bureau, Dayabay Economic and Technological Development Zone, Huizhou 516000, China
* Correspondence: yichending@hust.edu.cn; Tel.: +86-27-8754-3156

**Abstract:** China has achieved success in implementing the rural revitalization strategy and promoting the development of new urbanization. However, there are still many problems in the research and implementation on urban–rural integration development, such as insufficient research at the township level, unclear recognition of development patterns, and disconnection from land-use planning. Therefore, taking Hanchuan city in the Wuhan metropolitan area as a case study, this research constructs a comprehensive evaluation system of urban–rural integration development based on both on the current and potential level of development, and identifies the spatial characteristics and patterns in the study area. This study found that: (1) The comprehensive evaluation result of urban–rural integration development in Hanchuan City shows that a high level of development units are mainly distributed in the northeast and southwest, and gradually decreases from the northeast and southwest, indicating that towns in the central area are relatively weakly driven by the radiation of the surrounding growth poles. (2) Xiannvshan Street, Makou Town, Chenhu Town, and Xinyan Town with the highest comprehensive evaluation values were selected as the centers of urban–rural integration development in four directions. (3) Four typical patterns of urban–rural integration development, which are town gathering, agro-tourism interaction, industry-trade driven, and agricultural service, are derived by the gravity model and classification assignment method according to their respective centers. (4) According to the urban–rural integration development patterns, land-use strategies such as centralization for promoting linkage level of towns, differentiation for arranging various resources and infrastructures, and demonstration for optimization of experience to the whole area are proposed in a targeted manner. This study has important implications for the preparation and implementation of urban–rural integration development and provides effective planning guidance for promoting social equity and accessibility of facilities in the metropolitan area.

**Keywords:** urban–rural integration development; evaluation system; gravity model; spatial pattern; land-use planning implications

**Citation:** Lu, S.; Huang, Y.; Wu, X.; Ding, Y. Evaluation, Recognition and Implications of Urban–Rural Integration Development: A Township-Level Analysis of Hanchuan City in Wuhan Metropolitan Area. *Land* **2023**, *12*, 14. https://doi.org/10.3390/land12010014

Academic Editors: Bindong Sun, Tinglin Zhang, Wan Li, Chun Yin and Honghuan Gu

Received: 27 November 2022
Revised: 14 December 2022
Accepted: 19 December 2022
Published: 21 December 2022



## 1. Introduction

Urban–rural integration is considered to be the ideal state of development between urban and rural areas [1], which means that elements freely penetrate and interact with each other, and make the urban and rural develop together [2]. To overcome the negative impact of the long-term urban–rural dual system, the European Commission formulated the integration principle from the perspective of European spatial development in 1999 [3]. Nowadays, urban–rural integration has become an important trend in social and economic development for both the developed and developing countries [4,5].

Since 1949, China has implemented the urban–rural development concept of "supporting the industry with agriculture and nurturing cities with the countryside" [6], and gradually formed the situation of urban–rural dual division, leading to the problem of unbalanced urban–rural development over the past five decades. In the new era, China has paid great attention to the integrated development of urban and rural areas, such as continuously increasing the support for the rural areas and formulating a series of policies and systems to promote rural revitalization and urban–rural development. In general, China's urban–rural relationship has gone through a process from dual segmentation to overall planning and integration [7]. However, in 2020, the urban–rural income ratio in developed countries such as the UK and Canada was close to 1, and the urban–rural income ratio in low-income countries in Africa such as Uganda was only about 2.3, while the urban–rural income ratio gap in China was as high as 2.56 [8]. The unbalanced and insufficient development between urban and rural areas is still an important feature of the current urban–rural relationship in China.

Theories regarding the interactive relationships between urban and rural mainly include Utopian socialism, Marx and Engels' urban–rural relationship theory, "Garden city", "Organic evacuation" theory and "Desakota" model [9–12]. They emphasize balanced development and deny excessive bias towards urban and rural areas. From the perspective of urban–rural integration development, current research mainly focuses on measuring the level of urban–rural integration [13], evolving characteristics [14], influencing factors [15,16], classification of villages in the context of urban–rural integration [17], and summarizing excellent case experiences [18,19]. In addition, related studies have concluded that there are flows of people, goods, capital, information, and technology [20] between urban and rural areas. Mayer et al. focused on rural entrepreneurs who established links between urban and rural areas and investigated the role of their entrepreneurial activities in improving economic relations between urban and rural areas [21]. However, existing studies have paid less attention to the planning responses to the integrated urban–rural development, especially on how to scientifically delineate the scope of integrated urban–rural development zones and how to propose targeted optimization strategies for different development patterns. In particular, the problems of insufficient research at the county implementation level, unclear guidance and strategy of development patterns, and disconnect from spatial planning are particularly prominent.

To fill these gaps, this research selects Hanchuan City, located within the Wuhan metropolitan area, as the study area. Then, the comprehensive evaluation system which combines the current and potential level of urban–rural integration development together is constructed. The centers of urban–rural integration development are selected according to the comprehensive level of urban–rural integration development at township level, and the urban–rural integration development zones are delineated by using the gravity model. Finally, different types of urban–rural integration development patterns are identified, and corresponding strategies for urban–rural integration development planning guidance are proposed. The innovation points of this research include the following two main aspects. Firstly, this study constructs a comprehensive evaluation system which is made up of socio-economic and land-use-related indicators, to quantitatively assess the urban–rural integration development at township level. By using the gravity model, the scope of urban–rural integration development zones are delineated. Secondly, different leading patterns of urban–rural integration development zones delineated. Furthermore, urban–rural land-use strategies, such as centralization, differentiation and demonstration, are proposed according to the recognized patterns. This research could provide a direct basis for the preparation and implementation of urban–rural integration development, thus leading to better social equity and accessibility to facilities in the metropolitan area.

## 2. Materials and Methods

### 2.1. Study Area

Hanchuan City is located in the Wuhan metropolitan area in the middle reaches of the Yangtze River, with flat and low-lying terrain, low terrain complexity, and a strong agricultural foundation. Hanchuan City is a key county-level city for comprehensive agricultural development in the Jianghan Plain. Hanchuan City consists of two township-level subdistricts and 24 towns, with a total area of 1659.91 km$^2$. The urban land area accounts for about 1.89% and the rural settlement area accounts for about 9.21%. In 2020, the GDP of Hanchuan City was 64.566 billion Yuan, the resident population was about 0.9 million, and the urbanization rate was about 60.66 %. Figure 1c shows the administrative division and land use of Hanchuan City.

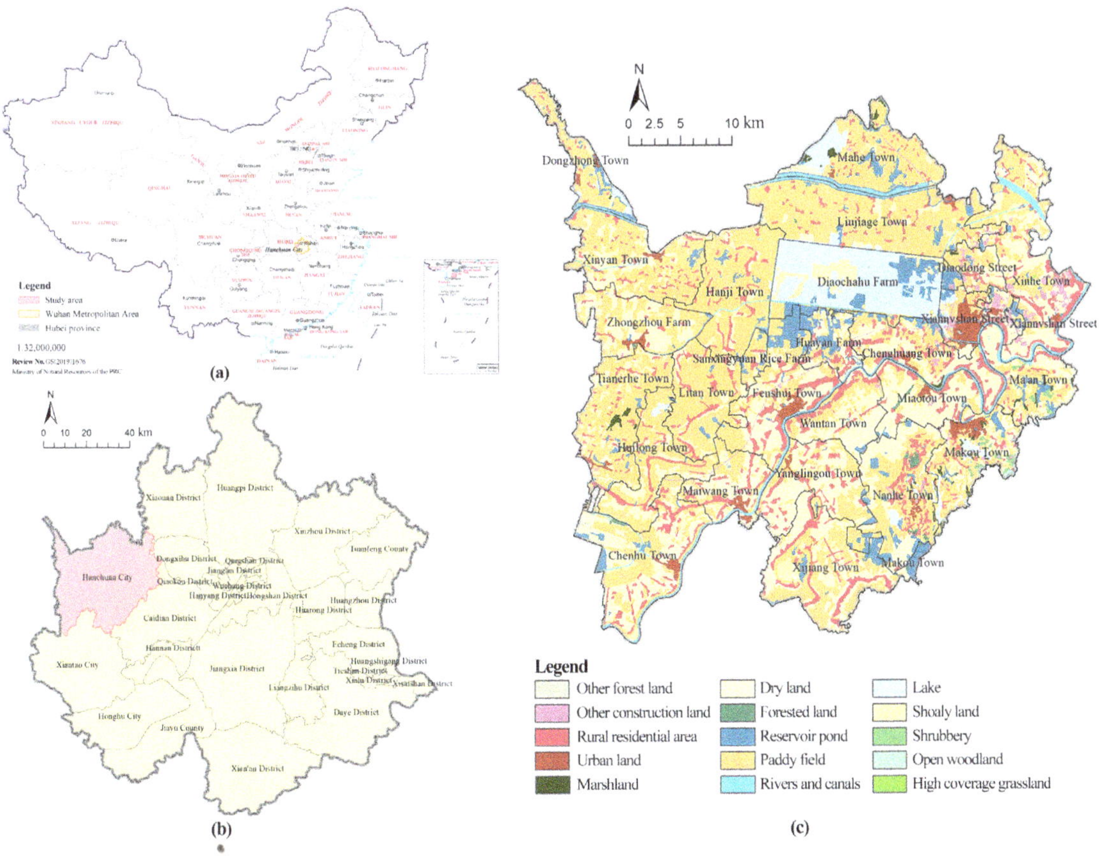


**Figure 1.** Study area. (**a**) Locations of the study area in China. (**b**) Locations of the study area in the Wuhan metropolitan area. (**c**) Land use of the study area.

### 2.2. Data Sources

The data used in this study mainly include socio-economic statistics, land-use data, and built environment data. Among these data, socio-economic statistics are mainly from the Xiaogan Statistical Yearbook 2020, and the China County Statistical Yearbook. The land-use data of Hanchuan is the LUCC data of 2020, provided by the Data Center for Resources and Environmental Sciences, Chinese Academy of Sciences (http://www.resdc.cn, accessed on 16 September 2022), which is now one of the most commonly used data sets for conducting related studies [22]. Spatial data of the built environment, such as road networks and transportation hubs, were obtained from the Hanchuan Planning Bureau.

*2.3. Methods*

In this study, the comprehensive evaluation of urban–rural integration development was a weighted combination of the current and potential levels of integration. The evaluation of current integration development level was carried out from five dimensions, including urban–rural demographic, spatial, economic, social and ecological integration. The evaluation of the potential integration development level was carried out from five dimensions, including location and transportation, town scale, economic level, facility construction, and characteristic resources. Based on the comprehensive evaluation results, towns with the highest integration level were selected as the centers of the urban–rural integration development zones in each direction within the municipal area. The gravity model was used to delineate the scope of the urban–rural integration development zones and identify the urban–rural integration development patterns. Finally, the corresponding planning and guiding strategies for urban–rural integration development were proposed in a targeted manner (Figure 2).

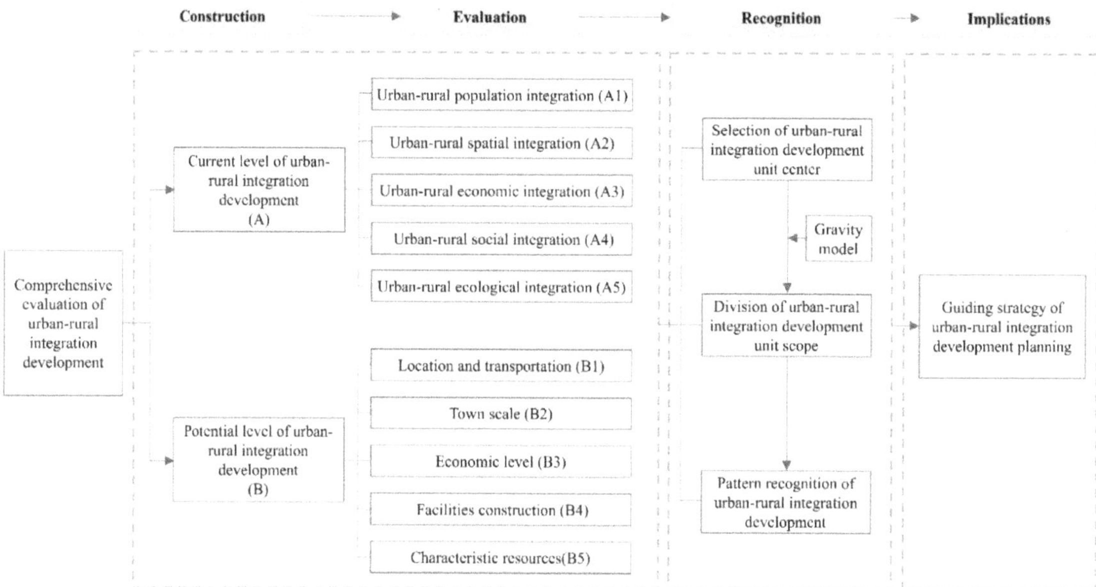


**Figure 2.** Research framework of this research.

2.3.1. Evaluation Model of the Current Integration Development Level

The level of current urban–rural integration development measures the degree of interconnectedness of natural, spatial, economic, and social elements between urban and rural areas [23]. In this study, the linear weighted sum method was used to calculate the current urban–rural integration development level score of each town [24]. The calculation formula is as follows:

$$L_i = \sum_{j=1}^{n} W_j X_{ij} \tag{1}$$

where $L_i$ is the score of the level of current urban–rural integration development of the $i$th town; $W_i$ and $X_{ij}$ are the weights and standardized values of the $j$th indicator of the $i$th town, respectively; and $n$ is the number of towns.

(1) Construction of the Indicator System

This study fully considers the calculability of indicators, the accessibility of data, and combines the current situation and demands of urban–rural development to construct the evaluation system of urban–rural integration development. As shown in Table 1, one target is the level of urban–rural integration development, and five criteria are urban–rural population integration, spatial integration, economic integration, social integration and ecological integration. The 18 indicators are introduced to conduct the measurements and calculation (Table 1).

**Table 1.** Evaluation system of the current urban–rural integration development level.

| Target | Criterions | Indicators | Unit | Method of Calculation | Index Attribute |
|---|---|---|---|---|---|
| Evaluation of the current Urban–rural Integration Development (A) | Urban–rural population integration (A1) [25] | Level of population urbanization (A01) | % | (Number of urban population/Total population) × 100% | Positive |
| | | Rate of urban and rural population going out (A02) | % | (Number of outworkers/Total population) × 100% | Negative |
| | | Proportion of agricultural labor (A03) | % | Number of labor force in (primary industry/Total labor force) × 100% | Negative |
| | Urban–rural spatial integration (A2) [26] | Urban–rural road network density (A04) | km/km$^2$ | (Road length/Total land area) × 100% | Positive |
| | | Urban–rural public service facilities density ratio (A05) | % | (Density of rural public service facilities/Density of urban public service facilities) × 100% | Positive |
| | | Level of land urbanization (A06) | % | (Area of built-up area/Total land area) × 100% | Positive |
| | Urban–rural economic integration (A3) [27] | Ratio of non-agricultural output value to agricultural output value (A07) | % | (Non-agricultural output value/Agricultural output value) × 100% | Positive |
| | | Per capita GDP (A08) | Yuan | GDP/Total population | Positive |
| | | Rural per capita net income (A09) | Yuan | - | Positive |
| | | Output value of industrial enterprises above designated size (A10) | Billion yuan | - | Positive |
| | Urban–rural social integration (A4) [28] | Every ten thousand people have the number of primary school teachers (A11) | People | Number of elementary school teachers/10,000 people | Positive |
| | | Number of students in urban–rural primary schools (A12) | People | - | Positive |
| | | Number of health technicians per thousand population in urban–rural areas (A13) | People | Number of health technicians/Thousands of people | Positive |
| | | Number of medical beds per thousand population (A14) | Beds/1000 people | Number of beds in urban–rural medical institutions/Thousands of people | Positive |
| | Urban–rural ecological integration (A5) [29] | Urban–rural garbage harmless disposal rate (A15) | % | (Amount of harmless garbage disposal/Amount of urban–rural garbage generation) × 100% | Positive |

(2) Data Normalizations

The units and attributes among indicators in the evaluation system are different and cannot be directly weighted and superimposed, it is necessary to standardize the evaluation indicators into a uniform manner. In this study, all data are processed by using the standardized method of extreme differences [30], which is calculated as follows:

Positive indicators:

$$Z_{ij} = \frac{X_{ij} - minX_{ij}}{maxX_{ij} - minX_{ij}} \quad (2)$$

Negative indicators:

$$Z_{ij} = \frac{maxX_{ij} - X_{ij}}{maxX_{ij} - minX_{ij}} \quad (3)$$

where $Z_{ij}$ is the standardized value of the $j$th indicator in the $i$th sample; $X_{ij}$ is the $j$th indicator value in the $i$th sample; $maxX_{ij}$ is the maximum value of the $j$th indicator in the $i$th sample; and $minX_{ij}$ is the minimum value of the $j$th indicator in the $i$th sample.

(3) Weight Calculation

The standardized indicator value of each town is passed for KMO and Bartlett's sphericity test [31]. The linear combination coefficients and the coefficients in the composite score model is calculated based on the principal component matrix. The weights of each indicator are distributed from 0 to 1, resulting in the final weight of each indicator.

### 2.3.2. Evaluation Model of Potential Urban–Rural Integration Development Level

Based on the standardization of data and calculation of weights, the potential development level of each town is weighted and summed, i.e., the weighted sum of the standardized values of each indicator multiplied by its weight is calculated [32]. The calculation formula is:

$$S_i = \sum_{j=1}^{n} w_j P_{ij} (j = 1, 2, 3, \cdots n) \quad (4)$$

where $S_i$ is the potential development score of the $i$th town, $P_{ij}$ is the standardized indicator of the $i$th town, $w_j$ is the weight of indicator $j$, and $n$ is the number of indicators in the evaluation system.

(1) Construction of the Indicator System

The hierarchical model of the analytic hierarchy process (AHP) is used to construct the evaluation system of potential urban–rural integration development into a structure of "1 + 5 + 18", which consists of 1 target, 5 criteria and 18 indicators. The target is the potential level of urban–rural integration development of each town. The criteria include five aspects: location and transportation, town scale, economic level, facility construction, and characteristic resources. The 18 indicators include specific indicators such as population size, township construction scale, industrial output value, and per capita net income of farmers (Table 2).

**Table 2.** Evaluation system of the potential level of urban–rural integration development.

| Target | Criterions | Indicators | Scoring Standard |
|---|---|---|---|
| Urban–rural integration development potential (B) | Location and Transportation (B1) [33] | Is the town located around economically developed areas (B01) | 5 points for proximity to large cities, 3 points for the periphery of the central city, 2 points for being located in the development axis, and 1 point for other areas |
| | | Distance of the town from transportation hubs and roads (B02) | High-speed railway station, high-speed exit, rail transit station, national highway, provincial highway, other roads are assigned 5, 4, 4, 3, 2, 1 |
| | Town scale (B2) [34] | Whether the township is a key development township (B03) | National key town assignment points 5, provincial model towns, central town assignment points 3, general town assignment points 1 |
| | | Population size (B04) | Total resident population size |
| | | Township construction scale (B05) | Township built-up area land size |
| | Economic level (B3) [35] | Industrial output (B06) | The output value of industrial enterprises above the scale |
| | | Number of Industrial Parks (B07) | Current and planned number of industrial parks |
| | | Per capita net income of farmers (B08) | - |
| | | Agricultural production (B09) | Average grain land yield |

**Table 2.** *Cont.*

| Target | Criterions | Indicators | Scoring Standard |
| --- | --- | --- | --- |
| | Facility Construction (B4) [36] | Educational Facilities (B10) | Number of primary and secondary schools |
| | | Medical Facilities (B11) | Number of medical beds per 1000 population |
| | | Garbage disposal (B12) | Township waste disposal rate |
| | Characteristic resources (B5) [37] | Natural Waters (B13) | Lakes, rivers, wetlands, assigned 5, 3, 2, respectively |
| | | Geographical system (B14) | Forest land, water area, arable land, and town are assigned 5, 3, 2, and 1 point, respectively. |
| | | Brand Resources (B15) | National well-known trademarks and provincial famous trademarks are assigned 5 and 3 points, respectively |
| | | Tourism Resources (B16) | National and provincial scenic spots are assigned 5 and 3 points, respectively |
| | | Tangible Cultural Heritage (B17) | According to the heritage level, the national level is assigned 5 points, the provincial level is assigned 3 points, the municipal level is assigned 1 point |
| | | Intangible Cultural Heritage (B18) | According to the level of intangible cultural heritage, the national level is assigned 5 points, the provincial level is assigned 3 points, and the municipal level is assigned 1 point. |

(2) Weight Calculation

The data for calculating the indicators are normalized using the forward processing method in Section 2.3.1. Hierarchical analysis and the Delphi method are used to determine the weights of each evaluation indicator [38]. The hierarchical structure model is established by using Yaahp software, and the weights of each indicator are finally calculated by establishing the relative importance comparison matrix of indicators after the consistency test.

#### 2.3.3. The Comprehensive Evaluation System of Urban–rural Integration Development

The evaluation results of the current and potential urban–rural integration development levels are weighted and summed, and the weights are calculated specifically using the expert scoring method. The revised scores of the current and potential development, and the final comprehensive evaluation results are obtained by adopting the combined opinions of several experts [39]. The specific formula is as follows:

$$T = a_1 L_i + a_2 S_i \tag{5}$$

where $T$ is the comprehensive evaluation result of urban–rural integration development; $L_i$ and $S_i$ are the evaluation results of the current and potential urban–rural integration development level, respectively; and $a_1$ and $a_2$ are the weights of the current and potential urban–rural integration development level, respectively.

#### 2.3.4. The Division of Urban–rural Integration Development Zones

Towns with the highest overall score are selected as the centers of the urban–rural integration development zones. The gravity model is used to calculate the gravity value between each center and the surrounding towns [40]. The higher the gravity value, the closer to the centers. In the gravity model formulation, the mass parameter and the distance parameter are key factors that affect the model's results. The specific formula is as follows:

$$I_{ij} = \frac{M_i M_j}{D_{ij}^2} \tag{6}$$

where $I_{ij}$ is the gravity between center $i$ and town $j$; $M_i$ and $M_j$ denote the comprehensive evaluation results of center $i$ and town $j$; and $D_{ij}$ denotes the road distance between $i$ and $j$.

# 3. Results

## 3.1. Comprehensive Evaluation of Urban–Rural Integration Development

### 3.1.1. The Current Level of Urban–Rural Integration Development

The current level of most towns in Hanchuan City is lower than the average. The towns with high levels form an axis in the direction from northeast to southwest. The indicator weights of each township in Hanchuan are derived by principal component analysis, and the current level of urban–rural integration development of each town is finally calculated. The development level of urban–rural integration in Hanchuan is 0.35, which can be judged that Hanchuan as a whole is in the primary stage of urban–rural integration. Among them, 7 out of all the 26 towns, such as Xiannvshan Street and Xinhe Town, have a higher value than the average value of Hanchuan City, accounting for 26.92%. While the other 19 towns, such as Miaotou Town and Dongzhong Town, are lower than the overall average value, accounting for 73.08%. Therefore, Hanchuan City still needs to continue to promote the development of urban–rural integration.

The natural breakpoint method was used to classify the current urban–rural integration development value of 26 towns of Hanchuan City into four levels. The first category is the towns with the highest level of urban–rural integration development, accounting for 3.84% of the total towns, such as Xiannvshan Street, which is the central urban area of Hanchuan City, with a current level of urban–rural integration development far ahead of other towns. The second category is the towns with a higher level of urban–rural integration development, including Makou Town, Chenghuang Town, Fenshui Town, Chenhu Town, Xinhe Town, and Diaodong Street, accounting for 23.08% of the total. The third category is towns with a lower current level of urban–rural integration development, including eight towns, such as Maiwang Town, Tianerhe Town, and Huilong Town, accounting for 30.77% of the total towns. The fourth category is towns with the lowest current level of urban–rural integration development, including 11 townships such as Xinyan Town, Dongzhong Town, and Mahe Town, accounting for 42.31% of the total. From the viewpoint of spatial distribution, the first and second categories of towns form a spatial axis in the direction of "northeast–southwest", which may provide strong support for determining the spatial development pattern of Hanchuan City (Figure 3).

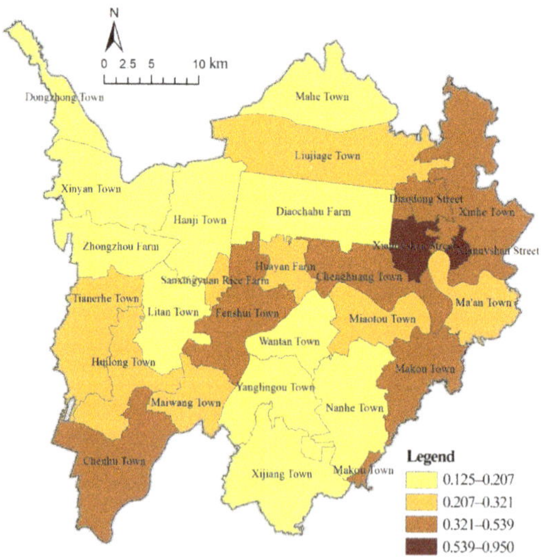


**Figure 3.** Spatial distribution of current urban–rural integration development level of each town in Hanchuan City.

3.1.2. Potential Level of Urban–Rural Integration Development

In order to specifically display the potential urban–rural integration development of each town, this study standardized the calculation results and uniformly set the highest value of potential urban–rural integration development of towns up to 100. According to the described method, the standardized scores of the potential urban–rural integration development of 26 towns were calculated, and the natural breakpoint method in ArcGIS software 10.0.2 was applied to classify the potential level of towns into three categories, which were high, medium and low development towns, respectively. Among them, the high development towns have comprehensive scores between 66.51–100, with a strong potential development level. The comprehensive scores of towns with medium development are between 34.68–66.51, which shows outstanding potential development in a certain way. The comprehensive scores of towns with low development are below 34.68, indicating weak development potentials.

The potential urban–rural integration development level of towns in Hanchuan City has a pyramidal structure from high to low. The townships with high, medium, and low development potentials account for 15.39%, 26.92%, and 57.69% of the total number of towns, respectively, and the overall balance is at a low level. The spatial distribution of towns with high development potential is concentrated in the eastern part of Hanchuan City adjacent to Wuhan City, with more obvious location advantages. The spatial distribution of towns with medium development potentials is more fragmented, either relying on the advantage of close contacts with urban areas or the rich characteristic resources. The low development towns are mainly located in the central and western parts of Hanchuan City. These towns are mostly traditional agricultural towns, which lack service cores to drive the overall development, and have no characteristic advantageous resources to use. They are not strongly connected with the urban areas (Figure 4).

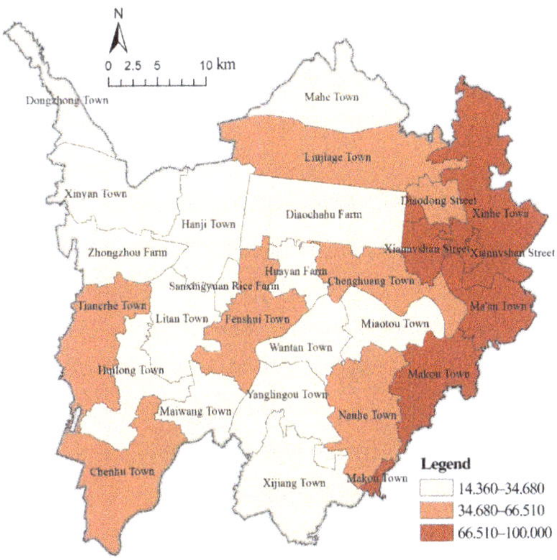


**Figure 4.** Spatial distribution of potential urban–rural integration development in each town of Hanchuan City.

3.1.3. The Comprehensive Evaluation of Urban–Rural Integration Development

The results of current and potential urban–rural integration development levels are weighted and summed to calculate the comprehensive evaluation results. The weights of the current and potential urban–rural integration development levels were determined to be 0.4 and 0.6, respectively, resulting in the revised current and potential scores (Table 3). The natural breakpoint method was used to classify the comprehensive evaluation results into four categories (Figure 5).

**Table 3.** Ranking of the comprehensive evaluation results of urban–rural integration development.

| Town | Urban–Rural Integration Level Evaluation Revision Score | Urban–Rural Integration Potential Evaluation Revision Score | Overall Score | Ranking |
|---|---|---|---|---|
| Xiannvshan Street | 0.380 | 0.418 | 0.798 | 1 |
| Makou Town | 0.213 | 0.410 | 0.624 | 2 |
| Xinhe Town | 0.216 | 0.354 | 0.570 | 3 |
| Chenhu Town | 0.165 | 0.278 | 0.443 | 4 |
| Ma'an Town | 0.104 | 0.304 | 0.408 | 5 |
| Chenghuang Town | 0.153 | 0.220 | 0.373 | 6 |
| Fenshui Town | 0.154 | 0.199 | 0.353 | 7 |
| Diaodong Street | 0.162 | 0.174 | 0.336 | 8 |
| Nanhe Town | 0.081 | 0.243 | 0.324 | 9 |
| Tianerhe Town | 0.097 | 0.216 | 0.313 | 10 |
| Miaotou Town | 0.128 | 0.107 | 0.235 | 11 |
| Maiwang Town | 0.118 | 0.107 | 0.225 | 12 |
| Xinyan Town | 0.073 | 0.145 | 0.218 | 13 |
| Mahe Town | 0.075 | 0.141 | 0.216 | 14 |
| Dongzhong Town | 0.083 | 0.128 | 0.210 | 15 |
| Liujiage Town | 0.093 | 0.116 | 0.210 | 16 |
| Huilong Town | 0.104 | 0.101 | 0.205 | 17 |
| Diaochahu Farm | 0.062 | 0.138 | 0.200 | 18 |
| Huayan Farm | 0.109 | 0.081 | 0.190 | 19 |
| Xijiang Town | 0.068 | 0.122 | 0.190 | 20 |
| Yanglingou Town | 0.080 | 0.089 | 0.169 | 21 |
| Sanxingyuan Rice Farm | 0.102 | 0.061 | 0.163 | 22 |
| Wantan Town | 0.081 | 0.069 | 0.150 | 23 |
| Hanji Town | 0.054 | 0.090 | 0.144 | 24 |
| Zhongzhou Farm | 0.060 | 0.079 | 0.139 | 25 |
| Litan Town | 0.053 | 0.060 | 0.113 | 26 |

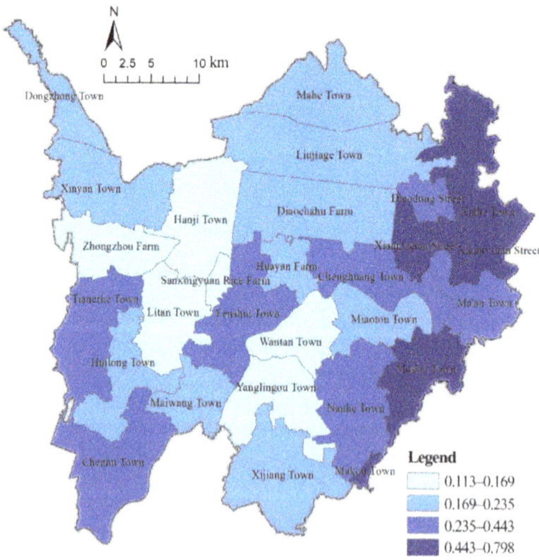


**Figure 5.** Spatial distribution of the comprehensive evaluation results of urban–rural integration development.

The spatial patterns of the comprehensive evaluation results of urban–rural integration development in Hanchuan city gradually decrease from the northeast and southwest to the central area. The eastern part of Hanchuan is adjacent to Wuhan, which is the core city of the Wuhan metropolitan area. The urban area of Hanchuan is also located in the northeastern part of the county, which is influenced by the radiation of Wuhan city and the central urban area of Hanchuan, thus the overall level of the central urban area and its surrounding townships is high and gradually decreases from the central urban area outward. Although some of the towns in southwestern Hanchuan are far away from Wuhan and Hanchuan's central urban area, such as Chenhu Town, they have the advantage of being close to the central urban area of Xiantao City, thus leading to a strong industrial base and a high level of integrated urban–rural development. In contrast, the towns in the central region are relatively weakly driven by the radiation of the surrounding growth poles (such as the central urban area of Hanchuan and Wuhan City adjacent to the eastern part of Hanchuan, and the central urban area of Xiantao City adjacent to the southwestern part).

### *3.2. Divisons of Urban–Rural Integration Development Zones*

#### 3.2.1. The Centers of Urban–Rural Integration Development Zones

According to the central place theory [41], regional network structure theory [42], and the territorial spatial master planning, the 26 towns in the whole area of Hanchuan city are roughly divided into four regions: northeast, southeast, southwest, and northwest. Towns with the highest combined scores of the current and potential development levels were selected as urban–rural integration development centers in each of the four regions. Therefore, Xiannvshan Street, Makou Town, Chenhu Town and Xinyan Town were selected as the centers of the northeast, southeast, southwest and northwest of Hanchuan City, respectively. These four centers drive the hinterland of surrounding towns to integrate development (Figure 6).

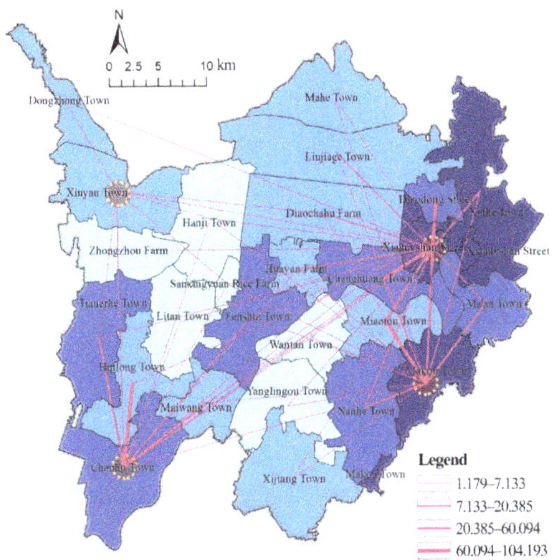


**Figure 6.** Spatial distribution of urban–rural integration development centers.

#### 3.2.2. Zones of Urban–Rural Integration Development

The gravity model was used to calculate the gravity value between the centers and surrounding towns. A larger gravity value means that the town is more closely connected to the selected center. In terms of the setting of quality parameters, the comprehensive evaluation score of each town was taken as the quality, which reflects the strength of a town's urban–rural integration development more comprehensively. In terms of the setting of the distance parameter, the road distance between the center and each town was taken as the distance parameter because Hanchuan is mainly road traffic. The gravity values between the other 22 townships and the four centers were obtained (as shown in Table 4 and Figure 6), and the spatial extent of the four urban–rural integration development zones were divided according to the principles of similarity and consistency of town development and their environment (Figure 7).

**Table 4.** Gravity values between 22 towns and four centers.

| Town | Xiannvshan Street | Makou Town | Chenhu Town | Xinyan Town |
|---|---|---|---|---|
| Xinhe Town | 60.09 | 8.79 | 1.82 | 1.07 |
| Ma'an Town | 18.77 | 49.3 | 1.08 | 0.6 |
| Chenghuang Town | 104.19 | 19.2 | 2.5 | 0.83 |
| Fenshui Town | 7.13 | 4.6 | 7.84 | 1.04 |
| Diaodong Street | 93.73 | 8.74 | 1.43 | 0.87 |
| Nanhe Town | 9.66 | 37.49 | 0.86 | 0.37 |
| Tianerhe Town | 1.85 | 1.09 | 23.13 | 1.91 |
| Miaotou Town | 38.96 | 66.21 | 1.05 | 0.44 |
| Maiwang Town | 2.79 | 1.46 | 25.74 | 0.5 |
| Mahe Town | 3.66 | 1.33 | 0.4 | 0.25 |
| Dongzhong Town | 1.18 | 0.63 | 0.79 | 3.78 |
| Liujiage Town | 14.63 | 3.04 | 0.66 | 0.39 |
| Huilong Town | 1.59 | 0.9 | 93.73 | 0.82 |
| Diaochahu Farm | 20.38 | 4.49 | 1.05 | 1.34 |
| Huayan Farm | 12.35 | 4.23 | 2.34 | 1.01 |
| Xijiang Town | 2.47 | 2.65 | 0.91 | 0.15 |
| Yanglingou Town | 3.26 | 3.27 | 0.75 | 0.16 |
| Sanxingyuan Rice Farm | 2.84 | 1.33 | 3.2 | 3.62 |
| Wantan Town | 5.05 | 5.09 | 1.81 | 0.18 |
| Hanji Town | 3.23 | 1.26 | 1.62 | 4.54 |
| Zhongzhou Farm | 1.37 | 0.66 | 1.7 | 5.6 |
| Litan Town | 1.19 | 0.7 | 4.67 | 1.04 |

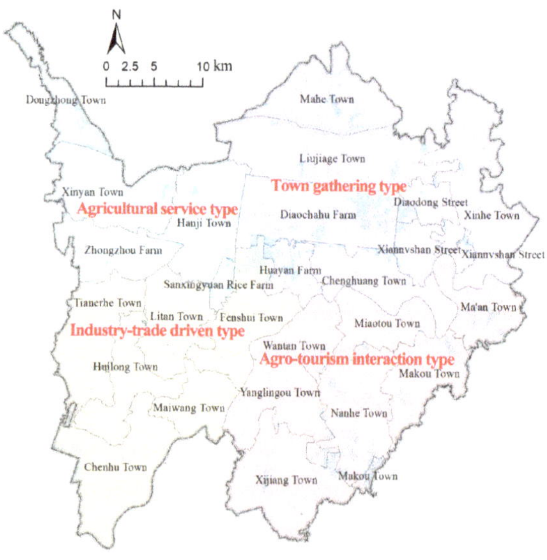


**Figure 7.** Spatial distribution of the urban–rural integration development pattern.

### *3.3. Recognition of Urban–Rural Integration Development Pattern*

According to the evaluation of the potential development, the multifunctional evaluation of each town within different zones is conducted based on five elements: location and transportation, town scale, agricultural production, non-farm economy, characteristic resources, and so on [43,44]. Using the categorical assignment method, the scores of each element were assigned to seven grade intervals, with a maximum value of eight and the minimum value of two. The scores of the towns within the four development zones were calculated by horizontally comparing the maximum value of each town's score, comprehensively considering the leading industries of each town and the leading functions of each development zone. The urban–rural integration pattern of Hanchuan City was finally classified into four types: town gathering, agro-tourism interaction, industry-trade driven, and agricultural service (Figure 7).

#### 3.3.1. The Pattern of the Town Gathering Type

The town gathering type has more comprehensive industrial advantages. Xiannvshan Street and Xinhe Town focus on secondary and tertiary industries with strong economic strength; Chenghuang Town and Liujiage Town have developing manufacturing and modern service industries while stabilizing agriculture of the traditional agricultural towns, such as Diaodong Street, Mahe Town and Huayan Farm. From the spatial distribution of the pattern, these eight towns are located in the dense area of the eastern part in Hanchuan City, among which Xiannvshan Street, Diaodong Street and Xinhe Town are partly located within the central urban area. Thus, this type was summarized as the town gathering type pattern by combining the industrial base and spatial distribution.

#### 3.3.2. The Pattern of the Agro-Tourism Interaction Type

The leading functions of the towns within the agro-tourism interaction type are characteristic resources and agricultural production. Through sorting the resource endowment, it was found that Nanhe Town, Yanglingou Town, Ma'an Town, and other towns have good agricultural landscapes or planting bases to develop rural tourism. From the spatial distribution, it can be seen that this type of integration development zone is located to the south of the Hanjiang River and close to the Caidian District of Wuhan City, leading to good tourism location advantages, and can link Huanglong Lake of Ma'an Township, Tianyu Lake of Makou Town, Nanhe Ancient Ferry in Nanhe Town and other scenic spots to form the Hanan agricultural tourism leisure resort area.

#### 3.3.3. The Pattern of the Industry-Trade Driven Type

Most of the towns' leading industries within the industry-trade driven type are secondary industries, and the urban functions are mainly a non-agricultural economy. In particular, Chenhu Town and Fenshui Town are, respectively famous for their metal products and pharmaceutical packaging, with strong industrial foundations. The surrounding Huilong Town, Maiwang Town, and Chenhu Town have both industrial divisions and cooperation, which is good for forming industrial clusters. While Tianerhe Town can be used as a product trading distribution center by its geographical advantages at the junction of the two cities. Though Litan Town is a traditional agricultural town, it can provide raw materials and labor for industrial development. Therefore, this pattern was summarized as the industry-trade driven type.

#### 3.3.4. The Pattern of the Agricultural Service Type

Towns within the agricultural service type are located in the vicinity of the Jianghan Plain, and their leading function is agricultural production. The traditional agricultural town of Xinyan is the unit center. The leading industries are high-quality rice cultivation, special aquaculture and fine processing industry of agricultural products. Among them, Xinyan Town and Hanji Town have the foundation of developing new modern agriculture; Dongzhong Town relies on a planting foundation to develop melon fruit and a flower industry; Zhongzhou Farm has the largest area of hybrid cotton base in Hanchuan. The last round of master planning also positioned the area as a grain production base, a special aquaculture base and an important fruit and vegetable production base of Hanchuan city. Thus, this pattern was summarized as the agricultural service type.

## 4. Implications for Urban–Rural Integration Development Planning

### 4.1. Centralization

Building a five-level hierarchy of "central urban area—subcentral town—key town—general town—new rural community" is needed to strengthen the agglomeration effect of the central urban area, sub-central town, and key town, to thus guide the orderly and efficient development of urban and rural areas and drive the construction of industries and facilities in general towns and central villages. Among them, the subcenter towns are supported by expanding the scale of construction, actively accepting the overflow of manufacturing functions from urban areas and neighboring cities in the region, developing employment- and technology-oriented labor-intensive industries, and improving public service facilities. As linkages between the central urban area, subcenter towns, general towns and key towns are important nodes and must act as the backbone of urban–rural integration in Hanchuan City by using favorable conditions such as transportation and natural resources to develop tourism, agricultural and sideline product processing industries to build a service center that radiates to the surrounding rural areas. Combining the analysis of the village distribution planning of each town and the territorial spatial master planning of the development conditions and development potential of villages, priority will be given to select settlements with good development conditions and build new rural communities in combination with resettlement.

### 4.2. Differentiation

Due to the great difference between the east and west directions of Hanchuan City, different guidance should be taken to promote the integrated development of urban and rural areas. The development paths and management of towns and villages in different zones should be planned in a focused and targeted manner. According to the differences in resource endowments and industrial development of each town, four different urban–rural integration development patterns are proposed, namely, town gathering type, agro-tourism interaction type, industry-trade driven type, and agricultural service type. Town gathering type pattern applies to the central urban areas of Hanchuan. Agro-tourism interaction type pattern applies to the areas with rich tourism resources. Industry-trade driven type pattern applies to the areas around key towns. Agricultural service type pattern applies to the areas with traditional agricultural production and poor industrial foundation and lack of infrastructure. Through the above classification, the development gap between the east and the west of Hanchuan city could be narrowed, leading to balanced development and striving to make up for the backwardness of the development of western villages and towns.

### 4.3. Demonstration

It is of great importance to build some demonstration areas, such as the central urban area of Hanchuan and Makou Town, to provide development guidance for other towns. The central urban area of Hanchuan has a good economic foundation and comprehensive functions. These areas focus on integrating the complex resource elements in the region to form an urban–rural ecosystem with a two-way flow of elements. Makou Town, relying on its strong industrial foundation and tourism resources, can serve as a dual demonstration area of agro-tourism interaction and industry-trade driven integration development pattern. It could build a modern rural community and give priority to the transformation from agriculture and rural areas to industry-trade towns. To narrow the gap between urban–rural development, these urban–rural integration demonstration zones could promote the two-way free flow of urban–rural production and the rational allocation of public resources. The demonstration zones should break the institutional drawbacks, complement the policy weaknesses to establish the urban–rural integration development system and policy system to provide a replicable typical experience for Hanchuan City.

## 5. Discussion

### 5.1. The Mechanism for Forming Urban–Rural Integration Development Patterns

Previous studies on urban–rural integration mainly focused on the evaluation of the development level of urban–rural integration [32], while less attention has been paid to what should be carried out after the evaluation. Therefore, the evaluation results cannot be well integrated with the subsequent planning implementation. In addition, in terms of evaluation contents, previous studies have often focused on measuring the current level of development, while less attention has been paid to the future development potential of urban–rural integration. Unlike previous studies, this study argues that the current and potential levels, patterns, and strategies of urban–rural integration development all differ within a city from the perspective of spatial heterogeneity [45].

At present, the county level urban–rural integration development is still in the stage of continuous exploration. Cultivating urban–rural integration development zone is an efficient way from the perspective of clusters. Based on the evaluation results of the current and potential level of urban–rural integration development, the gravity model was used to divide the study area into different urban–rural integration development zones with different development patterns. The formation of the four urban–rural integration development patterns is mainly the combinations of two forces, the internal and external of the city. For the internal forces, urban–rural integration development patterns are closely related to the spatial distribution of natural and human resources in the county, as well as the overall territorial spatial plan. For the external forces, the Wuhan metropolitan area, cannot be ignored as a driving force for its integrated urban–rural development. Hanchuan is adjacent to Wuhan, which is the core city of the Wuhan metropolitan area. According to the growth pole theory, Wuhan city is the first growth pole of the Wuhan metropolitan area and has a strong attraction effect on the population, economy and other factors in the surrounding areas [46]. Therefore, the integrated urban–rural development and the formation of specific patterns in Hanchuan are also subject to a larger role of the core city.

However, the specific role of the core cities in the metropolitan area in the urban–rural integration of the surrounding cities needs to be further discussed. For instance, when the core cities promote the development of urban and rural areas, and whether they promote the continuous reduction or expansion of the relative gap between urban and rural areas in the surrounding counties. The formation mechanism of their urban–rural integration development patterns may vary from cities outside the metropolitan area. There are likely to be some differences even in their internal urban–rural integration development patterns, except that the specific types of patterns are exactly the same as Hanchuan's urban–rural integration development patterns.

### 5.2. Limitations

China is actively promoting the development of new-type urbanization and urban–rural integration. To continually direct the efficient circulation of urban–rural elements and strengthen the urban–rural governance system, it should completely combine the resource advantages and potentials within each county and plan differentiated urban–rural integration development patterns. The results of this study can provide direct support for the preparation and implementation of township-level territorial spatial planning, and the layout planning of rural settlements in the county and the allocation of urban and rural infrastructure and public service facilities. This study builds an index system for evaluating urban–rural integration development level which is made up of socio-economic and land-use-related indexes; however, two main shortcomings exist. First, due to data limitations, this study was not based on a dynamic assessment [47] and did not consider the trends in the evolving level of urban–rural integration development over the study years. Secondly, due to some constraints, such as time and finance, a comprehensive survey on the degree of happiness of residents, or satisfaction with the current status of urban–rural integration development, or other subjective feelings were not considered. In the future, the evaluation model of urban–rural integration development level should be further improved by combining it with survey data and so on.

## 6. Conclusions

This study constructed a comprehensive evaluation system of urban–rural integration development based on the current and potential level of development, taking Hanchuan City, located in the Wuhan metropolitan area, as the case study area at the county level. After evaluating the comprehensive level of urban–rural integration development, the gravity model was used to scientifically delineate the scope of the urban–rural integration development zones. Four different types of urban–rural integration development patterns were identified: town gathering, agro-tourism interaction, industry-trade driven, and agricultural service. Finally, the corresponding optimization strategies were targeted. The comprehensive evaluation results and recognized patterns of urban–rural integration development were effectively connected with county-level and township-level territorial spatial planning, which could provide direct guidance for the determination of the county–town system and the development direction of each zone within the county. In addition, the methodological evaluation system of urban–rural integration development and pattern recognition established in this study is not only applicable to Hanchuan City within the Wuhan metropolitan area, but may also be of reference value to other counties or cities within or outside the metropolitan area.

**Author Contributions:** Conceptualization, S.L., Y.H. and Y.D.; methodology, X.W. and Y.D.; software, S.L., X.W. and Y.D.; validation, S.L. and Y.D.; formal analysis, X.W. and Y.D.; investigation, X.W., Y.H. and Y.D.; resources, S.L. and Y.H.; data curation, X.W. and Y.D.; writing—original draft preparation, S.L., X.W. and Y.D.; writing—review and editing, S.L. and Y.H.; visualization, X.W. and Y.D.; supervision, S.L. and Y.H.; project administration, S.L. and Y.H.; funding acquisition, S.L. and Y.H. All authors have read and agreed to the published version of the manuscript.

**Funding:** This research was funded by the National Natural Science Foundation of China (No. 41901390 and No. 51978299), the Natural Science Foundation of Hubei Province (No. 2021CFB012), the Social Science Foundation of Hubei Province (Post-Funded Project No. 203), and Key Program of the National Social Science Foundation of China (21AZD048).

**Data Availability Statement:** The data used in this study can be requested from the authors.

**Conflicts of Interest:** The authors declare no conflict of interest.

## References

1. Caffyn, A.; Dahlström, M. Urban-Rural Interdependencies: Joining up Policy in Practice. *Reg. Stud.* **2005**, *39*, 283–296. [CrossRef]
2. Liu, Y. Research on the Urban-Rural Integration and Rural Revitalization in the New Era in China. *Acta Geogr. Sin.* **2018**, *73*, 637–650. [CrossRef]
3. Overbeek, G. Opportunities for Rural-Urban Relationships to Enhance the Rural Landscape. *J. Environ. Policy Plan.* **2009**, *11*, 61–68. [CrossRef]
4. Liu, Y.; Li, Y. Revitalize the World's Countryside. *Nature* **2017**, *548*, 275–277. [CrossRef]
5. Simon, D. Co-Productive Tools for Transcending the Divide: Building Urban–Rural Partnerships in the Spirit of the New Leipzig Charter. *Land* **2021**, *10*, 894. [CrossRef]
6. Yao, Y.; Liang, M. Logic of Political Economics in the Development of Urban and Rural Integration–An Investigation on the Development of New China in the past 70 years. *Seeking Truth* **2019**, *46*(05), 11–18.
7. Liu, Y.; Lu, S.; Chen, Y. Spatio-Temporal Change of Urban-Rural Equalized Development Patterns in China and Its Driving Factors. *J. Rural Stud.* **2013**, *32*, 320–330. [CrossRef]
8. Li, S.; Chen, J.; Teng, Y. Rural Revitalization on the Path of Common Prosperity: Problems, Challenges and Suggestions. *J. Lanzhou Univ. (Soc. Sci.)* **2021**, *49*, 37–46.
9. Lovell, D.W. Socialism, Utopianism and the "Utopian Socialists". *Hist. Eur. Ideas* **1992**, *14*, 185–201. [CrossRef]
10. Bussard, R.L. The "dangerous Class" of Marx and Engels: The Rise of the Idea of the Lumpenproletariat. *Hist. Eur. Ideas* **1987**, *8*, 675–692. [CrossRef]
11. Batchelor, P. The Origin of the Garden City Concept of Urban form. *J. Soc. Archit. Hist.* **1969**, *28*(3), 184–200. [CrossRef]
12. Pribadi, D.O.; Pauleit, S. The Dynamics of Peri-Urban Agriculture during Rapid Urbanization of Jabodetabek Metropolitan Area. *Land Use Policy* **2015**, *48*, 13–24. [CrossRef]

13. Wang, Y.; Liu, Y.; Li, Y.; Li, T. The Spatio-Temporal Patterns of Urban-Rural Development Transformation in China since 1990. *Habitat Int.* **2016**, *53*, 178–187. [CrossRef]
14. Sun, Y.; Yang, Q. Study on Spatial–Temporal Evolution Characteristics and Restrictive Factors of Urban–Rural Integration in Northeast China from 2000 to 2019. *Land* **2022**, *11*, 1195. [CrossRef]
15. Yin, Z.H.; Choi, C.H. Does E-Commerce Narrow the Urban–Rural Income Gap? Evidence from Chinese Provinces. *Internet Res.* **2022**, *32*, 1427–1452. [CrossRef]
16. Requena, F. Rural–Urban Living and Level of Economic Development as Factors in Subjective Well-Being. *Soc. Indic. Res.* **2016**, *128*, 693–708. [CrossRef]
17. Tian, Y.; Qian, J.; Wang, L. Village Classification in Metropolitan Suburbs from the Perspective of Urban-Rural Integration and Improvement Strategies: A Case Study of Wuhan, Central China. *Land Use Policy* **2021**, *111*, 105748. [CrossRef]
18. Ye, Y.; Legates, R.; Qin, B. Coordinated Urban-Rural Development Planning in China. *J. Am. Plan. Assoc.* **2013**, *79*, 125–137. [CrossRef]
19. Chen, C.; LeGates, R.; Fang, C. From Coordinated to Integrated Urban and Rural Development in China's Megacity Regions. *J. Urban Aff.* **2019**, *41*, 150–169. [CrossRef]
20. Davoudi, S.; Stead, D. Urban–Rural Relationships: An Introduction and Brief History. *Built Environ.* **2006**, *122*, 25–27.
21. Mayer, H.; Habersetzer, A.; Meili, R. Rural-Urban Linkages and Sustainable Regional Development: The Role of Entrepreneurs in Linking Peripheries and Centers. *Sustainability* **2016**, *8*, 745. [CrossRef]
22. Liu, J.; Liu, M.; Deng, X.; Zhuang, D.; Zhang, Z.; Luo, D. The Land Use and Land Cover Change Database and Its Relative Studies in China. *J. Geogr. Sci.* **2002**, *12*, 275–282.
23. Chuanglin, F. Theoretical Foundation and Patterns of Coordinated Development of the Beijing-Tianjin-Hebei Urban Agglomeration. *Prog. Geogr.* **2017**, *36*, 15–24. [CrossRef]
24. Ma, L.; Liu, S.; Fang, F.; Che, X.; Chen, M. Evaluation of Urban-Rural Difference and Integration Based on Quality of Life. *Sustain. Cities Soc.* **2020**, *54*, 101877. [CrossRef]
25. Song, M.; Tao, W. Coupling and Coordination Analysis of China's Regional Urban-Rural Integration and Land-Use Efficiency. *Growth Change* **2022**, *53*, 1384–1413. [CrossRef]
26. Rao, C.; Gao, Y. Evaluation Mechanism Design for the Development Level of Urban-Rural Integration Based on an Improved TOPSIS Method. *Mathematics* **2022**, *10*, 380. [CrossRef]
27. Yan, J.; Chen, H.; Xia, F. Toward Improved Land Elements for Urban–Rural Integration: A Cell Concept of an Urban–Rural Mixed Community. *Habitat Int.* **2018**, *77*, 110–120. [CrossRef]
28. Qian, L.; Zhang, K.; Song, J.X.; Tang, W.Y. Regional Differences and Convergence of Urban-Rural Integration Development from the Perspective of Factor Flow. *J. Environ. Public Health* **2022**, *2022*, 2695366. [CrossRef]
29. Zhao, W.; Jiang, C. Analysis of the Spatial and Temporal Characteristics and Dynamic Effects of Urban-Rural Integration Development in the Yangtze River Delta Region. *Land* **2022**, *11*, 1054. [CrossRef]
30. Sun, Y.; Zhao, T.; Xia, L. Spatial-Temporal Differentiation of Carbon Efficiency and Coupling Coordination Degree of Chinese County Territory and Obstacles Analysis. *Sustain. Cities Soc.* **2022**, *76*, 103429. [CrossRef]
31. Malah, A.; Bahi, H. Integrated Multivariate Data Analysis for Urban Sustainability Assessment, a Case Study of Casablanca City. *Sustain. Cities Soc.* **2022**, *86*, 104100. [CrossRef]
32. Yang, Y.; Bao, W.; Wang, Y.; Liu, Y. Measurement of Urban-Rural Integration Level and Its Spatial Differentiation in China in the New Century. *Habitat Int.* **2021**, *117*, 102420. [CrossRef]
33. Bu, X.; Pu, L.; Shen, C.; Xie, X.; Xu, C. Study on the Spatial Restructuring of the Village System at the County Level Oriented toward the Rural Revitalization Strategy: A Case of Jintan District, Jiangsu Province. *Land* **2020**, *9*, 478. [CrossRef]
34. Zhang, B.; Zhang, J.; Miao, C. Urbanization Level in Chinese Counties: Imbalance Pattern and Driving Force. *Remote Sens.* **2022**, *14*, 2268. [CrossRef]
35. Long, H.; Zou, J.; Pykett, J.; Li, Y. Analysis of Rural Transformation Development in China since the Turn of the New Millennium. *Appl. Geogr.* **2011**, *31*, 1094–1105. [CrossRef]
36. Wang, Y.; Cao, X. Village Evaluation and Classification Guidance of a County in Southeast Gansu Based on the Rural Revitalization Strategy. *Land* **2022**, *11*, 857. [CrossRef]
37. Liu, S.; Ge, J.; Bai, M.; Yao, M.; He, L.; Chen, M. Toward Classification-Based Sustainable Revitalization: Assessing the Vitality of Traditional Villages. *Land Use Policy* **2022**, *116*, 106060. [CrossRef]
38. Jeong, J.S.; García-Moruno, L.; Hernández-Blanco, J.; Jaraíz-Cabanillas, F.J. An Operational Method to Supporting Siting Decisions for Sustainable Rural Second Home Planning in Ecotourism Sites. *Land Use Policy* **2014**, *41*, 550–560. [CrossRef]
39. Maliene, V.; Fowles, S.; Atkinson, I.; Malys, N. A Sustainability Assessment Framework for the High Street. *Cities* **2022**, *124*, 103571. [CrossRef]
40. Wu, J.; Yi, T.; Wang, H.; Wang, H.; Fu, J.; Zhao, Y. Evaluation of Medical Carrying Capacity for Megacities from a Traffic Analysis Zone View: A Case Study in Shenzhen, China. *Land* **2022**, *11*, 888. [CrossRef]
41. Taylor, P.J.; Hoyler, M. Lost in Plain Sight: Revealing Central Flow Process in Christaller's Original Central Place Systems. *Reg. Stud.* **2021**, *55*, 345–353. [CrossRef]
42. Forsberg, G.; Lindgren, G. Regional Policy, Social Networks and Informal Structures. *Eur. Urban Reg. Stud.* **2015**, *22*, 368–382. [CrossRef]

43. Reimer, B. A Sample Frame for Rural Canada: Design and Evaluation. *Reg. Stud.* **2002**, *36*, 845–859. [CrossRef]
44. Li, Y.; Long, H.; Liu, Y. Spatio-Temporal Pattern of China's Rural Development: A Rurality Index Perspective. *J. Rural Stud.* **2015**, *38*, 12–26. [CrossRef]
45. Mowl, G.; Barke, M.; King, H. Exploring the Heterogeneity of Second Homes and the 'Residual' Category. *J. Rural Stud.* **2020**, *79*, 74–87. [CrossRef]
46. Zheng, Y.; Tan, J.; Huang, Y.; Wang, Z. The Governance Path of Urban–Rural Integration in Changing Urban–Rural Relationships in the Metropolitan Area: A Case Study of Wuhan, China. *Land* **2022**, *11*, 1334. [CrossRef]
47. Sánchez-Zamora, P.; Gallardo-Cobos, R.; Ceña-Delgado, F. Rural Areas Face the Economic Crisis: Analyzing the Determinants of Successful Territorial Dynamics. *J. Rural Stud.* **2014**, *35*, 11–25. [CrossRef]

*land*

*Article*

# Does Urban Agglomeration Discourage Entrepreneurship in China? Micro-Empirical Evidence from China

**Wan Li [1], Bindong Sun [2,*], Shuaishuai Han [3] and Xiaoxi Jin [4]**

1 Business School, Zhengzhou University, Zhengzhou 450001, China
2 Research Center for China Administrative Division, Institute of Eco-Chongming, Future City Lab, The Center for Modern Chinese City Studies, School of Urban and Regional Science, East China Normal University, Shanghai 200241, China
3 Key Research Institute of Yellow River Civilization and Sustainable Development & Collaborative Innovation Center of Yellow River Civilization Provincial Co-construction, Henan University, Kaifeng 475001, China
4 School of Urban and Regional Science, East China Normal University, Shanghai 200241, China
* Correspondence: bdsun@re.ecnu.edu.cn

**Abstract:** As the net effect of agglomeration on entrepreneurship depends on the trade-off between positive and negative effects, urban agglomeration can either promote or discourage entrepreneurial activity in theory. However, there is an unexpected shortage of empirical confirmations on this potential cause-and-effect relationship. Our study strives to fill this empirical gap by providing credible evidence whether agglomeration, measured by the urban density or population, increases the probability of individuals being self-employed. Based on the China Labor-Force Dynamic Survey of 2012, 2014, and 2016, we find that big cities fail to facilitate individuals to start or run their own businesses. Further analyses illustrate that the entrepreneurs in large cities can be easily tempted by a wider range of salaried opportunities and are generally exposed to high fixed costs and intense competition. In contrast, entrepreneurship in large cities is of high reward. These results serve as direct evidence of the co-existence of agglomeration diseconomies and economies. This also suggests the direction of government policy in large cities, which is to alleviate, as much as possible, the negative impact on entrepreneurs.

**Keywords:** agglomeration economies; agglomeration diseconomies; entrepreneurship; self-employment; agglomeration cost

**Citation:** Li, W.; Sun, B.; Han, S.; Jin, X. Does Urban Agglomeration Discourage Entrepreneurship in China? Micro-Empirical Evidence from China. *Land* **2023**, *12*, 145. https://doi.org/10.3390/land12010145

Academic Editor: Maria Rosa Trovato

Received: 16 November 2022
Revised: 8 December 2022
Accepted: 12 December 2022
Published: 1 January 2023



## 1. Introduction

Entrepreneurship, the essence of which is creative destruction [1], does not only create employment and promote productivity, but also fundamentally affects cities' future evolution. A prominent phenomenon in China's entrepreneurial boom is the uneven geographical distribution of entrepreneurial activity. Beijing, Shanghai, Shenzhen, or other densely populated cities are often considered "pioneer cities of innovation and entrepreneurship in China" or "best cities of entrepreneurship in China". This is widely supported by the "China City Entrepreneurship Index" released by Renmin University of China (https://news.ruc.edu.cn/archives/126019, accessed on 14 December 2022) and the "Best Startup Cities in China" list issued by China's leading startup community (CYZONE) (https://www.cyzone.cn/article/132069.html, accessed on 14 December 2022). Given the highly spatial concentration of entrepreneurial activities, agglomeration economies are commonly considered as a starting point to understanding the generation and development of entrepreneurship [2–4]. Traditionally, cities with a large population or a high density have been regarded as "incubators" or "nurseries" for entrepreneurs [5,6]. Glaeser et al. (2010) also affirm that entrepreneurs in densely populated urban regions have the advantages of ready access to agglomerated local inputs, skills, ideas, and markets, among others.

However, there is a surprising lack of rigorous empirical evidence to test this assumed cause–effect relationship between urban agglomeration and entrepreneurship. There are a limited number of studies [7,8] that both used urban population size or density as their main variable of concern in the estimation of the effect of agglomeration on entrepreneurship and addressed endogeneity concerns. We consider that this absence of empirical verifications is due to the following issues. First, most researchers rarely question the positive effect of urban agglomeration. In previous empirical studies on the sources of entrepreneurship, while agglomeration has been covered, it has often been treated as a control variable [9–11]. Second, although these studies confirm that big/dense cities are friendlier to entrepreneurs, the causal relationship between these two variables remains questionable. The endogeneity problem has no easy treatment [12], as the main sources of endogeneity are sorting and potential omitting variables. We discuss these issues in more detail in the literature review section.

This paper, therefore, aims to provide a quantitative assessment on whether agglomeration, measured by urban density or population, increases the probability of individuals becoming entrepreneurs. With regard to entrepreneurship, there is no agreed measurement. We take respondents who claim they are self-employed as entrepreneurs, which is believed to be the most commonly used measurement of entrepreneurship [10]. Our study contributes to the literature in two important ways. First, it is one of the first quantitative attempts to establish the causal relationship between urban agglomeration and entrepreneurship. We use agglomeration, measured by urban density or population, as our focal variable and tackle the potential endogeneity problem by using a restricted subsample and two-stage least squares (TSLS) regressions. Our findings support the existence of agglomeration diseconomies and even suggest that the cost of agglomeration has surpassed its benefit in terms of entrepreneurship in China. Second, this paper identifies the source of our counter-intuitive finding, that is, why large cities fail to boost the possibility of being self-employed. Although previous studies have begun questioning the long-held positive effect of urban population size or density on entrepreneurship, there is still a shortage of evidence-based explorations in this area [7,13]. For an emerging market economy like China, which is in the process of institutional transformation and rapid urbanization, how to build and optimize urban entrepreneurial ecosystems is undoubtedly an issue worthy of attention in current research. The purpose of this paper is to explore the above-mentioned uneven geographical distribution of entrepreneurial activities from a relatively new perspective of urban agglomeration [14–16].

The paper is organized as follows. Section 2 lays out the theoretical background and progress on relevant empirical evidence. Section 3 describes the data and empirical strategy. Section 4 presents the econometric results and the final section concludes the paper.

## 2. Literature Review and Research Proposition

This section reviews the theoretical and empirical research on the relationship between urban agglomeration and entrepreneurship. Since it is a core topic in economic geography, there is a rich body of literature dedicated to agglomeration economies [3,5,12,14,17]. While it has long been established that the spatial concentration of firms and workers increases productivity, theoretically, the benefits of agglomeration accumulate faster initially, but eventually, costs prevail as population and density increase in cities [18,19]. Therefore, we next theoretically approach the effects of agglomeration on entrepreneurship from the benefit-cost perspective.

First proposed by Duranton and Puga (2004), agglomeration economies, or the benefits of agglomeration, are wildly widely recognized to stem from three sources: sharing, learning, and matching. Sharing means that the increased local outcomes of spatial concentration lie primarily in sharing indivisible facilities, input suppliers, industrial specialization, and risks, while learning suggests that the improvements in the local productivity of spatial agglomeration come largely from the generation, diffusion, and accumulation of knowledge. These two sources of agglomeration also motivate entrepreneurship, as the sharing and learning effects in large cities are accelerators for entrepreneurs [20–22]. However, the matching mechanism of agglomeration economies may not serve the same function when it comes to entrepreneurship. Specifically, the boost in local performance from urban agglomeration mainly lies in the improvement of either the quality or quantity of the matches between firms and workers. On one hand, this helps entrepreneurs find employees and partners easily and efficiently, thus encouraging entrepreneurial activity; on the other hand, a higher matching effect in large cities also implies it is easier to find a satisfactory job, meaning individuals tend to become salaried-job employees rather than risk-taking employers.

In addition to this matching effect, there are other often mentioned costs of agglomeration, such as high land/house prices or intense competition, which are expected to negatively affect entrepreneurship [23]. The high land/house price costs are commonly believed to have a direct negative impact on entrepreneurship. Induced by agglomeration, high land/house prices suggest office or store rent required is likely higher for entrepreneurs in larger cities. Moreover, high land/house prices also mean entrepreneurs need to offer high salaries to enable their employees to afford rent. As for the intense competition, while some scholars argue that it makes entrepreneurship more efficient [7,14], others believe that excessive competition can discourage entrepreneurs [7]. Other costs, such as congestion, pollution, and crime, do not directly affect the profits or costs of entrepreneurial activity, and are thus not further discussed in this paper.

Apart from the theoretical uncertainty, empirical studies on the impact of urban agglomeration on entrepreneurship are lacking. There are limited empirical papers devoted to this specific topic [7,8] and their findings are inconsistent. Specifically, considering Italian college graduates' work possibilities as entrepreneurs after graduation, Di Addario and Vuri (2010) found that young college graduates were discouraged from starting their entrepreneurial activity in the most densely populated provinces. However, Sato et al. (2012) found that a U-shaped relationship existed between population density and observed entrepreneurship in Japanese prefectures, and the impact of population density on observed entrepreneurship was positive in both small and large cities, while the impact was smaller (or even negative) in medium-sized cities. While there are empirical studies on entrepreneurship that include urban agglomeration as control variables, these studies do not generally discuss the endogeneity of agglomeration and arrive at varied findings [10,11,24]. Similarly, there are empirical-based studies that focus on the industrial structure within agglomerations to explore the impact of specialization and diversification on entrepreneurship [25,26]. We do not further discuss these two branches of literature here because their topic is beyond the scope of this paper.

There are, in fact, two critical challenges in empirically answering the question of whether urban agglomeration increases the probability of an individual becoming an entrepreneur. They are also the main endogeneity sources. The first challenge refers to addressing the sorting or self-selection effect [7,14,27,28]. Specifically, both risk-taking entrepreneurs and risk-averse employees prefer to relocate to large cities because of the greater availability of both entrepreneurial and employment opportunities there. This re-location influences both the population size and level of entrepreneurship of a city, thus leading to biased estimates of the impact of agglomeration on entrepreneurship. Moreover, it is difficult to determine whether this is an overestimate or underestimate. However, this self-selection or sorting effect may not introduce a heavy bias. According to Michelacci and Silva (2007) [29], entrepreneurship can be regarded as a local factor, given that entrepreneurs tend to start their business in the regions they were born.

The second challenge regards missing variables [7,30,31]. To some degree, it is impossible for any study to rule out the possibility of missing variables. Attributes such as the cultural atmosphere of entrepreneurship are likely to influence both the urban population and its entrepreneurship level but are difficult to fully capture. This can lead to biased and inconsistent estimates of urban agglomeration, and ultimately to the failure to establish the causal link between agglomeration and entrepreneurship. It is also worth noting that, while both studies deal with endogeneity using instrumental variables, neither pays special attention to the issue of self-selection.

Taken together, we make our research proposition as follows. It is difficult to conclude whether urban agglomeration promotes or discourages entrepreneurship, as the net effect of agglomeration on entrepreneurship depends on the trade-off between positive and negative influences. Notably, there is a good chance that agglomeration poses a disadvantage for entrepreneurship, with the potential disadvantages or agglomeration diseconomies being mainly embodied in alternative salaried opportunities, high land/house prices, and intense competition. Therefore, there is an urgent need for more empirical evidence to test this potential cause-and-effect relationship, while paying attention to endogeneity issues.

## 3. Data and Estimation Strategies

### 3.1. Data

Our main data source is the China Labor-Force Dynamic Survey (CLDS), which is a nation-wide database updated by Sun Yat-sen University every two years. The CLDS provides a representative image of China's workforce population and we focus on its 2012, 2014, and 2016 waves. Our sample consists of 11,551 working individuals (self-employer and employees), with a self-employment rate of about 17.98% (Table 1). The self-employment rate indicates that our data source is reliable, as it is consistent with the results of a sample survey of 1% of China's population. According to Wu et al. (2014) [32], which is based on the 2005 China's population sample survey (1/5 of a random subsample), the self-employment rate of the urban population was 13.1% in 2005. Generally, the self-employment rate is expected to remain stable and the rapid growth in 2014 mirrors the initiation of a policy on "mass entrepreneurship and innovation".

**Table 1.** Distribution of the sample between the self-employed and employees.

| Year | Self-Employer | | Employee | |
|---|---|---|---|---|
| | Number | Percentage (%) | Number | Percentage (%) |
| 2012 | 474 | 16.68 | 2368 | 83.32 |
| 2014 | 811 | 17.08 | 3938 | 82.92 |
| 2016 | 799 | 20.18 | 3161 | 79.82 |

Although there is no agreed measurement of entrepreneurship, self-employment is considered the most natural individual measure of entrepreneurship [10,33,34]. Hence, we start constructing the core explained variable Entrep, which is a dummy variable taking the value of 1 if the respondents state they are self-employed. Moreover, as a robust check, we also employ Active_Entrep, which is also a dummy variable taking the value of 1 if the respondents state that he or she was motivated to start a business based on taking advantage of a good business opportunity. The individuals who are self-employed as nannies or in odd jobs are dropped from the sample, as they are not really engaged in entrepreneurial activities. For further analysis, we also collect other information at the individual level (see Table 2).

Our core explanatory variable—urban agglomeration—is a proxy of urban density or population. The three CLDS waves considered in this study cover a total of 78 cities, providing a good national representation. A piece of supporting evidence is that the density and population distribution in our sample of 78 cities is similar to that of the national cities (Figure 1a,b). To be specific, the cities in our sample not only share a similar trend of density with all cities, but also have a wide population range, from 0.32 million (Yunfu) to 22.30 million (Shanghai). Moreover, these population-related data are all gathered from the 2010 Population Census of the People's Republic of China to ensure that the permanent population is considered. We also collect other city-level data from the China City Statistical Yearbook for the following analysis (Table 2).

**Table 2.** Variable definitions and summary statistics.

| Variable | Definition | Obs. | Mean | Std. Dev. |
|---|---|---|---|---|
| Entrep | Self-employed or not (1 = yes; 0 = no) | 9883 | 0.170 | 0.375 |
| Age | Age of the respondent (years) | 9883 | 40.475 | 10.492 |
| Male | Gender (1 = male; 0 = female) | 9883 | 0.554 | 0.497 |
| Edu_year | Years of schooling (years) | 9883 | 12.129 | 3.465 |
| Married | Marital status (1 = married; 0 = single) | 9883 | 0.821 | 0.384 |
| Income (ln) | Total income over the past year (Yuan, ln) | 9883 | 10.395 | 0.731 |
| Local_hukou | Possess a local hukou or not (1 = yes; 0 = no) | 9883 | 0.774 | 0.418 |
| Party | Being party member or not (1 = yes; 0 = no) | 9883 | 0.170 | 0.375 |
| Density (ln) | Population density (ln) | 78 | 9.602 | 0.487 |
| Population (ln) | Number of permanent population (ln) | 78 | 14.375 | 1.046 |
| Land area (ln) | Area of construction land (ln) | 78 | 4.772 | 1.063 |
| GDP_pop (ln) | GDP per capita (ln) | 78 | 10.329 | 0.534 |
| Coll_pop (ln) | Percentage of the total population with university education (ln) | 78 | −2.437 | 0.588 |
| Gov_gdp (ln) | Share of government expenditure (subtract expenditure on education and technology) in GDP (ln) | 78 | 7.249 | 0.418 |
| Ter_gdp (ln) | Share of tertiary sector output to GDP (ln) | 78 | 3.76 | 0.25 |
| Internet_pop (ln) | Number of international internet users per capita (ln) | 78 | −1.755 | 0.605 |

(ln) refers to the log-transformation of the data.

### 3.2. Estimation

To investigate whether urban agglomeration increases the probability of individuals being self-employed, we run the following logit regression:

$$Entrep = \alpha + \beta \ln(Density\ or\ Population) + \sum r_i ind_i + \sum \delta_j city_j + \varepsilon, \quad (1)$$

where, as previously discussed, *Entrep* is a dummy variable indicating whether the respondent works as a self-employed entrepreneur, Density and Population are two continuous proxies for urban agglomeration, and $ind_i$ and $city_j$ are vectors of control variables at individual and city levels ($i$ signifies different individuals, and $j$ stands for different cities), respectively. Specifically, $ind_i$ includes respondents' age (Age), gender (Male), years of schooling (Edu_year), marital status (Married), income (Income), possessing a local hukou or not (Local_hukou), and being a party member or not (Party), while $city_j$ includes the area of constructed land (Land_area), GDP per capita (GDP_pop), city's average level of education (Coll_pop), share of government expenditure in GDP (Gov_gdp), share of tertiary sector output to GDP (Ter_gdp), and number of internet users per capita (Internet_pop). These control variables were primarily sourced from entrepreneurship and urban agglomeration studies [7,8,10,12,35,36]. We also include industry, province, and year fixed effects in the specification.

Although we focus on the three waves of the CLDS survey, we still employ a (pooled) cross-sectional strategy rather than panel regression. The main reason lies in the fact that a panel-based identification requires variation in the entrepreneurial status of individuals between 2012 and 2016 for weighing the impact of agglomeration on entrepreneurship. However, the entrepreneurs who have entered and exited the market are only around 200 during the study period (2012–2014). Therefore, our individual-level variables, including the core explained variables, cover three years (2012, 2014, and 2016), while all city-level variables and the core explaining variables are only for 2010. The definition and statistical information for all variables are outlined in Table 2.

To address the two empirical challenges mentioned above, we adopt two approaches. Our solution for the self-selection issue is to identify a subsample of only respondents that have not moved across counties since the age of 14. In this way, we rule out the risk-taking entrepreneurs and risk-averse employees who prefer to relocate among cities to some extent. Regarding potential missing variables, apart from adding region and year dummy variables, we use a TSLS regression with an historical instrumental variable, that is, density or population in 1953, the data coming from the first census conducted in China.

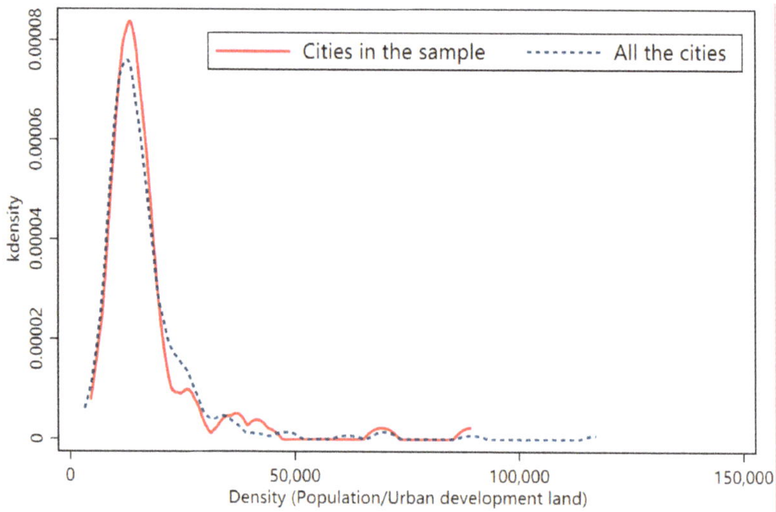


(**a**)

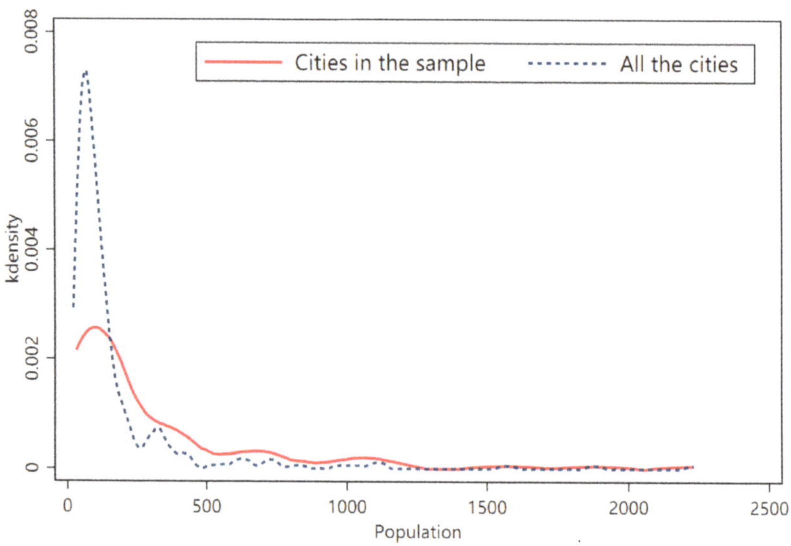


(**b**)

**Figure 1.** Kernel density estimate of urban density (**a**) and population (**b**).

Since the groundbreaking work of Ciccone and Hall (1996) [37], historical instrumental variables have become common practice in the study of agglomeration economies. This instrument can satisfy both the relevance and exogenous requirements. As for relevance, the population in 1953 shaped today's population. As shown in Table 3, the Kleibergen-Paap rk Wald F (KP F) statistic confirms the relevance of our instrument. As for exogeneity, the spatial pattern of China's population has changed dramatically in response to various broadly population-oriented projects, such as the well-known Shangshan Hsia-hsiang and The Third Front programs [38,39]. At the same time, entrepreneurship has also been deeply transformed by three decades of planned economy. Hence, the 1953 population should have no direct effect on current entrepreneurship.

**Table 3.** Baseline identifications under different approaches.

| Y: *Entrep* | (1) MLE | (2) Sorting | (3) IV | (4) MLE | (5) Sorting | (6) IV |
|---|---|---|---|---|---|---|
| Density (ln) | 0.567 *** | 0.583 *** | 0.517 ** | | | |
| | (0.094) | (0.097) | (0.132) | | | |
| Population (ln) | | | | 0.567 *** | 0.583 *** | 0.517 ** |
| | | | | (0.094) | (0.097) | (0.132) |
| Age | 1.070 ** | 1.033 | 1.024 | 1.070 ** | 1.033 | 1.001 |
| | (0.030) | (0.043) | (0.023) | (0.030) | (0.043) | (0.002) |
| Age$^2$ | 0.999 * | 1.000 | 1.000 | 0.999 * | 1.000 | 1.000 |
| | (0.000) | (0.000) | (0.000) | (0.000) | (0.000) | (0.000) |
| Male | 1.465 *** | 1.810 *** | 1.388 *** | 1.465 *** | 1.810 *** | 1.008 |
| | (0.115) | (0.232) | (0.107) | (0.115) | (0.232) | (0.005) |
| Edu_year | 0.895 *** | 0.874 *** | 0.922 *** | 0.895 *** | 0.874 *** | 0.922 *** |
| | (0.013) | (0.022) | (0.015) | (0.013) | (0.022) | (0.015) |
| Married | 1.588 *** | 1.781 *** | 1.288 ** | 1.588 *** | 1.781 *** | 0.987 |
| | (0.208) | (0.368) | (0.154) | (0.208) | (0.368) | (0.154) |
| Local_hukou | 0.628 *** | 0.639 *** | 0.775 ** | 0.628 *** | 0.639 *** | 0.775 ** |
| | (0.089) | (0.105) | (0.080) | (0.089) | (0.105) | (0.015) |
| Income (ln) | 1.806 *** | 1.691 *** | 1.341 *** | 1.806 *** | 1.691 *** | 1.341 *** |
| | (0.188) | (0.194) | (0.091) | (0.188) | (0.194) | (0.091) |
| Party | 0.391 *** | 0.375 *** | 0.620 *** | 0.391 *** | 0.375 *** | 1.002 |
| | (0.063) | (0.062) | (0.062) | (0.063) | (0.062) | (0.006) |
| Land area (ln) | 0.696 *** | 0.731 ** | 0.673 ** | 1.228 | 1.254 * | 1.322 ** |
| | (0.091) | (0.094) | (0.114) | (0.153) | (0.155) | (0.212) |
| GDP_pop (ln) | 0.498 ** | 0.760 | 1.125 | 0.498 ** | 0.760 | 1.646 ** |
| | (0.158) | (0.238) | (0.296) | (0.158) | (0.238) | (0.296) |
| Coll_pop (ln) | 0.962 | 0.710 | 0.704 | 0.962 | 0.710 | 0.704 |
| | (0.240) | (0.194) | (0.189) | (0.240) | (0.194) | (0.189) |
| Gov_gdp (ln) | 0.588 ** | 1.098 | 0.867 | 0.588 ** | 1.098 | 0.867 |
| | (0.152) | (0.336) | (0.214) | (0.152) | (0.336) | (0.214) |
| Ter_gdp (ln) | 1.948 | 0.877 | 1.448 | 1.948 | 0.877 | 3.317 *** |
| | (0.954) | (0.424) | (0.823) | (0.954) | (0.424) | (0.823) |
| Internet_pop (ln) | 0.974 | 1.223 | 1.011 | 0.974 | 1.223 | 0.923 |
| | (0.237) | (0.299) | (0.206) | (0.237) | (0.299) | (0.206) |
| Industry FE | YES | YES | YES | YES | YES | YES |
| Province FE | YES | YES | YES | YES | YES | YES |
| Year FE | YES | YES | YES | YES | YES | YES |
| KP F stat | - | - | 9.961 | - | - | 9.961 |
| Observations | 9883 | 4266 | 3485 | 9883 | 4266 | 3485 |
| LL | −3163 | −1425 | 510.7 | −3163 | −1425 | 510.7 |
| Pseudo R$^2$ | 0.297 | 0.293 | / | 0.297 | 0.293 | / |

Odds ratio (OR) coefficients above 1 indicate an increased occurrence of the event and vice versa. Standard errors adjusted for clustering at city level are between parentheses. * $p < 0.1$, ** $p < 0.05$, *** $p < 0.01$.

## 4. Empirical Evidence

### *4.1. Baseline Results of Urban Agglomeration on Entrepreneurship*

Table 3 reports the logit specification results of urban agglomeration on entrepreneurship for different proxy variables and econometric approaches. Columns (1)–(3) use the proxy variable Density, while columns (4)–(6) use Population as a proxy. Each column group uses the maximum likelihood estimate (MLE), sorting, and instrumental variable (IV) strategies, respectively. MLE is the most commonly used estimation strategy in logistic regression. Sorting refers to our adoption of subsamples that have not moved across counties since the age of 14 to tackle the potential sorting problem. IV implies estimation using TSLS with the 1953 density or population as IVs to address the missing variable concern.

Moreover, in the IV specification, the KP F statistic of columns (3) and (6) is close to 10, which is above the 15% maximal IV size (8.96) in the Stock–Yogo weak instrument test. This confirms the relevance of our instrumental variables.

Turning to our focal variable, the odds ratios for both Density and Population are below 1; that is, all else being equal, the higher the population density or the larger the population, the less chance individuals have of becoming self-employed. This finding is robust to three different strategies and the odds ratios are of roughly the same magnitude. In other words, big cities fail to incentivize individuals to start or run their own business, and our concerns about self-selection and omitted variables do not make a significant difference. The magnitude and significance of Density and Population are the same. As this is an inevitable result when considering Land area as a control variable, we only employ Density as the proxy for urban agglomeration in the following.

The results of the other controls are in line with expectations. At the individual level, male, married, and higher income individuals are more likely to be entrepreneurs, while individuals with high education, who hold a local residence, and are party members are less likely to engage in an entrepreneurial venture. This is consistent with Cejudo García et al. (2020) [40]. At the city level, the roles of these variables are rather mixed. In general, government intervention is harmful to individual entrepreneurship, while the average education level in the city does not affect whether an individual chooses to be self-employed.

Someone may argue that self-employment is not an appropriate measure for entrepreneurship, for many people are pushed into self-employment. In order to tackle this potential issue, we select the self-employed entrepreneur who claims that their motivation is to take advantage of a good business opportunity, as explained variables (Table 4). According to Table 4, all else being equal, the odds ratios for Density or Population are again below 1. In other words, for active self-employed entrepreneurs, big cities still play a negative role.

**Table 4.** Robust check with active entrepreneur.

| Y: *Active_Entrep* | (1)<br>MLE | (2)<br>Sorting | (3)<br>IV |
|---|---|---|---|
| Density (ln) | 0.423 *** | 0.332 *** | 0.517 ** |
| | (0.131) | (0.106) | (0.132) |
| Individual-level control variables | YES | YES | YES |
| City-level control variables | YES | YES | YES |
| Industry FE | YES | YES | YES |
| Province FE | YES | YES | YES |
| Year FE | YES | YES | YES |
| KP F stat | - | - | 9.961 |
| Observations | 5818 | 3640 | 3485 |
| LL | −1023 | −585.5 | 510.7 |
| Pseudo $R^2$ | 0.285 | 0.300 | / |

Odds ratio (OR) coefficients above 1 indicate an increased occurrence of the event and vice versa. Standard errors adjusted for clustering at city level are between parentheses. ** $p < 0.05$, *** $p < 0.01$. The individual-level control variables and city-level control variables in columns (1) and (2) are the same as in Table 3.

### *4.2. Potential Explanations for the Negative Impact of Agglomeration*

As shown above, the probability of becoming an entrepreneur decreases as urban density increases. This result is robust for controlling for a wide range of individual- and city-level features and after correcting for the two potential endogenous sources of agglomeration. Here, we explore the potential explanations for our counter-intuitive finding from three aspects. Additionally, as the endogeneity problem is largely insensitive according to the benchmark regressions, in the follow-up specifications, we do not specifically target endogeneity to avoid MLE non-convergence. This is always the case, especially when the sample size is small. The sample size is reduced in many of the specifications in this section.

#### 4.2.1. Matching Effect

It is believed that an increase in density or population can increase the probability of finding a match and improves the quality of matches. This translates into easier access to a satisfying job in dense or large cities; as a result, individuals tend to be wage-earning employees rather than risk-taking employers.

To verify this reasoning, we first divide our sample into high and low groups based on the availability of employment opportunities to check whether the magnitude and significance of Density differ. Next, we replace the explained variables with the job satisfaction to examine differences in match quality. Particularly, the availability of employment opportunities is measured by the total number of workers in the 2008 Industrial Census, while job satisfaction (ranging from 1 to 5 for strongly dissatisfied to strongly satisfied, respectively) is derived from the CLDS questionnaire.

We find that the coefficient on Density is significantly lower than 1 in column (1) but insignificant in column (2) of Table 5. This implies that the negative effect of agglomeration is stronger in cities with a higher availability of employment opportunities. Meanwhile, according to column (3), higher urban density is indeed associated with higher job satisfaction among employees. However, the increase in density has no significant effect on employer job satisfaction, based on column (4). This reflects the high quality of matches in large cities, where employees are more likely to find desirable jobs. In short, the negative effect of density can be explained by the fact that the densest markets are better at matching quantity with quality, thus creating a trade-off for entrepreneurship.

**Table 5.** Agglomeration economies and entrepreneurship: testing matching effect.

| | (1) | (2) | (3) | (4) |
|---|---|---|---|---|
| | Y: *Entrep* | | Y: *Job Satisfaction* | |
| | **High** | **Low** | **Employee** | **Employer** |
| Density (ln) | 0.207 *** | 1.199 | 1.188 ** | 1.003 |
| | (0.116) | (0.619) | (0.087) | (0.078) |
| Individual-level control variables | YES | YES | YES | YES |
| City-level control variables | YES | YES | YES | YES |
| Industry FE | YES | YES | YES | YES |
| Province FE | YES | YES | YES | YES |
| Year FE | YES | YES | YES | YES |
| Observations | 6753 | 2725 | 8135 | 1645 |
| LL | −2071 | −989.6 | −8393 | −1779 |
| Pseudo $R^2$ | 0.276 | 0.337 | 0.0360 | 0.0434 |

High signifies that the availability of jobs in this subsample is higher than the 50th percentile for the sample cities, while Low indicates it is below the 50th percentile. The individual-level control variables in columns (1)–(4) are the same as in Table 3. The city-level control variables in columns (1) and (2) are the same as in Table 3, but columns (3) and (4) include only GDP_pop and Land Area. OR coefficients are shown. Standard errors adjusted for clustering at the city level are between parentheses. ** $p < 0.05$, *** $p < 0.01$.

#### 4.2.2. High House Price

As a non-tradable resource, land and housing prices are bound to increase with density, which can impose a high fixed cost on entrepreneurs and raise entrepreneurship entry barriers. Hence, it is generally agreed that high land/housing costs are a strong discouragement to entrepreneurship. Additionally, a side-effect of the high land/prices is that entrepreneurs typically have to pay high salaries to make house rent affordable for their employees. This may make the cost of labor additionally hinder entrepreneurship.

To explore whether this is the case, we split the sample into two high-low groups based on average house prices and salary in the city (Table 6). The coefficient on Density with high house prices is significantly below 1 but does not show significance for the subsample of low house prices. This empirically confirms the discouraging effect of house prices on entrepreneurship. In terms of salary, although the coefficient on Density with low labor cost is greater than 1, it is insignificant. In fact, none of the coefficients on density are significant when grouped by salary (columns (3) and (4)). In other words, the dampening effect of labor costs is not verified. Overall, high housing prices in big cities tend to discourage individuals from entrepreneurship.

**Table 6.** Agglomeration and entrepreneurship: testing the effect of house price and salary.

| Y: *Entrep* | (1) | (2) | (3) | (4) |
|---|---|---|---|---|
| | House Prices | | Salaries | |
| | High | Low | High | Low |
| Density (ln) | 0.644 * | 0.686 | 0.605 | 1.235 |
| | (0.166) | (0.373) | (0.232) | (0.364) |
| Other control variables | YES | YES | YES | YES |
| Industry FE | YES | YES | YES | YES |
| Province FE | YES | YES | YES | YES |
| Year FE | YES | YES | YES | YES |
| Observations | 6442 | 3082 | 6471 | 3066 |
| LL | −1977 | −1125 | −1986 | −1115 |
| Pseudo $R^2$ | 0.292 | 0.303 | 0.275 | 0.325 |

High signifies that the house price or salary in this subsample is higher than the 50th percentile for the sample cities, while Low indicates it is below the 50th percentile. The other control variables in columns (1)–(4) are the same as in Table 3. OR coefficients are shown. Standard errors adjusted for clustering at the city level are between parentheses. * $p < 0.1$.

#### 4.2.3. Intense Competition

As for competition, it is widely accepted that there exists higher competition in larger markets. This is also confirmed by our empirical examination in column (1) of Table 7. Taking the employer's personal perception of intense business competition in the past year (Fierce, ranges 1 to 5 for free of competition to fierce competition, respectively) as the explanatory variable, an ordered logistic regression shows that the higher the density, the more intense perceived competition is.

**Table 7.** Agglomeration economies and entrepreneurship: testing the effect of competition.

| | (1) | (2) | (3) |
|---|---|---|---|
| | Y: *Fierce Competition* | Y: *Entrep* | |
| | | High | Low |
| Density (ln) | 2.006 *** | 0.734 * | 0.562 |
| | (0.511) | (0.118) | (0.346) |
| Other control variables | YES | YES | YES |
| Industry FE | YES | YES | YES |
| Province FE | YES | YES | YES |
| Year FE | YES | YES | YES |
| Observations | 999 | 3117 | 6476 |
| LL | −1159 | −1137 | −1967 |
| Pseudo $R^2$ | 0.059 | 0.296 | 0.299 |

High signifies that the density of firms in this subsample is higher than the 50th percentile for the sample cities, while Low indicates it is below the 50th percentile. The other control variables in columns (1)–(3) are the same as in Table 3. OR coefficients are shown. Standard errors adjusted for clustering at the city level are between parentheses. * $p < 0.1$, *** $p < 0.01$.

More importantly, fierce competition may invariably increase the difficulty of starting a business, which in turn discourages over-thinking entrepreneurs. Hence, we measure the degree of competition using the density of firms from the 2008 Industrial Census and divide the sample into high and low competition groups. As shown in Table 7, the subsample with a high competition degree has a regression coefficient on density significantly below 1 (column (2)), while the coefficient is insignificant for a low competition degree (column (3)). Therefore, the fierce competition in big cities does, as expected, drive individuals away from becoming employers.

### 4.3. Reward for Entrepreneurs in Large Cities

Based on Sections 4.1 and 4.2, we can conclude that large cities fail to encourage individuals to start or run their own businesses because entrepreneurs in large cities can easily be tempted by a wider range of salaried opportunities and face high fixed costs and intense competition. However, these findings can easily be translated into misleading policy, that is, limiting or restricting individuals from engaging in entrepreneurship in large cities. In fact, if a firm or entrepreneur can survive high housing prices and fierce competition in large cities, they can expect to reap significant rewards. In this sense, it is worth encouraging entrepreneurship in big cities.

To find out whether this is true, we respectively take the gross profit of firms, number of employees, and operational income of entrepreneurs as dependent variables and observe the coefficients on density. Table 8 confirms that, as urban density increases, firms and entrepreneurs indeed perform better. This can serve as a friendly reminder that the firms and entrepreneurs surviving in large cities are productive and do receive rewards.

**Table 8.** Rewards for entrepreneurs in large cities.

| | (1) | (2) | (3) |
|---|---|---|---|
| | Y: *Gross Profit of the Firm (Million RMB)* | Y: *Number of Employees* | Y: *Operational Income of Entrepreneurs* |
| Density (ln) | 0.006 * | 0.340 * | 0.500 *** |
| | (0.003) | (0.184) | (0.125) |
| Other control variables | YES | YES | YES |
| Industry FE | YES | YES | YES |
| Province FE | YES | YES | YES |
| Year FE | YES | YES | YES |
| Observations | 1082 | 253 | 457 |
| $R^2$ | 0.161 | 0.356 | 0.299 |

Other control variables in columns (1)–(3) are the same as in Table 3. Standard errors adjusted for clustering at the city level are between parentheses. * $p < 0.1$, *** $p < 0.01$.

## 5. Discussion

Theoretically, it is difficult to draw conclusions on whether urban agglomeration promotes or hinders entrepreneurship. Based on our empirical examination, we find that, all else being equal, the higher the population density or the larger the population, the less chance individuals have of becoming self-employed. This baseline result is in line with Di Addario and Vuri (2010), who found that young Italian university graduates were reluctant to start their entrepreneurial activities in the most densely populated provinces. A U-shaped relationship was found in Japanese prefectures [8], but it could not be confirmed in our study (the square items of population and density are not statistically significant).

As for the reasons why large cities fail to encourage individuals to start their own businesses, we empirically find that entrepreneurs in large cities can easily be tempted by a wider range of salaried opportunities. This is against the suggestion of van Oort and Bosma (2013), who argue that the matching effect in large cities can help entrepreneurs find employees and partners easily and efficiently, thus encouraging the development of startups. Our findings lend more support to the idea that this matching effect makes it easier for individuals to find a satisfying job, which makes individuals tend to be employees with salaried jobs rather than risk-taking employers. Moreover, we find that entrepreneurs in large cities face intense competition. This is consistent with the Di Addario and Vuri (2010) argument about excessive competition.

Moreover, incredible rewards can be expected if a firm or entrepreneur can survive high housing prices and fierce competition in large cities. We do find firms in large cities are more likely to make better profits and hire more employees, and entrepreneurs can earn higher incomes. This may explain why big cities have traditionally been seen as pioneering cities for entrepreneurship [41–43].

## 6. Conclusions

Cities with large or dense populations have traditionally been treated as entrepreneurial "incubators" or "nurseries" [44,45]. However, there is a surprising lack of rigorous empirical evidence to test this assumed cause–effect relationship between agglomeration economies and entrepreneurship. Based on the 2012, 2014, and 2016 CLDS waves, this paper tries to fill this empirical gap using credible specifications. We find that large cities fail to boost individuals to start or run their own businesses, and this primary finding is robust in correcting the two potential endogeneity sources of agglomeration. Further analyses illustrate that entrepreneurs in large cities can be easily tempted by a wider range of salaried opportunities and are largely exposed to high fixed costs and intense competition. Additional examinations find that firms in larger cities yield better profits and hire more employees, and entrepreneurs are more likely to have higher incomes.

These findings lead to critical implications for boosting entrepreneurship. First, our baseline finding is a timely reminder that the cost of agglomeration has even outweighed its benefit in terms of entrepreneurship in China. It can be further deduced that China's cities may be experiencing deviations from their optimal sizes. The most prominent agglomeration diseconomy is excessive housing prices, which pose a serious obstacle to entrepreneurship.

Second, our findings should not simply be reduced to the idea that we should limit or restrict individuals from engaging in entrepreneurship in large cities. Although a high density is a strong discouragement for individuals becoming entrepreneurs, the survivors in large cities can always expect significant rewards. According to our empirical examination, firms in large cities are more likely to make better profits and hire more employees, and entrepreneurs can earn higher incomes. The key message we aim to deliver is that entrepreneurs in large cities suffer from many disadvantages and mitigating these vulnerabilities is a top priority.

Third, targeted government policies to mitigate agglomeration diseconomies can focus on the following aspects. As entrepreneurs in large cities can be easily tempted by a wider range of salaried opportunities and are largely exposed to high fixed costs and intense competition, policymakers in large cities could at least nurture the culture of self-employment, reduce taxes for entrepreneurs, and encourage legitimate competition.

**Author Contributions:** Conceptualization, W.L., B.S., S.H. and X.J.; Data curation, X.J.; Formal analysis, W.L.; Funding acquisition, W.L., B.S. and S.H.; Supervision, B.S.; Validation, S.H.; Writing—original draft, W.L.; Writing—review & editing, B.S. All authors have read and agreed to the published version of the manuscript.

**Funding:** This research was funded by National Natural Science Foundation of China grant number 41901184, 42071210, Fundamental Research Funds for the Central Universities grant number 2022ECNU-XWK-XK001, Humanities and Social Science Youth Foundation of Ministry of Education of China grant number 22YJC790037, and Philosophy and Social Science Annual Project Youth Foundation of Henan Province of China grant number 2022CJJ130.

**Institutional Review Board Statement:** Not applicable.

**Informed Consent Statement:** Not applicable.

**Data Availability Statement:** Not applicable.

**Conflicts of Interest:** The authors declare no conflict of interest.

## References

1. Schumpeter, J. *The Theory of Economic Development*; Harvard University Press: Cambridge, MA, USA, 1934.
2. Capello, R. Entrepreneurship and Spatial Externalities: Theory and Measurement. *Ann. Reg. Sci.* **2002**, *36*, 387–402. [CrossRef]
3. Glaeser, E.L.; Rosenthal, S.S.; Strange, W.C. Urban Economics and Entrepreneurship. *J. Urban Econ.* **2010**, *67*, 1–14. [CrossRef]

4. Rosenthal, S.S.; Strange, W.C. Small Establishments/big Effects: Agglomeration, Industrial Organization and Entrepreneurship. In *Agglomeration Economics*; Glaeser, E.L., Ed.; University of Chicago Press: Chicago, IL, USA, 2010; pp. 277–302.
5. Duranton, G.; Puga, D. Micro-Foundations of Urban Agglomeration Economies. In *Handbook of Regional and Urban Economics*; Henderson, J.V., Thisse, J.F., Eds.; Elsevier: Amsterdam, The Netherlands, 2004; Volume 4, pp. 2063–2117.
6. Hoover, E.M.; Vernon, R. *Anatomy of a Metropolis: The Changing Distribution of People and Jobs within the New York Metropolitan Region*; Harvard University Press: Cambridge MA, USA, 2013.
7. Di Addario, S.; Vuri, D. Entrepreneurship and Market Size. The Case of Young College Graduates in Italy. *Labour Econ.* **2010**, *17*, 848–858. [CrossRef]
8. Sato, Y.; Tabuchi, T.; Yamamoto, K. Market Size and Entrepreneurship. *J. Econ. Geogr.* **2012**, *12*, 1139–1166. [CrossRef]
9. Reynolds, P.D.; Miller, B.; Maki, W.R. Explaining Regional Variation in Business Births and Deaths: U.S. 1976–1988. *Small Bus. Econ.* **1995**, *7*, 389–407. [CrossRef]
10. Glaeser, E.L. Entrepreneurship and the City. In *Entrepreneurship and Openness*; Edward Elgar Publishing: Cheltenham, UK, 2009.
11. Zheng, S.; Du, R. How Does Urban Agglomeration Integration Promote Entrepreneurship in China? Evidence from Regional Human Capital Spillovers and Market Integration. *Cities* **2020**, *97*, 102529. [CrossRef]
12. Combes, P.P.; Gobillon, L. The Empirics of Agglomeration Economies. In *Handbook of Regional and Urban Economics*; Henderson, J.V., Thisse, J.F., Eds.; Elsevier: Amsterdam, The Netherlands, 2015; Volume 5, pp. 247–348.
13. Nocke, V. A Gap for Me: Entrepreneurs and Entry. *J. Eur. Econ. Assoc.* **2006**, *4*, 929–956. [CrossRef]
14. Duranton, G.; Puga, D. The Economics of Urban Density. *J. Econ. Perspect.* **2020**, *34*, 3–26. [CrossRef]
15. Barkham, R.J. Entrepreneurial Characteristics and the Size of the New Firm: A Model and an Econometric Test. *Small Bus. Econ.* **1994**, *6*, 117–125. [CrossRef]
16. Audretsch, D.B.; Fritsch, M. The Geography of Firm Births in Germany. *Reg. Stud.* **1994**, *28*, 359–365. [CrossRef]
17. Combes, P.P.; Duranton, G.; Gobillon, L.; Puga, D. The Productivity Advantages of Large Cities: Distinguishing Agglomeration from Firm Selection. *Econometrica* **2012**, *80*, 2543–2594. [CrossRef]
18. Rauch, J.E. Productivity Gains from Geographic Concentration of Human Capital: Evidence from the Cities. *J. Urban Econ.* **1993**, *34*, 380–400. [CrossRef]
19. Henderson, J.V. The Sizes and Types of Cities. *Am. Econ. Rev.* **1974**, *64*, 640–656.
20. van Oort, F.G.; Bosma, N.S. Agglomeration Economies, Inventors and Entrepreneurs as Engines of European Regional Economic Development. *Ann. Reg. Sci.* **2013**, *51*, 213–244. [CrossRef]
21. Qian, H.; Zhao, C. Space-Time Analysis of High Technology Entrepreneurship: A Comparison of California and New England. *Appl. Geogr.* **2018**, *95*, 111–119. [CrossRef]
22. Buenstorf, G.; Klepper, S. Why Does Entry Cluster Geographically? Evidence from the US Tire Industry. *J. Urban Econ.* **2010**, *68*, 103–114. [CrossRef]
23. Combes, P.-P.; Duranton, G.; Gobillon, L. The Costs of Agglomeration: House and Land Prices in French Cities. *Rev. Econ. Stud.* **2018**, *86*, 1556–1589. [CrossRef]
24. Sun, B.; Zhu, P.; Li, W. Cultural Diversity and New Firm Formation in China. *Reg. Stud.* **2019**, *53*, 1371–1384. [CrossRef]
25. van Oort, F.G.; Stam, E. Agglomeration Economies and Entrepreneurship in the ICT Industry. ERIM Report Series Reference No: ERS-2006-016-ORG. 2006, pp. 1–24. Available online: https://ssrn.com/abstract=902745 (accessed on 15 November 2022).
26. Capozza, C.; Salomone, S.; Somma, E. Local Industrial Structure, Agglomeration Economies and the Creation of Innovative Start-Ups: Evidence from the Italian Case. *Entrep. Reg. Dev.* **2018**, *30*, 749–775. [CrossRef]
27. Andersson, M.; Lööf, H. Agglomeration and Productivity: Evidence from Firm-Level Data. *Ann. Reg. Sci.* **2011**, *46*, 601–620. [CrossRef]
28. Forslid, R.; Okubo, T. Spatial Sorting with Heterogeneous Firms and Heterogeneous Sectors. *Reg. Sci. Urban Econ.* **2014**, *46*, 42–56. [CrossRef]
29. Michelacci, C.; Silva, O. Why so Many Local Entrepreneurs? *Rev. Econ. Stat.* **2007**, *89*, 615–633. [CrossRef]
30. Zheng, L.; Zhao, Z. What Drives Spatial Clusters of Entrepreneurship in China? Evidence from Economic Census Data. *China Econ. Rev.* **2017**, *46*, 229–248. [CrossRef]
31. Catalini, C. Microgeography and the Direction of Inventive Activity. *Manag. Sci.* **2018**, *64*, 4348–4364. [CrossRef]
32. Wu, X.Y.; Wang, M.; Li, L.X. Did High Housing Price Discourage Entrepreneurship in China. *Econ. Res. J.* **2014**, *9*, 121–134. (In Chinese)
33. Harding, J.P.; Rosenthal, S.S. Homeownership, Housing Capital Gains and Self-Employment. *J. Urban Econ.* **2017**, *99*, 120–135. [CrossRef]
34. Blanchflower, D.G. Self-Employment in OECD Countries. *Labour Econ.* **2000**, *7*, 471–505. [CrossRef]
35. Barkham, R. Regional Variations in Entrepreneurship: Some Evidence from the United Kingdom: Regional Variations in Entrepreneurship. *Entrep. Reg. Dev.* **1992**, *4*, 225–244. [CrossRef]
36. Qin, N.; Kong, D. Human Capital and Entrepreneurship. *J. Hum. Cap.* **2021**, *15*, 513–553. [CrossRef]
37. Ciccone, A.; Hall, R.E. *Productivity and the Density of Economic Activity*; Nber Working Paper, No: 4313; American Economic Association: Nashville, TN, USA, 1993.
38. Bernstein, T.P. *Up to the Mountains and Down to the Villages: The Transfer of Youth from Urban to Rural China*; Yale University Press: New York, NY, USA, 1977.

39. Naughton, B. The Third Front: Defence Industrialization in the Chinese Interior. *China Q.* **1988**, *115*, 351–386. [CrossRef]
40. Cejudo García, E.; Cañete Pérez, J.A.; Navarro Valverde, F.; Ruiz Moya, N. Entrepreneurs and Territorial Diversity: Success and Failure in Andalusia 2007–2015. *Land* **2020**, *9*, 262. [CrossRef]
41. Acs, Z.J.; Glaeser, E.L.; Litan, R.E.; Fleming, L.; Goetz, S.J.; Kerr, W.; Klepper, S.; Rosenthal, S.S.; Sorenson, O.; Strange, W.C. Entrepreneurship and Urban Success: Toward a Policy Consensus. 2008. Available online: https://papers.ssrn.com/sol3/papers.cfm?abstract_id=1092493 (accessed on 15 November 2022).
42. Bosma, N.; Sternberg, R. Entrepreneurship as an Urban Event? Empirical Evidence from European Cities. *Reg. Stud.* **2014**, *48*, 1016–1033. [CrossRef]
43. Florida, R.; Adler, P.; Mellander, C. The City as Innovation Machine. *Reg. Stud.* **2017**, *51*, 86–96. [CrossRef]
44. Duranton, G.; Puga, D. Nursery Cities: Urban Diversity, Process Innovation, and the Life Cycle of Products. *Am. Econ. Rev.* **2001**, *91*, 1454–1477. [CrossRef]
45. Fan, D.; Su, Y.; Huang, X. Nursery City Innovation: A CELL Framework. *Public Adm. Rev.* **2022**, *82*, 764–770. [CrossRef]

land

MDPI

Article

# Optimal Regional Allocation of Future Population and Employment under Urban Boundary and Density Constraints: A Spatial Interaction Modeling Approach

David Jung-Hwi Lee [1] and Jean-Michel Guldmann [2,*]

1 Tennessee Department of Transportation, Long Range Planning Division, Nashville, TN 37243, USA
2 Department of City and Regional Planning, The Ohio State University, Columbus, OH 43210, USA
* Correspondence: guldmann.1@osu.edu

**Abstract:** This paper develops an optimization modeling framework to select strategies of land development and population and employment densities for a growing metropolitan area. The modeling core involves a non-linear commuting model, which accounts for spatial structure variables and is empirically estimated by Tobit regression. This commuting model is then embedded into a non-linear optimization model that allocates increments in the population and employment (activities) to available land, while minimizing the total future commuting costs under various combinations of land expansion boundaries and population and employment densities. The resulting minimum cost surface is approximated via polynomial regression and combined with land development and congestion cost functions to derive the overall optimal strategy. These models are estimated and calibrated with data from the Census Transportation Planning Package (CTPP) and Auditor's property database, and are applied to the Fredericksburg metropolitan area, Virginia. The results demonstrate that the optimal development densities are very sensitive to the congestion cost function. A land development strategy that allows for limited sprawl might be a smart policy to reduce both regional vehicle mile travel (VMT) and related congestion and pollution.

**Keywords:** population location and density; employment location and density; commuting spatial interactions; urban boundary; land availability; cost minimization

**Citation:** Lee, D.J.-H.; Guldmann, J.-M. Optimal Regional Allocation of Future Population and Employment under Urban Boundary and Density Constraints: A Spatial Interaction Modeling Approach. *Land* **2023**, *12*, 433. https://doi.org/10.3390/land12020433

Academic Editors: Bindong Sun, Tinglin Zhang, Wan Li, Chun Yin and Honghuan Gu

Received: 15 December 2022
Revised: 30 January 2023
Accepted: 30 January 2023
Published: 7 February 2023



## 1. Introduction

### 1.1. Historical Overview of Urban Modeling

Accurately predicting the spatial pattern of population and economic activity is necessary for developing successful regional plans and policies. Computer-based urban simulation models originated in the U.S. in the 1950s' metropolitan transportation studies and used geographic accessibility concepts. However, the attempts to build large-scale urban models failed over the next 15 years. After Lowry [1] introduced a comprehensive spatial interaction model called "The Lowry Model" to simulate location patterns of residential and commercial/service activities for a given pattern of basic (export manufacturing) employment locations, while accounting for accessibility, a renaissance in urban modeling took place, based on spatial interaction modeling (SIM), with initial formulations using the gravity model. During the 1960~1970s, the focus of SIM was primarily on population and activities. The 1980~1990s witnessed efforts at integrating land-use and transportation modeling. More recently, comprehensive models have involved environmental modeling, and the advent of the digital era, advances in computer technology, sophisticated spatial analysis methods, and the availability of big data, combined with geographic information systems (GIS), have generated new urban models.

In order to provide an appropriate background for this research, we critically review the literature on SIM, the relationship between SIM and planning optimization models, and the costs of urban sprawl and congestion. We then summarize the shortcomings of past research and outline the goals of this research.

### 1.2. Spatial Interaction Modeling: Structure and Variables

Spatial interaction modeling (SIM) represents various models that explain and predict spatial flows, including residence–workplace commuting, shopping travel, inter-city travel, migration, tourism, commodity flows, financial transactions, and various telecommunication forms. SIM ranges from the standard gravity model (GM), reflecting Newtonian physics, to entropy models to discrete spatial choice models. The basic GM is formulated as follows:

$$T_{ij} = kR_i W_j / D_{ij}{}^{\alpha} \quad (1)$$

In this equation, $T_{ij}$ is the flow from origin i to destination j, $R_i$ and $W_j$ are the measures of the sizes of the origin i and destination j, respectively, $D_{ij}$ is the distance between them, and $\alpha$ is a positive parameter that represents the distance friction. The $R_i$ and $W_j$ variables are proxies for the abilities of the origin to generate flows and of the destination to attract them. Generalized versions of Equation (1) include several variables that characterize both the origin and destination, and several friction factors. This model has been termed as unconstrained SIM. The estimation of (1), subject to the given total outflows for all the origins and given total inflows for all the destinations, is termed as constrained or entropy SIM. The focus here is on the unconstrained case. These models consider aggregate flow data (e.g., the number of commuters between the origins and destinations). Another interpretation of SIM is related to discrete choice models (e.g., multi-nomial logit models), using disaggregate data at the level of the individual decision maker. Anas [2] argues that the gravity and discrete choice models are two equivalent views of the same problem. For reviews of the theory and applications of SIM, one can refer to the work of Sen and Smith [3], who discuss the theoretical foundations and practical applications of gravity models to commuting, and Nijkamp and Ratajczak [4], who review the relevance of gravitational principles in regional science and spatial economics, and address their application to trade flow analysis.

The above SIM approach suffers from the problem of independence from irrelevant alternatives. SIM models have been improved by incorporating variables that represent the effects of the spatial structure, thus eliminating the estimation bias of the friction parameters. Fotheringham [5] introduced a competing destination (CD) factor that measures the accessibility of any destination j to all (or a subset of) the other destinations. If the effect of CD is negative, competition to attract flows can be detected among the destinations; the closer destination j is to the other destinations, the smaller the flow terminating at j. If the effect of CD is positive, agglomeration effects can be observed among the neighboring destinations (e.g., a set of different brand stores within a shopping mall). Another approach to accounting for the spatial structure involves the intervening opportunities (IO) factor [6], which measures the accessibility of an origin to destinations located between the origin and the destination. IO measures the absorbing effects on the originating flow. Gitlesen and Thorsen [7] present an application of the CD concept to commuting modeling in Norway, while accounting for discontinuities in the road network.

Sirmans [8] is among the first to highlight the importance of incorporating various socio-economic determinants into SIM models, including cost, gender, race, income, age, and education and outlines the following points: (1) cost variables are expected to have a negative influence on commuting flows; (2) age variables are expected to have a negative influence on commuting flows, due to increased costs; (3) the higher the education level, the higher the commuting flows; (4) income variables are expected to have a positive influence on commuting flows; (5) race variables (percentages of minorities) are expected to have a negative influence on commuting flows. The results also point out that the determinants of commuting vary across gender.

Sandow [9] shows that women commute shorter distances than men. Sermons and Koppelman [10] also show that the presence of children, the occupation of the male worker in a household, and the last change in the female worker's workplace are important determinants of gender differences in commuting behavior. Prashker et al. [11] investigated various factors that influence an individual's choice of residence location, using a logit model, and showed the importance of area characteristics and commuting distance in selecting a residential location, with significant differences between genders. They also showed that commuting distance becomes less important with increasing income, education, and car ownership. O'Kelly et al. [12], using Irish data, show that commuting trip length varies by occupation and gender. Lin et al. [13] reviewed the impacts of socio-economic factors on commuting.

Another SIM research stream is the relationship between housing prices/locations, employment centers' locations, and commuting. Kim et al. [14] developed an empirical model to show how housing prices, wages, and commuting times affect joint residential and workplace location choice. They show that residents trade lower housing costs for lower wages, and higher housing costs for higher wages. Wu [15] analyzed the impacts of employment and housing development on commuting in the Silicon Valley region and indicated that housing affordability and land-use patterns are important determinants of residential location choices and commuting flows, and that accessibility, local government expenditures, land availability, and ethnic background are important determinants of the spatial distribution of employment. Glenn et al. [16] show that commuting flows result not only from wage differentials and distances, but also from a spatial mismatch between the types of jobs and the categories of workers. Ahrens and Lyons [17], using a gravity model with Irish data, show that rising housing rents lead to longer commutes. Sohn [18] examined how commuting patterns reveal urban structures (where jobs and housings are located), by including locational variables (distance from the city center) for the origins and destinations of commuting flows in a modified gravity model.

### 1.3. Planning Optimization Models and Spatial Interaction Modeling

There have been various research efforts to design normative models for delineating more efficient urban patterns, including convex programming models that embed spatial interaction models within activity-allocation frameworks [19]. Kim [20] further expands this approach by adding alternative transportation systems. Some important works in this line of research include [21–25]. Prastacos' POLIS model maximizes total locational surplus and combines the allocation of employment and a multi-modal transportation system. It is a programming formulation of the Lowry model and incorporates the location of basic employment with data from the San Francisco region.

Barber [26] uses the Lowry model reformulated in matrix form by Garin [27] to develop a linear goal programming model, which determines the basic employment distribution that minimizes deviations from target zonal populations. Using the newly distributed population, the model then estimates zone-specific service and retail employment. Barber [28] develops a linear programming model to allocate the future growth in basic employment to minimize total travel time. Basic employment, and hence basic land-use requirements, is the control variable, whereas service employment and population and their land-use requirements are not. The objective function reflects total travel time for all work- and home-based service trips.

More recent land-use allocation and transportation simulation models include those proposed by Ma et al. [29] and Samani et al. [30]. These models include gravity-based components of the ITLUP (integrated transportation and land-use planning) model, initially developed by Putnam [31]. Some recent studies focus on the relationship between accessibility and the spatial structure of economic activities. Wu et al. [32] evaluate the relationship between the spatial structure of medical resources and the accessibility of medical facilities in different traffic analysis zones and shed light on the potential optimal solutions for the spatial allocation and efficient utilization of medical services. Zhang et al. [33] scrutinize the relationship between the spatial pattern of roadway networks and the quality of business environments. Finally, one should mention the recent stream of multi-objective land-use optimization models, exemplified by [34,35]. However, these models do not incorporate the transportation system or transportation interactions (e.g., commuting), and built-up areas are not differentiated in terms of internal land use. The focus of these models is on sustainability and ecological protection.

### 1.4. Sprawl versus Compact City: Cost Assessment

The effect of urban sprawl on commuting patterns seems to be controversial. Some argue that sprawl has negative consequences for commuting, with longer commutes and congestion [36]. Ewing et al. [37] find no relationship between sprawl and commuting time. Weber and Sultana [38], using 1990 and 2000 Census Transportation Planning Package (CTPP) data for Birmingham, Alabama, differentiate workplace sprawl from residential sprawl, and examine the impact of employment sprawl on the commuting of White and Black workers. Their results show that workplace sprawl reduces commuting distances for those who commute to the sprawling areas and suggest that workers may be able to reduce their commutes as more workplaces relocate to suburban areas. However, workplace sprawl may increase commutes for those who may not be able to adjust their residential location. Variables found to influence commuting length include race, income, mode of transportation, location, population and household density, employment density, homeownership, and time leaving home for work [39].

Dunphy and Fisher [40] report that increasing density decreases the number of daily trips per person, but also assert that high density causes more congestion and pollution. Levinson and Kumar [41] test the influence of residential density on commuting patterns, and conclude that density is an important explanatory variable, with noticeable negative effects on the speed and distance of trips. They use 1980/1990 U.S. Census data and 1990/1991 Nationwide Personal Transportation Survey (NPTS) data. Auto travel time is negatively related to density below a density threshold (10,000 persons per square mile) and positively so above this threshold. O'Toole [42] indicates that there is no consensus about how much compact development reduces total driving and he suggests that the benefits of compact development are often likely to be overstated and its costs understated. The costs of compact development include loss of property rights, reduced geographic mobility, higher housing costs and lower home-ownership rates, higher taxes or reduced urban services to subsidize compact development, increased traffic congestion, and reduced economic mobility. Cambridge Systematics, Inc. [43] reports that congestion would be clearly a major result of a compact development plan and estimates that doubling densities from an average of 3000 people per square mile to an average of 7000 people would reduce per capita driving by less than 15 percent, but would still lead to a 100 percent increase in total vehicle travel miles. Without new road/highway expansion to accommodate this increased demand, there would be a large increase in regional congestion. Stevens [44] conducted a meta-regression analysis of the results of 46 studies to derive a clearer understanding of the influence of compact development on driving, and found a generally small, although significant, reduction in driving.

Air pollution has also been analyzed in the context of the sprawl/compactness debate. Emrath and Liu [45] show that the vehicle miles travelled (VMT) declines as the compactness of subdivisions increases, but with less efficient speeds. However, on balance, $CO_2$ emissions still tend to be lower in more compact developments. Stone [46], using data on 45 large U.S. metropolitan areas, shows that sprawling areas are associated with more ozone exceedances than more spatially compact metropolitan regions. Schindler and Caruso [47] develop a theoretical monocentric urban model to analyze the trade-off between traffic-based pollutant emissions and pollution exposure. Solving the model with parameters drawn from the literature, they find that emissions increase with sprawl and exposure increases with compactness, underscoring the difficulty in assessing compactness net benefits. Finally, Zhang et al. [48] show that there is a significant correlation between urban development patterns and $PM_{2.5}$ concentrations.

### 1.5. Summary and Research Goals

Although spatial interaction modeling of commuting has been the subjects of much urban research, there have not been many planning/optimization applications to residential and employment allocations that incorporate SIM. Both Barber's and Prastacos' models are essentially programming formulations of the Lowry model, where only basic employment is a control variable, with residential allocations automatically derived. In addition, few commuting models have incorporated SIMs and spatial structure factors. It is clear that improved SIMs should incorporate spatial structure effects in order to avoid the misspecification of conventional gravity models and that planning/optimization models should also consider land development costs and congestion/pollution costs, in addition to commuting costs.

Given the above shortcomings, the goals of this research are as follows:

1. Develop a new SIM for commuting trip distribution, based on Tobit regression estimation [49] and including spatial structure variables measured by competing destinations (CD) [5] and intervening opportunity (IO) [6] factors. It is expected that incorporating these factors will better represent commuting behavior and commuting costs.
2. Using the Tobit commuting SIM, develop a new commuting cost minimization model that simultaneously allocates target increments in the population and employment to geographical units across a city or metropolitan area under various scenarios of (a) population and employment densities (land consumption per resident and per employee) and (b) land availability in each geographical unit, as determined by the growth boundaries and environmental constraints. The results of this optimization include a minimum commuting cost surface, which is then to be estimated by polynomial regression, with the densities as independent variables.
3. Combining the polynomial commuting cost model with estimated land development cost models and synthetic congestion cost models, develop a total cost minimization model to determine the optimal densities under various growth boundary scenarios and various parametric assumptions for the congestion cost functions.
4. Use data on a specific U.S. metropolitan area to test the feasibility of the above-methodological goals. This would be a proof-of-concept goal, but is not intended to provide an actual plan for the local authorities of this metropolitan area.

## 2. Data and Methods

### 2.1. Overview of the Study Area

The Fredericksburg Area Metropolitan Planning Organization (FAMPO) was established in 1992, in accordance with Federal regulations, stating that "a metropolitan planning organization (MPO) shall be designated for each urbanized area with a population of more than 50,000 individuals." To be classified as an urbanized area, a central place and any contiguous areas must have a density of at least 1000 persons per square mile. Based on the 1990 Census, an urbanized area consisting of the city of Fredericksburg and portions of Spotsylvania and Stafford counties met this threshold. FAMPO chose to expand its boundaries to include the three jurisdictions in their entirety. The Planning District 16 (George Washington Regional Commission—GWRC) in Virginia deals with FAMPO jurisdictions of two additional rural counties, King George and Caroline. For convenience, the terms "FAMPO region" and "George Washington Region" are used interchangeably in this paper. The location of FAMPO within Virginia is indicated in Figure 1.

The FAMPO region, because of its proximity to the rapidly growing suburbs of the Washington, D.C., metropolitan area (to the north) and the Richmond-Petersburg metropolitan area (to the south), is the fastest growing region in Virginia, with a 2006 population of 310,000 persons, nearly a third more than in 2000. The projections suggest that an additional 250,000 people will be living in FAMPO by 2035. As a result, the region is experiencing the growing pains related to sprawl, traffic safety and congestion. FAMPO's central location and proximity to expanding employment opportunities has encouraged the significant migration of new residents, both to fill local jobs and to seek affordable housing and rural and lower density suburban lifestyles.

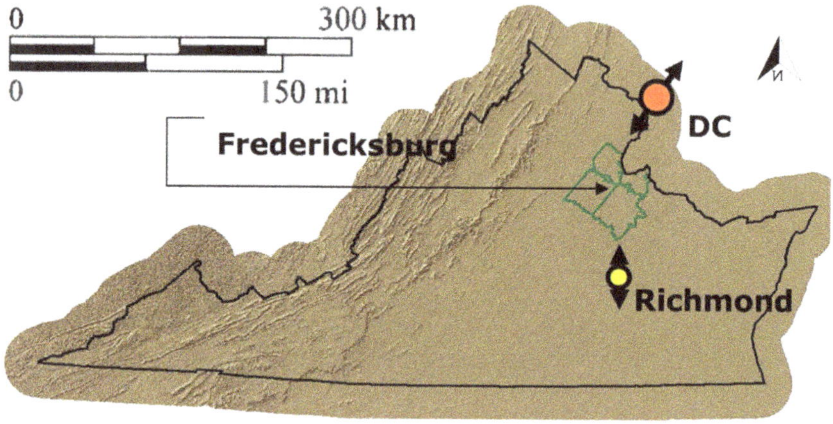


**Figure 1.** State of Virginia and Location of Study Area (FAMPO) (green lines).

### *2.2. Data Sources*

#### 2.2.1. CTPP 2000

Most of the data are drawn from the 2000 Census Transportation Planning Package (CTPP), a set of special tabulations prepared for transportation planners, based on data from the decennial census. CTPP data are downloadable from the following website: CTPP Data—Transportation.org. It is the only Census product that summarizes data by place of work and provides information on travel flows between homes and workplaces. It provides summary tabulations for traffic analysis zones (TAZs) and other small geographic areas. The CTPP is divided into the following three parts: Part 1 includes residence-based data, summarizing worker and household characteristics; Part 2 includes place-of-work data; and Part 3 data includes data on commuting flows from residences to workplaces. The geographical unit of analysis in this research is the TAZ, and there are 188 TAZs in FAMPO. The year 2000 was the last year when the CTPP was produced by the Bureau of the Census, in collaboration with the Bureau of Transportation Statistics, using data from the Long Form survey (16% sampling). This decennial survey was cancelled by the U.S. Congress and replaced by the Annual Community Survey (ACS), with a sampling rate of 3%. Data derived from the ACS are more uncertain, hence the choice of the CTPP 2000 data. It is, however, important to emphasize that the goal of this paper is to present a new planning methodology and to use data to demonstrate its feasibility, and not to produce a plan to be used by FAMPO.

The 2000 population and employment distributions are mapped in Figures 2 and 3. The highest population concentrations are located in the north of Stafford County (A); south of Route 3 and west of the I-95 interstate highway (B); and around the city of Fredericksburg, the center of the region (C). There are three employment clusters, which are as follows: the CBD of Fredericksburg (D); the Quantico military base located in the north of Stafford County (E); and Dahlgren, the site of a U.S. naval base located at the eastern corner of King George County (F).

Population and employment distribution across the FAMPO region are summarized by jurisdiction in Table 1. Two urban counties, Spotsylvania (37.5%) and Stafford (38.4%), function as major residential areas and also, together with the city of Fredericksburg, provide most of the regional employment (31.1% Spotsylvania; 31.8% Stafford; and 23.2% Fredericksburg).

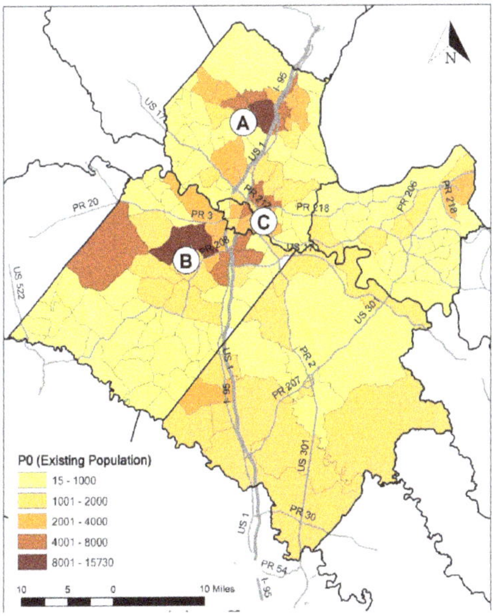


**Figure 2.** FAMPO Population in 2000.

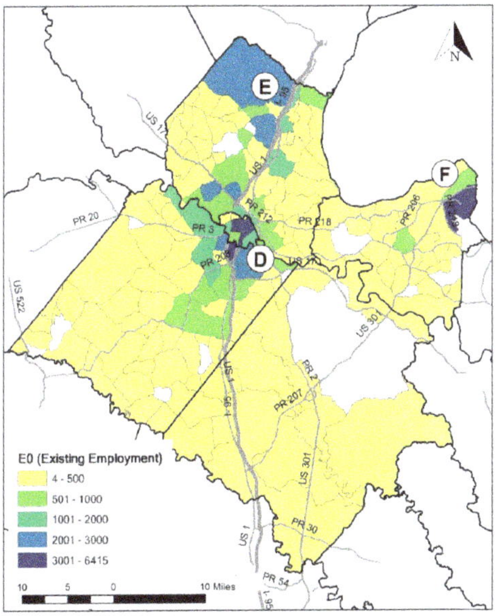


**Figure 3.** FAMPO Employment in 2000.

Zone-to-zone and jurisdiction-to-jurisdiction flows for 2000 are summarized in Table 2. Fredericksburg displays high interactions with the other jurisdictions (18.91%), while Spotsylvania has the highest level of internal flows (23.62%). A high share of the people living in Stafford work there (17.08%).

**Table 1.** FAMPO Region Population and Employment in 2000.

| Jurisdiction | Population | % | Employment | % |
|---|---|---|---|---|
| Caroline | 22,120 | 9.2% | 1945 | 2.3% |
| Fredericksburg | 19,275 | 8.0% | 19,760 | 23.2% |
| King George | 16,805 | 7.0% | 9912 | 11.6% |
| Spotsylvania | 90,405 | 37.5% | 26,521 | 31.1% |
| Stafford | 92,460 | 38.4% | 27,059 | 31.8% |
| Total | 241,065 | 100.0% | 85,197 | 100.0% |

**Table 2.** Commuting Flows (Number of Commuters) in 2000.

| | | | Flow | % |
|---|---|---|---|---|
| FAMPO | | Internal | 7125 | 10.61 |
| | | TAZ-to-TAZ | 60,023 | 89.39 |
| Jurisdiction | Caroline | Internal | 1261 | 1.88 |
| | | Jurisdiction-to-Jurisdiction | 203 | 0.30 |
| | Fredericksburg | Internal | 4056 | 6.04 |
| | | Jurisdiction-to-Jurisdiction | 12,699 | 18.91 |
| | King George | Internal | 4314 | 6.42 |
| | | Jurisdiction-to-Jurisdiction | 3173 | 4.73 |
| | Spotsylvania | Internal | 15,863 | 23.62 |
| | | Jurisdiction-to-Jurisdiction | 6377 | 9.50 |
| | Stafford | Internal | 11,469 | 17.08 |
| | | Jurisdiction-to-Jurisdiction | 7733 | 11.52 |
| Total Flow | | | 67,148 | 100% |

2.2.2. Property Data

Real estate data have been collected from the planning departments of local governments and combined into a regional data set to maintain consistency and comparability. The collected assessment data all apply to 2006. The following variables are available for each record that characterizes a parcel: parcel I.D.; land use; total land value; total building value; total property value; size (acre); jurisdiction; TAZ I.D. Average values per parcel for size, land value, building value, and property value by jurisdiction are provided in Table 3. These data can be obtained from the following Tax Assessor Offices:

Real Estate Taxes | Fredericksburg, VA—Official Website (fredericksburgva.gov);
Stafford County, VA (staffordcountyva.gov);
Assessment Office, Spotsylvania County, VA; Real Estate, Caroline County, VA;
Real Estate, King George County, VA (kinggeorgecountyva.gov).

**Table 3.** Average Parcel Data in 2006.

| Jurisdiction | Residential Property | | | | Workplace Property | | | |
|---|---|---|---|---|---|---|---|---|
| | Average Size (Acre) | Average Property Value (USD) | Average Land Value (USD) | Average Building Value (USD) | Average Size (Acre) | Average Property Value (USD) | Average Land Value (USD) | Average Building Value (USD) |
| Caroline | 2.3658 | 220,462 | 47,598 | 162,618 | 5.3395 | 576,642 | 144,972 | 304,321 |
| Fredericksburg | 0.3525 | 260,831 | 53,914 | 178,740 | 1.2749 | 945,372 | 388,820 | 506,586 |
| King George | 3.2624 | 245,225 | 72,054 | 167,082 | 4.5400 | 517,810 | 196,227 | 271,619 |
| Spotsylvania | 1.3127 | 147,640 | 65,580 | 82,059 | 14.0769 | 3,717,248 | 3,330,981 | 386,267 |
| Stafford | 1.1375 | 372,153 | 102,621 | 269,531 | 3.3182 | 1,241,485 | 479,788 | 761,687 |
| FAMPO | 1.4254 | 247,782 | 77,390 | 167,696 | 8.3923 | 2,309,510 | 1,809,006 | 481,376 |

## 2.3. Variables

### 2.3.1. Dependent Variable

The TAZ-to-TAZ flow table from CTPP 2000 Part 3 includes a large number of zero values. The number of potential commuting connections (all records) is 35,344 (188 × 188), including 188 intra zonal flows. Among these connections, 90.2% (31,875) have zero flow values. Tables 4 and 5 present the descriptive statistics for these flows. Records with zero flows embody useful information, and therefore cannot be discarded in statistical analyses.

**Table 4.** Descriptive Statistics for Non-Zero Flows in 2000.

| Variable | N | Mean | Standard Deviation | Minimum | Maximum |
|---|---|---|---|---|---|
| F: flow | 3469 | 19.35 | 34.17 | 4.00 | 990.00 |
| P: population | 3469 | 2332.27 | 2659.69 | 15.00 | 15,730.00 |
| E: employees | 3469 | 1683.79 | 1687.31 | 4.00 | 6415.00 |
| D: distance | 3469 | 10.51 | 6.97 | 0.42 | 41.07 |

**Table 5.** Descriptive Statistics for Zero Flows in 2000.

| Variable | N | Mean | Standard Deviation | Minimum | Maximum |
|---|---|---|---|---|---|
| F: flow | 31875 | 0 | 0 | 0 | 0 |
| P: population | 31875 | 1167.99 | 1765.41 | 0 | 15,730.00 |
| E: employees | 31875 | 319.25 | 734.31 | 0 | 6415.00 |
| D: distance | 31875 | 18.76 | 9.25 | 0.70 | 50.53 |

### 2.3.2. Independent Variables

The potential independent variables have been directly drawn from CTPP Parts 1 and 2 or are derived from these primary variables. These variables can be grouped as follows: Group A: residence-based variables (CTPP 2000 Part 1); Group B: workplace-based variables (CTPP 2000 Part 2); Group C: impedance variable; Group D: spatial structure variables.

Group A

The larger the population of a residential TAZ (P), the larger the commuting flow expected to originate from it. Gender differences have been shown to affect human behavior; therefore, the share of the male population (P_M_RES) is selected. Unemployment rates are also likely to have negative impacts on flows; therefore, the total (P_UNEMP_RES) and male (P_MUNEMP_RES) unemployment rates are selected. A high percentage of people driving alone to work implies more cars on the roads, and therefore larger commuting flows. High car-pooling rates can be expected to reduce flows; therefore, the following variables are selected:

- Percentage of workers driving alone from their residence (P_DA_RES);
- Percentage of workers carpooling from their residence (P_CP_RES);
- Percentage of male workers driving alone from their residence (P_MDA_RES);
- Percentage of male workers carpooling from their residence (P_MCP_RES).

Age variables, such as the percentages of residents aged 25 to 64 (P_AGE25_64) and of residents aged 65 + (P_AGE65PLUS), are likely to have positive and negative impacts on commuting flows, respectively. Employment occupation may have an effect on flows, although the direction of the effect is a priori unclear. The following variables are selected:

- Percentage of residents in sales or service occupations (P_OCC1_RES);
- Percentage of residents in clerical or administrative support occupations (P_OCC2_RES);
- Percentage of residents in manufacturing, construction, or maintenance occupations (P_OCC3_RES);
- Percentage of residents in professional, managerial, or technical occupations (P_OCC4_RES);
- Percentage of male residents in sales or service occupations (P_MOCC1_RES);

- Percentage of male residents in clerical or administrative support occupations (P_MOCC2_RES);
- Percentage of male residents in manufacturing, construction, or maintenance occupations (P_MOCC3_RES);
- Percentage of male residents in professional, managerial, or technical occupations (P_MOCC4_RES).

It has been argued that low-income minorities experience poor employment opportunities due to underprivileged accessibility. In order to test such effects, and particularly race impacts on travel patterns, the following variables are selected:

- Percentage of Hispanic or Latino residents (P_HIS_RES);
- Percentage of White residents (P_WHT_RES);
- Percentage of Black or African American residents (P_BLK_RES).

A higher share of White residents within a population, probably associated in part with higher income, is likely to produce larger commuting flows. The share of disabled people (P_DIS_RES) is likely to have a negative relationship with flows. Income and earnings are likely to have a positive effect on flows. To test these hypotheses, the following variables are selected:

- Percentage of resident households with an income of USD 75,000 or more in 1999 (P_HINC_RES);
- Median resident household income (MHI_RES);
- Percentage of resident workers with high earnings (USD 50,000+) in 1999 (P_HERN_RES);
- Percentage of resident workers below the poverty level in 1999 (P_POV_RES);

Home ownership is measured by the following variables:

- Percentage of households with self-owned housing (P_OWNSELF_RES);
- Percentage of households with owned housing with and without a mortgage (P_OWN_RES).

High vehicle availability is measured by the percentage of households with 3 or more vehicles (P_3VEH_RES). High housing occupancy rates are measured by the percentage of occupied housing units (P_OCCU_RES). Higher education rates are measured by the percentage of residents with a bachelor's degree or higher (HEDU_RES). These variables are also likely to have positive effects on commuting flows. Descriptive statistics for Group A variables across the 188 TAZs of FAMPO are presented in Table 6.

Group B

The level of employment (EMP) at the workplace (destination) is likely to have a positive impact on attracted flows. It is also likely that higher rates of full-time workers lead to higher flows. This is measured by the percentage of people who worked 40+ hours per week in 1999 at their workplace (P_Full_EMP). Vehicle availability is also likely to increase commuting flows, and it is measured by the percentage of people with 2 or more vehicles at their workplace (P_Veh2plus_EMP). Variables that are likely to have a negative relationship with flows include the following:

- Percentage of employees below the poverty level (P_BlwPov_EMP);
- Mean travel time (MTT_EMP);
- Percentage of workers with low earnings (P_LERN_EMP);
- Percentage of workers that carpool (P_CarPool_EMP).

The percentages of workers with high earnings (P_HEARN_EMP), driving alone (P_DA_EMP), and arriving at the morning peak period (P_AM7_10_EMP) are likely to have positive effects on commuting flows. Type-of-industry variables are likely to have mixed impacts and include the following:

- Percentage of workers in manufacturing (P_Mfg_EMP);
- Percentage of workers in wholesale trade (P_WhlTrd_EMP);

- Percentage of workers in retail trade (P_RetTrd_EMP);
- Percentage of workers in service industries (P_serv_EMP);
- Percentage of workers in public administration (P_Pub_EMP).

**Table 6.** Descriptive Statistics for Group A Variables.

| Variable | N | Mean | Median | Standard Deviation | Minimum | Maximum |
|---|---|---|---|---|---|---|
| P | 188 | 1282.26 | 625.00 | 1903.91 | 0.0 | 15,730.0 |
| P_DA_RES | 188 | 0.7601 | 0.7782 | 0.1498 | 0.0 | 1.000 |
| P_BLK_RES | 188 | 0.1525 | 0.1165 | 0.1421 | 0.0 | 0.674 |
| P_OCC1_RES | 188 | 0.1618 | 0.1627 | 0.0844 | 0.0 | 0.600 |
| P_OCC2_RES | 188 | 0.1858 | 0.1920 | 0.0914 | 0.0 | 0.455 |
| P_OCC3_RES | 188 | 0.2150 | 0.2000 | 0.1140 | 0.0 | 0.580 |
| P_OCC4_RES | 188 | 0.1769 | 0.1700 | 0.1002 | 0.0 | 0.495 |
| P_M_RES | 188 | 0.4981 | 0.4912 | 0.0918 | 0.0 | 0.984 |
| P_UNEMP_RES | 188 | 0.0214 | 0.0165 | 0.0256 | 0.0 | 0.150 |
| P_MUNEMP_RES | 188 | 0.0189 | 0.0000 | 0.0305 | 0.0 | 0.200 |
| P_CP_RES | 188 | 0.1480 | 0.1334 | 0.1031 | 0.0 | 0.700 |
| P_MDA_RES | 188 | 0.7538 | 0.7802 | 0.1866 | 0.0 | 1.000 |
| P_MCP_RES | 188 | 0.1517 | 0.1303 | 0.1305 | 0.0 | 1.000 |
| P_MOCC1_RES | 188 | 0.1232 | 0.1172 | 0.1031 | 0.0 | 0.667 |
| P_MOCC2_RES | 188 | 0.0837 | 0.0817 | 0.0696 | 0.0 | 0.400 |
| P_MOCC3_RES | 188 | 0.3279 | 0.3094 | 0.1731 | 0.0 | 1.000 |
| P_MOCC4_RES | 188 | 0.2047 | 0.2000 | 0.1375 | 0.0 | 0.695 |
| P_HIS_RES | 188 | 0.0185 | 0.0000 | 0.0309 | 0.0 | 0.192 |
| P_WHT_RES | 188 | 0.7876 | 0.8220 | 0.1699 | 0.0 | 1.000 |
| P_DIS_RES | 188 | 0.1429 | 0.1250 | 0.1028 | 0.0 | 0.600 |
| P_HINC_RES | 188 | 0.4013 | 0.4006 | 0.2163 | 0.0 | 1.000 |
| MHI_RES | 188 | 54,720.88 | 52,675.00 | 18,921 | 0 | 109,770 |
| P_HERN_RES | 188 | 0.2071 | 0.2032 | 0.1140 | 0.0 | 0.5052 |
| P_POV_RES | 188 | 0.0218 | 0.0112 | 0.0331 | 0.0 | 0.250 |
| P_OWNSELF_RES | 188 | 0.1881 | 0.1667 | 0.1385 | 0.0 | 1.000 |
| P_OWN_RES | 188 | 0.7895 | 0.8546 | 0.2137 | 0.0 | 1.125 |
| P_3VEH_RES | 188 | 0.3406 | 0.3354 | 0.1614 | 0.0 | 0.769 |
| P_OCCU_RES | 188 | 0.9131 | 0.9486 | 0.1428 | 0.0 | 1.000 |
| HEDU_RES | 188 | 0.0384 | 0.0341 | 0.0381 | 0.0 | 0.388 |

Descriptive statistics for these variables are presented in Table 7.

**Table 7.** Descriptive Statistics for Group B Variables.

| Variable | N | Mean | Median | Standard Deviation | Minimum | Maximum |
|---|---|---|---|---|---|---|
| EMP | 188 | 453.1755 | 75.0000 | 964.605 | 0.0000 | 6415.0 |
| P_Full_EMP | 188 | 0.3321 | 0.3099 | 0.2435 | 0.0000 | 1.0000 |
| P_Veh2Plus_EMP | 188 | 0.7334 | 0.8265 | 0.2895 | 0.0000 | 1.0000 |
| P_BlwPov_EMP | 188 | 0.0364 | 0.0108 | 0.0698 | 0.0000 | 0.6667 |
| MTT_EMP | 188 | 24.712 | 25.000 | 15.369 | 0.0000 | 102.3 |
| P_LERN_EMP | 188 | 0.3681 | 0.3631 | 0.2498 | 0.0000 | 1.0000 |
| P_CarPool_EMP | 188 | 0.0883 | 0.0809 | 0.0923 | 0.0000 | 0.4000 |
| P_Mfg_EMP | 188 | 0.0379 | 0.0000 | 0.0948 | 0.0000 | 0.7500 |
| P_WhlTrd_EMP | 188 | 0.0192 | 0.0000 | 0.0480 | 0.0000 | 0.4000 |
| P_RetTrd_EMP | 188 | 0.0864 | 0.0106 | 0.1413 | 0.0000 | 1.0000 |
| P_serv_EMP | 188 | 0.3687 | 0.3637 | 0.3014 | 0.0000 | 1.0000 |
| P_Pub_EMP | 188 | 0.0327 | 0.0000 | 0.0777 | 0.0000 | 0.4427 |
| P_Finan_EMP | 188 | 0.0561 | 0.0000 | 0.1341 | 0.0000 | 1.0000 |

Group C

Distances have been computed as Euclidian distances (miles) between the zone centroids. Table 8 provides descriptive statistics for the inter-TAZ distances (D).

**Table 8.** Descriptive Statistics for Inter-TAZ Distances.

| Variable | N | Mean | Median | Standard Deviation | Minimum | Maximum |
|---|---|---|---|---|---|---|
| D | 35,344 | 17.947 | 17.380 | 9.379 | 0.420 | 50.530 |

Group D

The intervening opportunity (IO) and the competing destinations (CD) factors are based on employment. The CD factor measures the accessibility of destination $j$ to other destinations in the neighborhood of $j$, while the IO factor measures the accessibility of origin $i$ to other origins in the neighborhood of $i$. The following three different types of IO factors have been proposed by Guldmann [50]: the IO circle, IO sector, and IO corridor. In this research, the IO circle, as originally used by [6], is retained. The neighborhoods for the IO and CD factors of a given TAZ are circles of a 10-mile radius centered on the centroid of the TAZ. A higher IO factor is expected to reduce outgoing commuting flows (negative relationship), while the CD factor could have either negative or positive effects on commuting flows. A positive effect suggests the presence of agglomeration forces at the destination, and a negative one suggests the presence of competition forces. The IO and CD factors are defined mathematically as follows:

$$\text{IO} = \sum_k E_k d_{ik}^{\gamma}, \rightarrow k \neq i \text{ and } k \in \text{ Neighborhood of TAZ } i \tag{2}$$

$$\text{CD} = \sum_l E_l d_{jl}^{\varepsilon}, \rightarrow l \neq j \text{ and } l \in \text{ Neighborhood of TAZ } j \tag{3}$$

In order to illustrate the computation of the IO and CD factors, one must consider Figure 4, with the origin TAZ 5 and destination TAZ 17. The neighborhood TAZs for TAZ 5, within a 10-mile radius, are {2,6,7,8}. Similarly, the neighborhood TAZs for TAZ 17 are {6,15,18}. The factors are computed as follows:

$$\text{IO}_{5,17} = \sum_{\substack{k \neq 5 \\ k \in \{2,6,7,8\}}} E_k d_{5,k}^{\gamma} \tag{4}$$

$$\text{CD}_{5,17} = \sum_{\substack{l \neq 17 \\ l \in \{6,15,18\}}} E_l d_{17,l}^{\varepsilon} \tag{5}$$

### 2.4. Statistical and Optimization Methodology

Spatial interaction models (SIMs) of commuting flows are estimated with, as explanatory variables, the population $P_i$ at the origin $i$, the employment $E_j$ at the destination $j$, several socio-economic variables characterizing either $i$ or $j$ ($X \ldots, Y \ldots$), the distance $d_{ij}$, and competing destinations ($\text{CD}_j$) and intervening opportunity ($\text{IO}_i$) variables that characterize the spatial structure. If $F_{ij}$ is the commuting flow between $i$ and $j$, a general SIM can be calculated as follows:

$$F_{ij} = f(P_i, E_j, d_{ij}, X, \ldots, Y, \ldots, \text{CD}_j, \text{IO}_i) \tag{6}$$

Tobin [49] analyzed household expenditures on durable goods, while taking into account the fact that expenditures cannot be negative. He proposed a regression method applied to data with censored values, which became known as the Tobit model. The basic dependent variable in this research, commuting flow, cannot be negative, and any examination of an actual flow matrix shows that many flows are equal to zero. The Tobit model is a reasonable approach to deal with this problem.

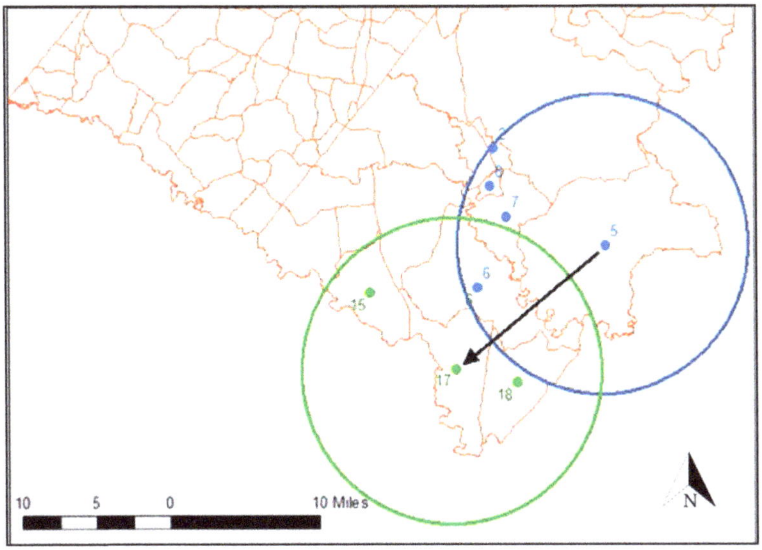


**Figure 4.** IO and CD Neighborhoods for TAZs 5 and 17.

The conceptual basis for using the Tobit model is based on resident worker (RW) utility maximization. One must assume that the RW has a choice among multiple origin–destination (O–D) trips (by virtue of residential and employment location decisions), and that only one O–D trip turns out to be positive, with all the others turning out to be negative at the utility maximum. These negative values are not observed and become zero values in terms of actual trips. The observed flows can then be viewed as the sums of these individual commuting decisions, and the zero flow values represent the censored unobserved negative values. Therefore, the latent variable of the Tobit model captures both positive and negative sums of commuting flows. With standard regression approaches with only positive observed values, the information embodied in zero flow observations is lost. Ordinary least squares (OLS) estimation applied to a truncated sample will be biased and inconsistent. The Tobit model allows the explicit inclusion of zero commuting observations. This is particularly important if there are large volumes of zero observations.

The latent variable In the Tobit is specified as follows:

$$\hat{F}_{ij} = \beta x_{ij} + \varepsilon \tag{7}$$

The actual flow $F_{ij}$ is defined as

$$F_{ij} = \begin{cases} \hat{F}_{ij} = \beta x_{ij} + \varepsilon & \text{if } \hat{F}_{ij} > 0 \\ 0 & \text{if } \hat{F}_{ij} \leq 0 \end{cases} \qquad \varepsilon \sim N(0, \sigma^2) \tag{8}$$

$\hat{F}_{ij}$ is the latent variable that represents the "desired" commuting flows, which can be negative. A Tobit linear commuting flow SIM can be formulated as follows:

$$F_{ij} = a_0 + \sum_i a_i X_i + \sum_j b_j X_j + \sum_{i,j} c_{ij} Z_{ij} \tag{9}$$

where $a_i$, $b_j$, and $c_{ij}$ are the parameters, $X_i$ and $X_j$ are the variables that characterize the origin $i$ and destination $j$, $Z_{ij}$ represents the impedance variables (e.g., distance, time and price) and $F_{ij}$ is the commuting flow.

$P_i^0$ and $E_j^0$ represent the existing (base year) population and employment in zones $i$ and $j$, respectively. The population and employment allocation problem involves the optimal allocation of total population and employment increments, $\Delta P_T$ and $\Delta E_T$, to all zones where land is available for a certain target year. $x_i$ and $z_j$ are the population and employment increments allocated to zones $i$ and $j$, and $\mathbf{Z}$ and $\mathbf{X}$ are the corresponding vectors. In addition, *ULP* is the population density (land area per new resident), *ULE* is the employment density (land area per new employee), and $LAND_i$ is the land available in zone $i$ for new residents and new employees. The parameters ULP and ULE uniformly apply to all the geographical units. However, the model could be easily modified to test for spatially varying density scenarios. If $C_{ij}$ is the fixed unit commuting cost between $i$ and $j$, a general total commuting cost minimization model can be as follows:

$$\text{Minimize } Z = \sum_{i,j} C_{ij} F_{ij} \tag{10}$$

This is subject to

$$\sum x_i = \Delta P_T \tag{11}$$

$$\sum z_j = \Delta E_T \tag{12}$$

$$F_{ij} \geq \hat{F}_{ij}(P_i^0 + x_i, E_j^0 + z_j, d_{ij}, X \ldots Y \ldots, \mathrm{CD}_j(\mathbf{Z}), \mathrm{IO}_i(\mathbf{X})) \tag{13}$$

$$ULP.xi + ULE.zi \leq LAND_i \tag{14}$$

$$F_{ij} \geq 0 \; x_i \geq 0 \; z_j \geq 0 \tag{15}$$

The objective (10) is to find the minimum total commuting cost. Constraint (11) guarantees that the sum of all increments in the population equals the total population increment, and constraint (12) ensures the same for employment. The Tobit constraint (13) defines the commuting flow between $i$ and $j$, and constraint (15) guarantees that $F_{ij} = 0$ when the right-hand side of constraint (13) is negative. ULP and ULE are the given parameters in the optimization model, but can be varied in the context of scenario analyses. Constraint (14) simply states that the land to be used for new residents and employees in zone I cannot exceed the land available. In the specific case of the FAMPO region with 188 zones (TAZs), this model has 36,097 variables and 35,911 constraints.

However, minimizing the total commuting costs cannot be the sole objective of the model. Other costs need to be considered. For instance, compact development is assumed to reduce pollution emissions by reducing driving and housing a higher percentage of people in multi-family and mixed-use developments at more central locations, reducing utility costs and utilizing more transit and fewer highways. Land development costs would be smaller in such developments. However, some argue that compact developments can be more costly than often estimated [42], because compact cities may increase emissions by increasing roadway congestion. These costs are often neglected. In addition, compact development may entail higher housing costs and lower homeownership rates, reduced geographic and economic mobility, higher taxes, reduced urban services, higher consumer costs, etc. A more general cost function can then be stated as follows:

$$\begin{aligned} TOTALCOST &= commutingcost(TCOM) \\ &+ landdevelopmentcost(LDC) \\ &+ congestioncost(TCON) \end{aligned} \tag{16}$$

All the costs in (16) were formulated in terms of both residential and employment densities, or, alternatively, in terms of unit land consumption per resident (ULP) and per employee (ULE). The higher the ULP and ULE, the lower the corresponding densities.

Commuting costs and land development costs increase, and congestion costs decrease with ULP and ULE. These cost curves, and the total cost curve are illustrated in Figure 5. The LDC would include all annualized land/building costs for both employment and the population. The TCOM represents the annual commuting cost. The TCON represents the congestion costs for both the population and employment. Within the given ranges of ULP and ULE, the total cost function (TCD) will point to the optimal ULP and ULE that minimize the total cost.

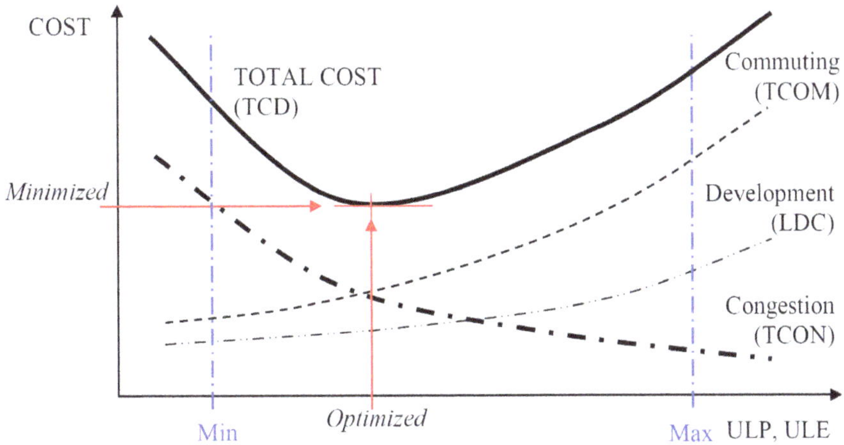


**Figure 5.** Commuting Cost, Land Development Cost, Congestion Cost, and Optimal Density.

## 3. Results

### *3.1. Tobit Regression*

The estimation of the final Tobit model with SAS$^{TM}$ procedure QLIM (qualitative and limited dependent variable models) is the outcome of a multi-step exploratory process. The first estimated model (Model 1) involved only three independent variables that appear in most gravity models, which are as follows: population (P), employment (EMP) and distance (D). All the coefficients turned out to be highly significant ($p < 0.0001$), with the expected positive sign for P and EMP, and the expected negative sign for D, and with $R^2 = 0.297$ and Pseudo-$R^2 = 0.042$. The Pseudo-$R^2$ is defined as follows:

$$R^2_{MF} = \frac{LRT}{LRT*} = \frac{(l_M - l_0)}{(l_{MAX} - l_0)} = 1 - \frac{l_M}{l_0}, \tag{17}$$

where *LRT* is the likelihood ratio statistic; $l_M$ is the log-likelihood value of the model; $l_0$ is the log-likelihood value if the non-intercept coefficients are restricted to zero; $l_{MAX}$ is the maximum possible likelihood. One can refer to [51] for a discussion on the Pseudo-$R^2$ for limited dependent variable models.

The second step was to add the spatial structure variables IO and CD to Model 1. Since the IO and CD factors involve additional parameters (exponents of distance) that need to be estimated, a grid sensitivity analysis was conducted to search for the optimal parameter set. For both $\gamma$ (IO factor) and $\varepsilon$ (CD factor), the range ($-2.0$, 0) was selected, as typical in the literature, with a 0.1 increment. Hence, 400 combinations of $\gamma$ and $\varepsilon$ values were evaluated in this sensitivity analysis. The combination of $\gamma = -0.1$ and $\varepsilon = -0.3$ yielded the highest log-likelihood, as well as the highest Pseudo-$R^2$. All the five variables of this model (Model 2) turned out to be very significant ($p < 0.0001$), with a negative sign for IO (as expected) and a positive sign for CD, pointing to agglomeration effects at the destination.

The third step was to add socio-economic variables, as listed in Groups A and B, to Model 2. The following variables improved in Model 2 (significance and sign; overall performance) and were retained in Model 3: P_DA_RES, P_BLK_RES, P_OCC1_RES,

P_OCC2_RES, P_OCC3_RES, P_OCC4_RES, P_Mfg_EMP, P_WhlTrd_EMP, P_RetTrd_EMP, P_Finan_EMP, P_serv_EMP, and P_Pub_EMP. However, none of the INCOME, GENDER and AGE-related variables turned out to be significant. For Model 3, $R^2$ = 0.355, and Pseudo-$R^2$ = 0.121.

The final model (Model 4) expands on Model 3 by introducing quadratic terms. The following significant quadratic terms were selected:

$$E2 = EMP * EMP \tag{18}$$

$$P2 = P * P \tag{19}$$

$$POPEMP = P * EMP \tag{20}$$

$$EMPCD = EMP * CD \tag{21}$$

Table 9 represents the parameter estimates of Model 4, with $R^2$ = 0.466, and Pseudo-$R^2$ = 0.239. Model 4 has stronger performance criteria than Model 3, pointing to the non-linear relationship between commuting flows and the variables P, EMP, D, and CD. The more workers that drive alone to their workplace (P_DA_RES), the higher the flow. The magnitude of this variable coefficient is relatively high (33.11). The share of Black citizens within a population (P_BLK_RES) also has a positive impact on flows. The Black population in the region is a highly educated and affluent middle-class community, hence its mobility and likely positive impact on flows. The occupation and industry variables display the expected signs. The more residents with sales or service (P_OCC1_RES) or clerical or administration (P_OCC2_RES) occupations, the larger the commuting flows. These occupations have stronger impacts than the other two occupations. The percentages of workers in manufacturing, wholesale trade, retail trade, fire, service, and public administration occupations all have positive impacts on commuting flows. Wholesale trade (P_WhlTrd_EMP) and public administration (P_Pub_EMP) have the largest coefficients, 90.08 and 89.72, respectively, followed by manufacturing (55.57), retail trade (45.22), finance industries (36.17), and service occupations (22.13). This result is consistent with the existence of large regional distribution centers, such as CVS and UPS, as well as government and military workers.

### *3.2. Minimizing Commuting Costs in the Allocation of Population and Employment*

#### 3.2.1. Scenarios

The control totals for population and employment for the horizon year 2035 were obtained from the Virginia Employment Commission (VEC) and the GWRC. The model allocates the total regional increments in the population and employment to the 188 TAZs of FAMPO, while assuming that the existing population and employment levels remain at their current locations. The existing and target population and employment levels are presented in Table 10.

Vacant land is made available for any future housing and employment development, except in physically, environmentally, and historically sensitive lands. Developed/developable land is delineated using a geographical information system (GIS) and is classified into the following five categories: existing residential developed land, existing commercial developed land, existing industrial developed land, undevelopable land, and vacant developable land. Vacant developable land is selected for possible further development expansion.

Each jurisdiction in FAMPO has primary settlement/growth area boundaries in its comprehensive long-range plan. These boundaries are not exactly the same as the urban growth boundaries (UGB) that control urban expansion into farm and forest lands in Portland, Oregon. One can also refer to [52] for information on rigid vs. flexible boundaries in the context of urban growth/development boundaries and an application for delineation. They are not used to control growth, but rather to define long-term city boundaries. However, this land-use control tool functions rather well in managing urban growth in the region. It has been observed that new housing developments built since 2000 have taken place around and within these boundaries. The following five land development scenarios were initially considered: L1—within UGB (35,535 acres); L2—within UGB + 1.0 mile (171,241 acres); L3—within UGB + 2.0 mile (252,508 acres); L4—within UGB + 3.0 mile (305,921 acres); L5: all developable land (503,412 acres). These scenarios are illustrated in Figure 6. After initial exploratory modeling, the scenarios L1 and L5 were discarded as too restrictive and too unconstrained, respectively. The Scenarios L2 and L4 were then retained as sufficiently contrasted scenarios to provide insights into the impact of land availability.

**Table 9.** Tobit Parameter Estimation of Model 4.

| Parameter | Estimate | Standard Error | t-Value | Approx Pr > \|t\| |
|---|---|---|---|---|
| Intercept | −80.267754 | 4.532873 | −17.71 | <0.0001 |
| P | 0.011133 | 0.000426 | 26.13 | <0.0001 |
| EMP | 0.027680 | 0.001218 | 22.73 | <0.0001 |
| D | −3.966485 | 0.158067 | −25.09 | <0.0001 |
| IO_E | −0.000324 | 0.000024329 | −13.33 | <0.0001 |
| CD_E | 0.000160 | 0.000035343 | 4.53 | <0.0001 |
| P_DA_RES | 28.859800 | 3.679491 | 7.84 | <0.0001 |
| P_BLK_RES | 9.602796 | 3.023634 | 3.18 | 0.0015 |
| P_OCC1_RES | 23.840650 | 6.515631 | 3.66 | 0.0003 |
| P_OCC2_RES | 12.881209 | 5.595655 | 2.30 | 0.0213 |
| P_OCC3_RES | 16.725051 | 4.991596 | 3.35 | 0.0008 |
| P_OCC4_RES | 17.346662 | 5.671378 | 3.06 | 0.0022 |
| P_Mfg_EMP | 36.628397 | 4.083194 | 8.97 | <0.0001 |
| P_WhlTrd_EMP | 53.338335 | 7.888539 | 6.76 | <0.0001 |
| P_RetTrd_EMP | 22.840645 | 3.075172 | 7.43 | <0.0001 |
| P_Pub_EMP | 43.681025 | 5.569463 | 7.84 | <0.0001 |
| P_Serv_EMP | 19.602157 | 1.794940 | 10.92 | <0.0001 |
| P_Finan_EMP | 21.147896 | 2.754040 | 7.68 | <0.0001 |
| P2 | −0.000000662 | $3.1856992 \times 10^{-8}$ | −20.78 | <0.0001 |
| E2 | −0.000002661 | 0.000000158 | −16.87 | <0.0001 |
| D2 | 0.048005 | 0.004843 | 9.91 | <0.0001 |
| POPEMP | 0.000002331 | 0.000000113 | 20.68 | <0.0001 |
| EMPCD | −0.000000109 | $1.807805 \times 10^{-8}$ | −6.03 | <0.0001 |
| *R*-square | 0.4657 | | | |
| Pseudo-$R^2$ | 0.2394 | | | |
| Log-likelihood | −20,466 | | | |

**Table 10.** Target Population and Employment.

| Jurisdiction | | 2035 Employment | 2035 Population |
|---|---|---|---|
| Caroline County | | 14,216 | 47,007 |
| Fredericksburg | | 43,679 | 29,852 |
| King George County | | 17,821 | 40,744 |
| Spotsylvania County | | 62,551 | 236,885 |
| Stafford County | | 69,574 | 238,208 |
| Total GWRC (PD 16) | | 207,841 | 592,696 |
| 2000 (Existing) | | Increments | |
| Employment | Population | ΔE | ΔP |
| 85,197 | 241,065 | 122,644 | 351,631 |

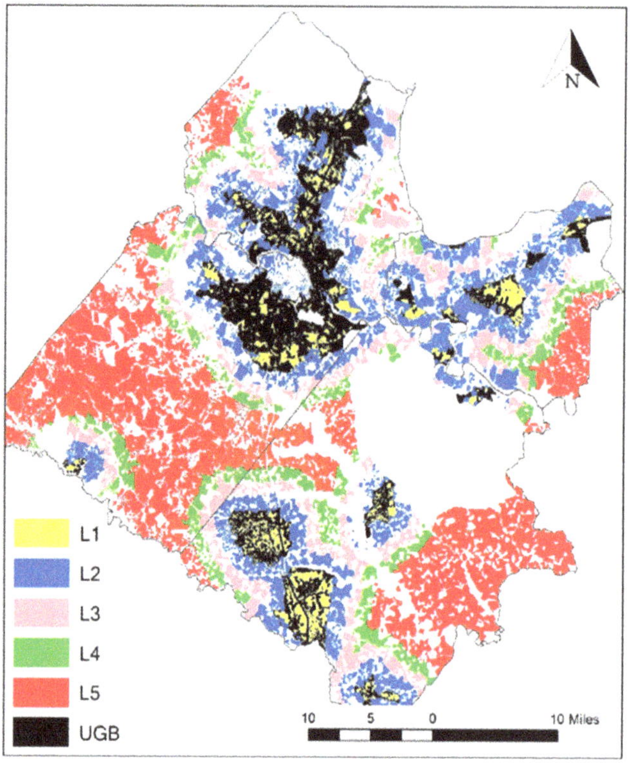


**Figure 6.** Land Development Scenarios with Buffers around UGBs.

While rural areas have higher average land consumption per resident (ULP) and employee (ULE), in the range of (2~3) acres, urban areas are characterized by denser developments in the range of (0.1~0.2) acres. All the areas of the FAMPO parcels of land currently occupied by the population and employment have been summed up at the TAZ level. Using the existing (2000) TAZ population and employment data, the following region-wide average densities have been derived: ULP = 0.429 acres and ULE = 0.223 acres. Using the increments $\Delta P$ and $\Delta E$ for population and employment, and the above density values, the total amount of land required by 2035 would be 178,199 acres. It is assumed that, in the future (target year 2035), the average ULP and ULE values will be smaller than the current values, and the following $9 \times 9$ grid of values is considered:

ULP: (0.10–0.50) by 0.05 increments
ULE: (0.05–0.25) by 0.025 increments

Except for the variables POP and EMP, which are endogenous to the optimization model, all the other variables of the Tobit model are also assumed to remain constant over time. It is also assumed that the share variables (P_DA_RES, P_BLK_RES, P_OCC1_RES, P_OCC2_RES, P_OCC3_RES, P_OCC4_RES, P_Mfg_EMP, P_WhlTrd_EMP, P_RetTrd_EMP, P_Pub_EMP, P_Serv_EMP, and P_Finan_EMP) remain constant and equal to their current values.

### 3.2.2. Model Formulation

The optimization model presented in Section 2.4 (Equations (10)–(15)) is adjusted as follows. First, the commuting travel costs are proxied by the total vehicle miles traveled

(VMT). If $D_{ij}$ is the distance between TAZs $i$ and $j$, the objective function to be minimized is as follows:

$$Z = \sum_i \sum_j D_{ij} F_{ij} \tag{22}$$

Second, Equation (13) is reformulated as Equation (23) by using the Tobit parameters presented in Table 9, in which $P_i^0$ is the existing 2000 population in TAZ $i$, $E_j^0$ is the existing 2000 employment level in TAZ $j$, $(P_i^0 + x_i)^2$ is the square of the final population (existing plus increment) in TAZ $i$; $(E_j^0 + z_j)^2$ is the square of the final employment level (existing plus increment) in TAZ $j$.

The model is a non-linear program because of the squares and products of the decision variables. The optimization is implemented with the general algebraic modeling system (GAMS) and with the non-linear solver CONOPT.

However, because of its non-linearity and non-convexity, this model only provides local optima, and a global optimum search is infeasible, due to the model size (36,097 variables and 35,911 constraints). In other words, the optimal solution obtained for any set of values for ULP and ULE may vary depending on the initial starting point of the optimization algorithm. Thus, the only way to deal with this problem is to repeatedly solve the model with the CONOPT solver, and to select the solution with the smallest objective function (OF) value. The results presented below are the outcomes of this process.

$$\begin{aligned}
F_{ij} \geq\; & -80.267754 + 0.011133 \cdot (P_i^0 + x_i) + 0.027680 \cdot (E_j^0 + z_j) - 3.966485 D_{ij} \\
& -0.000324 \cdot \left( \sum_{\substack{k \neq i \\ k \in Neighborhood_of_i}} (E_k^0 + z_k) D_{ik}^{-0.3} \right) + 0.000160 \cdot \left( \sum_{\substack{l \neq j \\ l \in Neighborhood_of_j}} (E_l^0 + z_l) D_{jl}^{-0.1} \right) \\
& +28.859800 \cdot (P_DA_RES)_i + 9.602796 \cdot (P_BLK_RES)_i \\
& +23.840650 \cdot (P_OCC1_RES)_i + 12.881209 \cdot (P_OCC2_RES)_i \\
& +16.725051 \cdot (P_OCC3_RES)_i + 17.346662 \cdot (P_OCC4_RES)_i \\
& +36.628397 \cdot (P_MFG_EMP)_j + 53.338335 \cdot (P_WhlTrd_EMP)_j \\
& +22.840645 \cdot (P_RetTrd_EMP)_j + 43.681025 \cdot (P_Pub_EMP)_j \\
& +19.602157 \cdot (P_Serv_EMP)_j + 21.147896 \cdot (P_Finan_EMP)_j \\
& -0.000000662 \cdot (P_i^0 + x_i)^2 - 0.000002661 \cdot (E_j^0 + z_j)^2 + 0.048005 D_{ij}^2 \\
& +0.000002331 \cdot (P_i^0 + x_i) \cdot (E_j^0 + z_j) - 0.000000109 \cdot (E_j^0 + z_j) \cdot \left( \sum_{\substack{l \neq j \\ l \in Neighborhood_of_j}} (E_l^0 + z_l) D_{jl}^{-0.1} \right)
\end{aligned} \tag{23}$$

### 3.2.3. Optimization Results

In order to contrast the VMT minimization results between high- and low-density scenarios, two sets of ULP and ULE pairs, including (1) high density: ULP = 0.10; ULE = 0.050 and (2) low density: ULP = 0.40; ULE = 0.200, were selected among the 9 × 9 grid of values and combined with the L2 and L4 land development scenarios.

The optimal TAZ-to-TAZ flows have been summarized into jurisdiction-to-jurisdiction flows, as presented in Table 11. The optimal values of the OF, total flows and average commuting distance are summarized in Table 12. The OF and the total flow decrease with more land available and a higher density, which indicates that a land-use policy that confines new developments within UGBs would significantly increase regional VMT. This large flow increase leads to increased congestion and air pollution costs, which are addressed in Section 3.3.4.

**Table 11.** Optimal Commuting Flows by Jurisdiction.

| **Scenario L2** | | | | To | | |
|---|---|---|---|---|---|---|
| **ULP = 0.400** | **ULE = 0.200** | **CR** | **FR** | **KG** | **SF** | **SP** |
| From | CR | 1000 | 3830 | 4849 | 7395 | 4223 |
| | FR | 0 | 961 | 33 | 528 | 802 |
| | KG | 9 | 2205 | 5531 | 2891 | 2369 |
| | SF | 0 | 4481 | 1575 | 10,991 | 3796 |
| | SP | 40 | 5855 | 1545 | 6215 | 10,695 |
| Scenario L2 | | | | To | | |
| ULP = 0.100 | ULE = 0.050 | CR | FR | KG | SF | SP |
| From | CR | 29 | 1582 | 382 | 967 | 1463 |
| | FR | 0 | 812 | 0 | 341 | 562 |
| | KG | 0 | 350 | 92 | 20 | 282 |
| | SF | 0 | 2016 | 0 | 3251 | 1322 |
| | SP | 0 | 2848 | 0 | 1417 | 3936 |
| Scenario L4 | | | | To | | |
| ULP = 0.400 | ULE = 0.200 | CR | FR | KG | SF | SP |
| From | CR | 136 | 288 | 1349 | 2304 | 1415 |
| | FR | 0 | 780 | 48 | 472 | 592 |
| | KG | 0 | 813 | 1057 | 1405 | 919 |
| | SF | 0 | 3836 | 1146 | 9045 | 2751 |
| | SP | 0 | 4886 | 1544 | 5245 | 8266 |
| Scenario L4 | | | | To | | |
| ULP = 0.100 | ULE = 0.050 | CR | FR | KG | SF | SP |
| From | CR | 2 | 7 | 0 | 0 | 23 |
| | FR | 0 | 763 | 0 | 357 | 549 |
| | KG | 0 | 0 | 17 | 0 | 0 |
| | SF | 0 | 1787 | 0 | 4030 | 1170 |
| | SP | 16 | 4038 | 0 | 2075 | 3837 |

CR: Caroline; FR: Fredericksburg; KG: King George; SF: Stafford; SP: Spotsylvania.

**Table 12.** Optimal Objective Function, Total Flow, and Average Commuting Distance.

| **Land Scenario** | | **Density Scenario** | |
|---|---|---|---|
| | | **ULP = 0.400 ULE = 0.200** | **ULP = 0.100 ULE = 0.050** |
| L2 | Objective function | 838,777 | 113,647 |
| | Total flows | 81,819 | 21,671 |
| | Average commuting distance | 10.25 | 5.24 |
| L4 | Objective function | 473,283 | 106,347 |
| | Total flows | 48,294 | 18,671 |
| | Average commuting distance | 9.80 | 5.70 |

Figure 7 displays the optimal TAZ-to-TAZ flows, while Figures 8 and 9 present the optimal allocations of the incremental population and employment, respectively. As more land becomes available, the incremental population and employment tend to move away towards rural areas. In addition, as density increases, more people and jobs are located closer to the urban cores. An interesting observation is that, for a given density scenario, when less land is available (e.g., L2 rather than L4), this requires higher levels of commuting flows. The L2 UGB requirement leads to more spatial separation of population and employment than under less restrictive land-use controls (L4). Thus, a tight UGB strategy does not appear to be the best way to reduce VMT, and the resulting congestion and air pollution. Another interesting observation is that moving from low to high density and from L2 to L4 leads to the southern TAZs being more and more disconnected in terms of flows, with most of the inter-TAZ and inter-jurisdictional flows concentrated in the northern TAZs. Compact development scenarios evidently result in the highest volumes of commuting flows. This could be a justification for further transit development and use. Most of the observed commuting flows involve private car transportation, and other modes of transportation are not explicitly considered, because their shares are currently very small.

| Total Flows (Average Commuting Distance) | | | Density Scenario | |
|---|---|---|---|---|
| | | | ULP=0.400 ULE=0.200 | ULP=0.100 ULE=0.050 |
| | | | *Low* | *High* |
| Land Scenario | L2 | *UGB 1 Mile* | 81,819 (10.25) | 21,671 (5.24) |
| | L4 | *UGB 3 Mile* | 48,294 (9.80) | 18,671 (5.70) |

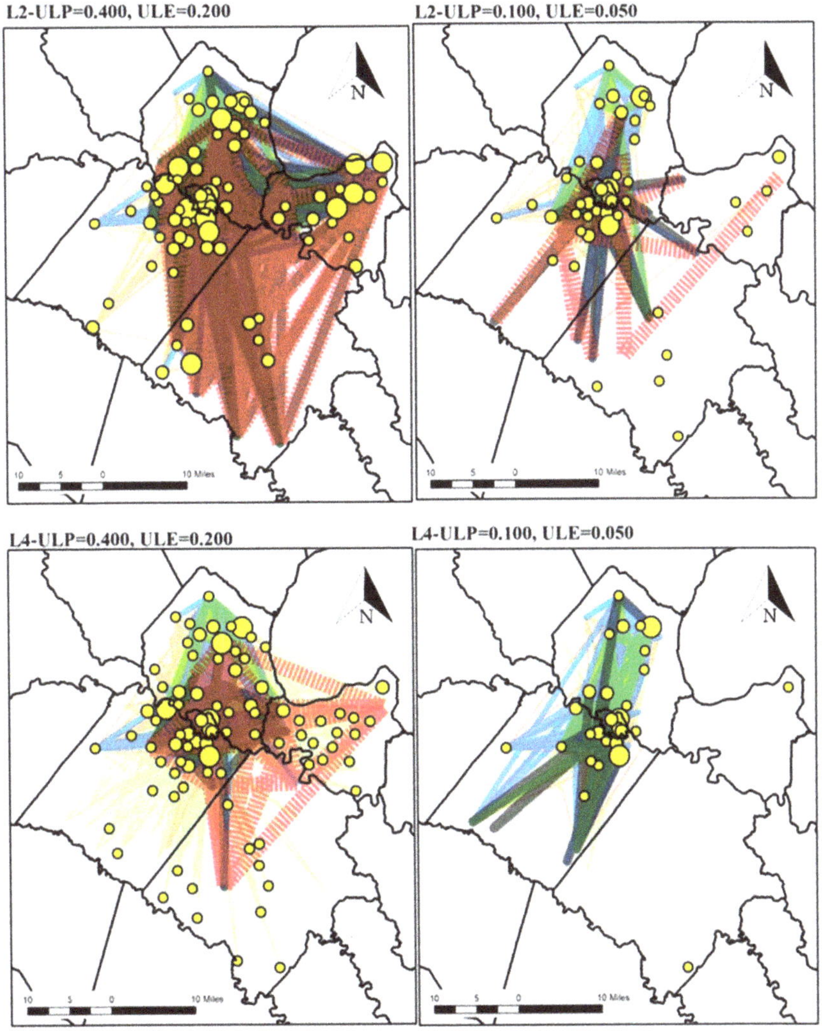


**Figure 7.** Optimal Zone–to–Zone Flows.

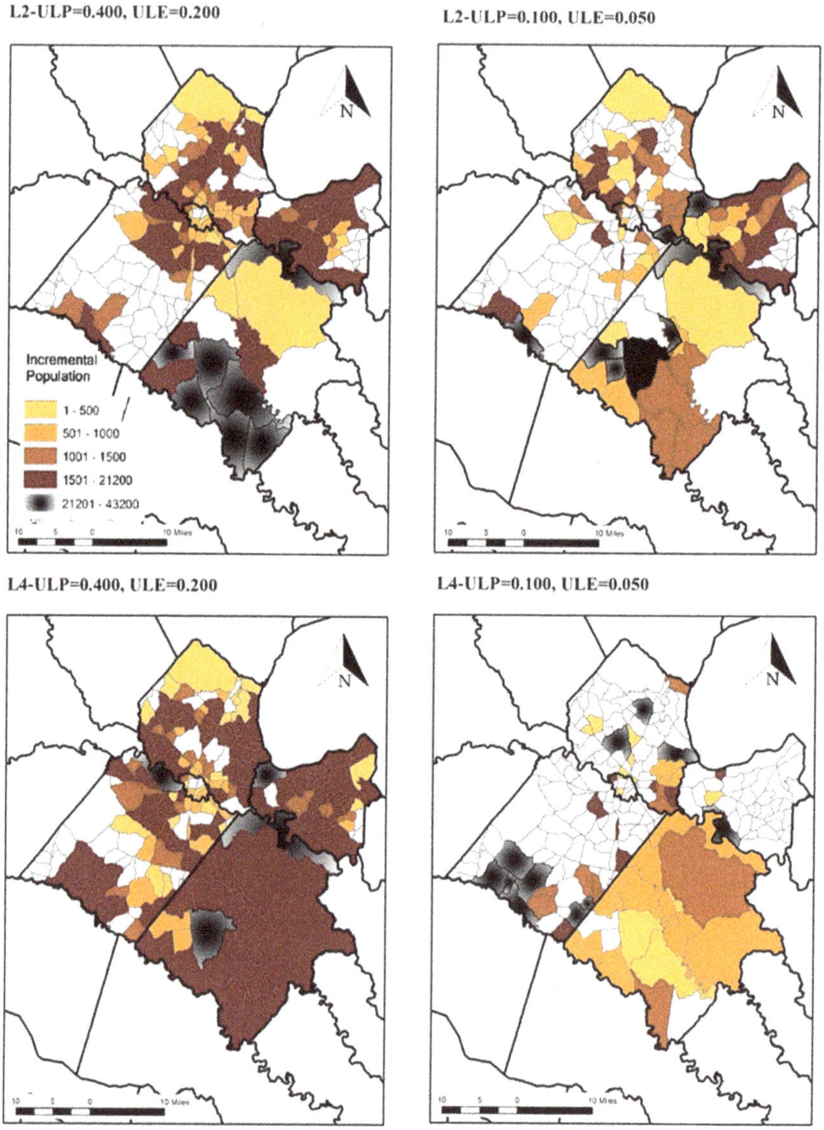


**Figure 8.** Optimal Allocation of Incremental Population.

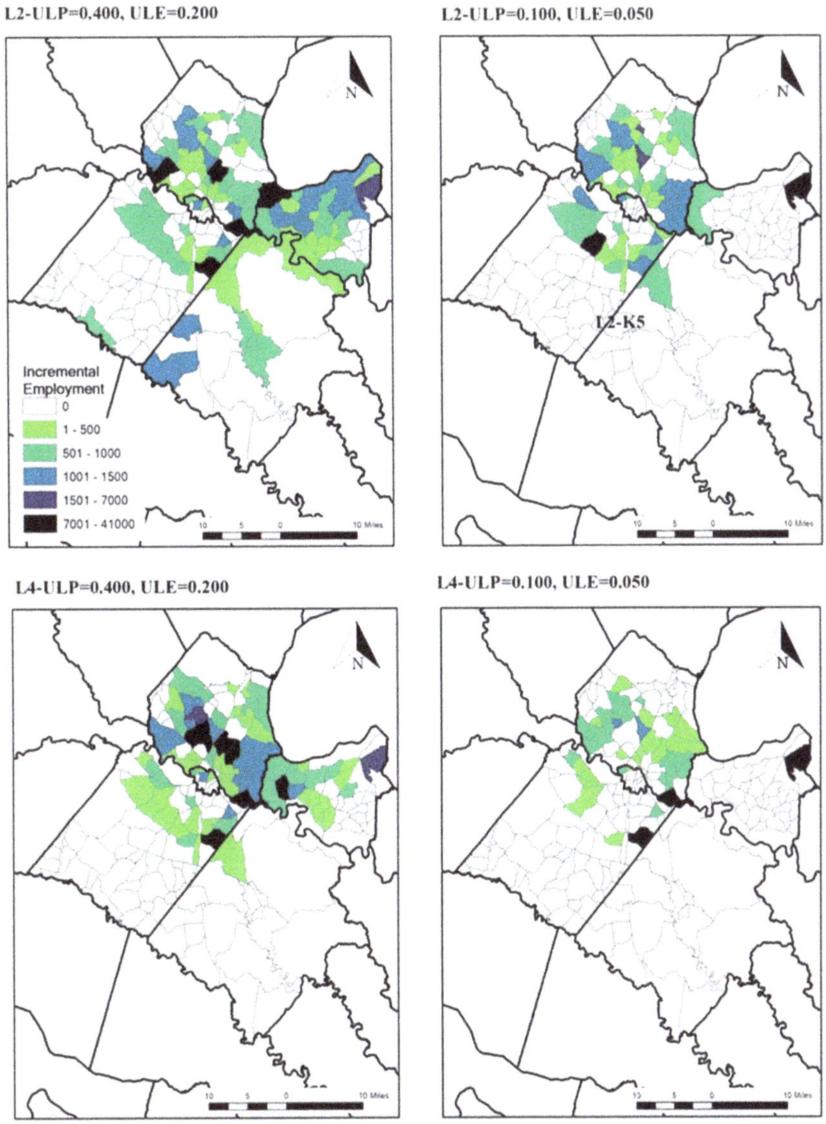


**Figure 9.** Optimal Allocation of Incremental Employment.

## 3.3. *Minimizing All Costs in the Allocation of Population and Employment*

### 3.3.1. Overview of Costs

The previous VMT-minimizing results demonstrate that density constraints are critical in determining the distribution of populations and employment, and therefore must be carefully considered. For any given land development strategy, the highest possible densities allow for minimizing commuting costs. However, commuting costs do not represent all urban and regional costs, which also include land development costs and congestion/pollution costs. The purpose of this section is to develop and optimize such an expanded cost function (TDC), taken as the sum of the commuting costs (TCOM), the total land development costs (LDC), and the total congestion costs (TCON).

### 3.3.2. Estimation of the Commuting Cost Surface

In Section 3.2.3, the total commuting cost (TCOM) was minimized under two density scenarios. However, it is now necessary to consider the variations in TCOM over a wider range of density values. A 9 × 9 grid analysis (ULP = 0.10~0.50 by 0.05 increments; ULE = 0.050~0.250 by 0.025 increments) of density scenarios is used to build up the relationship between the commuting cost and ULP and ULE. The general approach is to solve the VMT minimization model over these 81 density parameters (ULP, ULE) for land development strategies L2 and L4, and then to approximate the resulting cost surfaces through polynomial regression analysis. The normalized minimum VMT values for each pair of ULP and ULE are presented in Table 13. Normalized values provide a clearer picture of the variations in the minimum VMT. For Scenario L2, the maximum value (100) represents VMT = 930,741; for Scenario L4, the corresponding value is VMT = 560,790. VMT values under L2 are larger than those under L4. This is reasonable because L4 is less constraining (provides more location opportunities), which should lead to a lower VMT for any given set of (ULP, ULE) values.

**Table 13.** Minimum Commuting Costs for a 9 × 9 grid of ULP and ULE values.

| | | **Normalized** | | | | | | | | |
|---|---|---|---|---|---|---|---|---|---|---|
| Land Scenario L2 | | | | | | ULE | | | | |
| | | 0.050 | 0.075 | 0.100 | 0.125 | 0.150 | 0.175 | 0.200 | 0.225 | 0.250 |
| | 0.10 | 12.21 | 15.05 | 17.09 | 17.17 | 17.23 | 19.56 | 20.26 | 21.67 | 22.12 |
| | 0.15 | 15.06 | 16.24 | 17.56 | 18.32 | 20.61 | 22.26 | 22.99 | 23.18 | 23.44 |
| | 0.20 | 26.31 | 27.15 | 28.03 | 28.16 | 28.42 | 28.92 | 29.06 | 29.15 | 29.55 |
| | 0.25 | 33.14 | 39.99 | 40.34 | 41.25 | 41.32 | 42.06 | 44.21 | 45.06 | 45.30 |
| ULP | 0.30 | 40.62 | 46.95 | 53.61 | 57.55 | 59.47 | 61.18 | 62.55 | 64.59 | 66.17 |
| | 0.35 | 48.17 | 60.51 | 66.65 | 71.39 | 79.09 | 81.80 | 85.00 | 87.99 | 91.40 |
| | 0.40 | 53.13 | 64.22 | 70.87 | 78.52 | 80.79 | 85.27 | 90.12 | 100.00 | |
| | 0.45 | 63.43 | 76.50 | 80.40 | | | | | | |
| | 0.50 | | | | | | | | | |
| | | **Normalized** | | | | | | | | |
| Land Scenario L4 | | | | | | ULE | | | | |
| | | 0.050 | 0.075 | 0.100 | 0.125 | 0.150 | 0.175 | 0.200 | 0.225 | 0.250 |
| | 0.10 | 18.96 | 20.82 | 21.69 | 25.55 | 26.88 | 27.71 | 28.01 | 28.72 | 29.08 |
| | 0.15 | 23.10 | 23.41 | 26.63 | 29.99 | 32.82 | 36.03 | 36.15 | 36.26 | 36.39 |
| | 0.20 | 31.63 | 35.46 | 38.57 | 39.61 | 39.90 | 40.13 | 40.33 | 40.50 | 40.73 |
| | 0.25 | 39.40 | 40.17 | 41.09 | 41.72 | 42.26 | 42.71 | 43.14 | 43.56 | 43.97 |
| ULP | 0.30 | 44.12 | 46.64 | 51.31 | 53.13 | 53.93 | 54.71 | 55.55 | 56.36 | 57.20 |
| | 0.35 | 48.18 | 51.69 | 56.12 | 65.54 | 69.32 | 71.19 | 71.22 | 72.31 | 73.40 |
| | 0.40 | 49.13 | 56.64 | 64.89 | 71.43 | 80.24 | 83.10 | 84.40 | 85.64 | 87.05 |
| | 0.45 | 51.15 | 59.24 | 65.18 | 74.43 | 81.43 | 83.41 | 85.23 | 86.67 | 88.28 |
| | 0.50 | 55.52 | 63.55 | 72.26 | 80.09 | 88.76 | 92.80 | 95.45 | 97.63 | 100.00 |

Note: Red cells correspond to unfeasible solutions.

The normalized commuting cost surfaces are illustrated in Figure 10. The surface for L2 displays sudden drops at certain values of ULP and ULE, due to infeasibility. The VMT values increase as the ULP and ULE increase. As expected, the VMT is minimized when the ULP and ULE are the smallest. The surfaces suggest that the VMT is more sensitive to the ULP than to ULE. The relationship between VMT and ULP and ULE for each land development strategy is estimated using a third-order cubic polynomial regression analysis and the results are presented in Table 14.

As the models minimize the total commuter mile flows (Equation (22)), it is necessary to convert this quantity into the corresponding annual commuting cost, with the following equation:

$$TCOM = CPM \times ND \times \sum_i \sum_j d_{ij} F_{ij} \tag{24}$$

where *CPM* is the average commuting cost per person mile, and *ND* is the number of commuting days per year. *CPM* is estimated at USD 0.202 by dividing the 2005 purchases of cars and trucks and the spending on gas and oil (USD 988.2 billions), by the 2005 number of person travel miles (4884.557 billions of vehicle miles). In order to annualize the total commuting cost, the average number of workdays per year (ND) is calculated by assuming 2 weeks of vacation (10 days) and 10 days of federal holidays. Hence, ND = 240 days (50 weeks $\times$ 5 days–10 days).

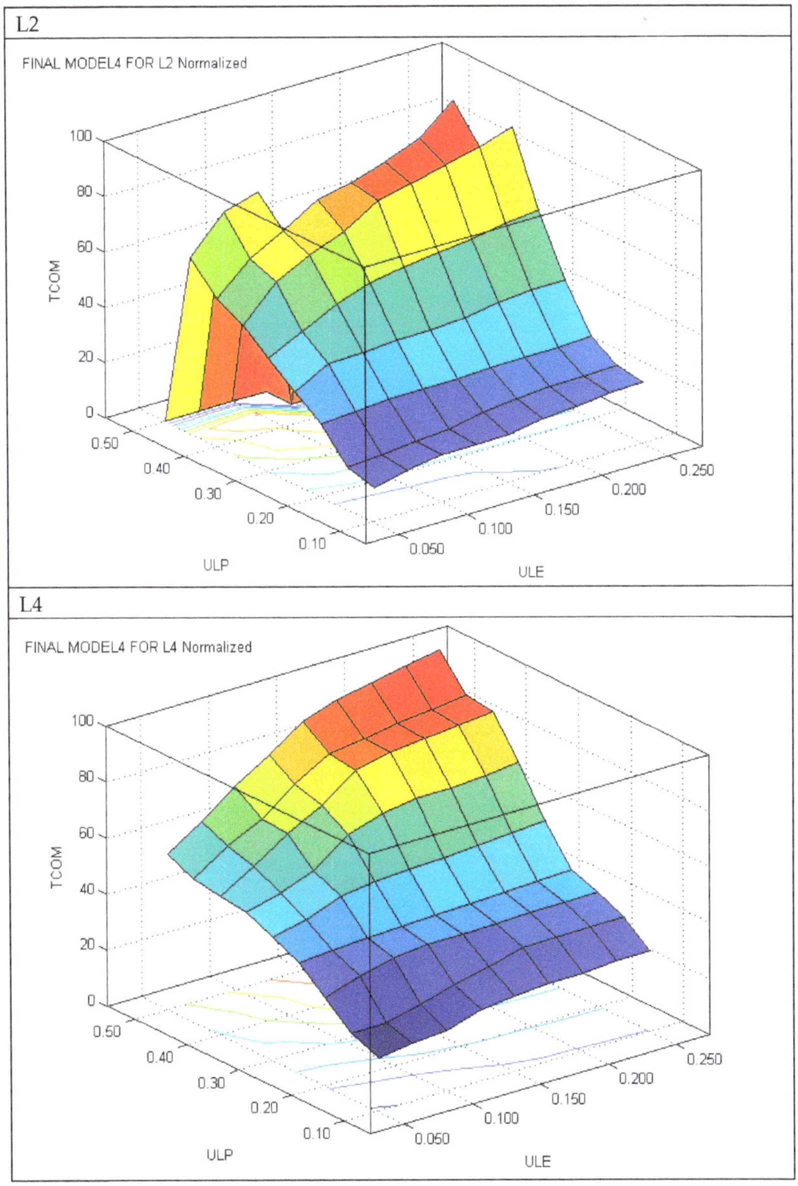


**Figure 10.** Commuting Cost Surface.

**Table 14.** Results of 3rd Order Regression of TCOM over ULP and ULE.

| Variables | Land Development Strategy L2 | L4 |
|---|---|---|
| Intercept | 188,109 (1.71) * | 100,497 (2.16) * |
| ULP | −3,410,307 (−3.34) ** | −377,350 (−1.07) |
| ULE | 2,816,147 (1.89) * | 117,457 (0.17) |
| ULP × ULP | 19,994,468 (5.65) ** | 3,921,962 (3.54) ** |
| ULE × ULE | −14,772,237 (−1.68) * | 2,136,213 (0.48) |
| ULP × ULE | −4,477,808 (−0.93) | 878,521 (0.52) |
| ULP × ULP × ULP | −26,345,984 (−6.61) ** | −5,919,163 (−5.02) ** |
| ULE × ULE × ULE | 31,053,141 (1.72) * | −3,265,161 (−0.35) |
| ULP × ULP × ULE | 24,799,246 (3.94) ** | 9,630,043 (4.90) ** |
| ULP × ULE × ULE | −6,222,188 (−0.62) | −13,204,059 (−3.36) * |
| $R^2$ | 0.987 | 0.983 |

( ) t-statistics; * significant at 90% level, two-tailed test; ** significant at 99% level, two-tailed test.

#### 3.3.3. Estimation of Land Development Costs

In order to develop land development cost functions for residence and workplace locations, with the ULP and ULE as determinants, parcel-level property values and developed acres data are drawn from the 2006 (4th quarter) Real Estate database that is used by local governments for tax assessment (Section 2.2.2). These data are aggregated at the TAZ level in order to match them with population and employment data. Land development cost functions, which involve acreage, population, and employees, are constructed as follows:

Total property value (land + building) = f (residential acreage; population) (25)

Total property value (land + building) = g (commercial + industrial + retail + office acreage; employees) (26)

The following three functional forms have been considered: (1) linear–linear; (2) log–log; and (3) log–linear. The log–log specification resulted in the highest $R^2$. However, because the ULP and ULE are the basic variables in the commuting flow function TCOM, the land development cost functions were re-estimated, in log–log form, with ULP and ULE as determinants, together with population P_2006 and employment E_2006. As the densities involve the ratios of acreages to population or employment, the same information is embodied in the new formulations. Furthermore, the exponents of P_2006 and E_2006 must be equal to 1 to avoid scale effects with regard to these variables. This homogeneity allows the estimated functions to be applied to any increment in the population and employment. The new regression results are presented in Table 15.

**Table 15.** Regression Results for Land Development Cost Functions.

| Land Development Cost (Residential) | | | | Land Development Cost (Employment) | | | |
|---|---|---|---|---|---|---|---|
| Intercept | 11.218 | 262.90 (<0.0001) | $R^2$ 0.85 | Intercept | 10.810 | 93.39 (<0.0001) | $R^2$ 0.78 |
| LN(P_2006) | 1.000 | Infty (<0.0001) | | LN(E_2006) | 1.000 | Infty (<0.0001) | |
| LN(ULP) | 0.014 | 0.34 (0.7311) | | LN(ULE) | 0.502 | 10.40 (<0.0001) | |
| RESTRICT | 16.899 | 2.37 (0.0173) | | RESTRICT | 98.442 | 5.01 (<0.0001) | |

Therefore, the total cost of development for the increments $\Delta P$ and $\Delta E$ are as follows:

$$\text{Population}: \ \text{LDCP} = e^{11.218} \cdot \Delta P \cdot ULP^{0.014} \quad (27)$$

$$\text{Employment}: \ \text{LDCE} = e^{10.810} \cdot \Delta E \cdot ULE^{0.502} \quad (28)$$

In order to annualize these costs, the equivalent annual cost (EAC) formula obtained from the Board of Governors of the Federal Reserve System is used, with the following equation:

$$EAC = (\text{Asset Price} \times IR)/(1 - (1 + IR)\hat{}(-N)) \quad (29)$$

IR = average mortgage interest rate over 1997~2006 = 6.71% = 0.0671
N = number of periods = 30 years (normal mortgage payment period)

The annualized functions (27) and (28) are adjusted with the multiplier 0.078.

### 3.3.4. Congestion Cost Synthetic Functions

High-density, compact cities can entail traffic congestion costs, reduced urban services, air pollution and noise costs, etc. These costs can be assumed to decrease as density decreases. However, they are not easily statistically estimated due to the lack of necessary data. Here, these costs are assumed to be related to population, employment, and land consumptions, with the following synthetic functional forms:

$$TCONP = K1 * \Delta P * (ULP^{**} - b) \quad (30)$$

$$TCONE = K2 * \Delta E * (ULE^{**} - d) \quad (31)$$

where b, d, K1 and K2 are positive parameters.

Consistent with the theories, these costs decline with increasing ULP and ULE values (decreasing densities). The parameters K1 and K2 impact the height of the cost curves, whereas b and d impact their steepness.

### 3.3.5. Total Development Cost Minimization

Therefore, the annualized total development cost, TDC, is as follows:

$$TDC = TCOM + LDCP + LDCE + TCONP + TCONE \quad (32)$$

The allocated population and employment increments, ΔP and ΔE, are fixed. They are implicit in the commuting cost function, and explicit in the other functions, where they serve as given parameters. Hence, each of the cost components is only a function of the inverse densities ULP and ULE. Therefore,

$$TDC = TDC\,(ULP, ULE) \quad (33)$$

For instance, TDC with land strategy L2 is as follows:

$$\begin{aligned} TDC &= 0.202 \times 240 \times (188109 - 3410307 \times ULP + 2816147 \times ULE \\ &+ 19994468 \times ULP^2 - 14772237 \times ULE^2 - 4477808 \times ULP \times ULE \\ &- 26345984 \times ULP^3 + 31053141 \times ULE^3 + 24799246 \times ULP^2 \times ULE - 6222188 \times ULP \times ULE^2) \\ &+ \frac{(e^{11.218} \times \Delta P \times ULP^{0.014} \times 0.0671)}{(1-(1+0.0671)^{-30})} + \frac{(e^{10.810} \times \Delta E \times ULE^{0.502} \times 0.0671)}{(1-(1+0.0671)^{-30})} \\ &+ K1 \times \Delta P \times ULP^{-b} + K2 \times \Delta E \times ULE^{-d} \end{aligned} \quad (34)$$

A similar function for land strategy L4 is also easily formulated, but is not presented here. The optimal values ULP* and ULE* that minimize TDC depend upon the values of the congestion cost function parameters K1, K2, b, and d. The optimal ULP* and ULE* are obtained over the following grid of values for K1, K2, b, and d:

K1 = 0.1, 0.3, 0.5;
K2 = 0.1, 0.3, 0.5;
b = 1.0, 3.0, 5.0;
d = 1.0, 3.0, 5.0.

The optimal values of the development densities (ULP and ULE) are obtained by solving a simple two-variable optimization problem over a grid of 81 combinations of values for K1, K2, b, and d. In addition, the upper and lower bounds of the ULP and ULE are included in the model to be consistent with the bounds used in estimating the TCOM functions. The model is as follows:

$$\text{Minimize TDC (ULP, ULE)} \tag{35}$$

s.t.

$$0.100 \leq \text{ULP} \leq 0.500 \tag{36}$$

$$0.050 \leq \text{ULE} \leq 0.250 \tag{37}$$

For a given set of K1, K2, b, and d values, the optimal ULP and ULE values that minimize the total cost TDC are presented in Tables 16 and 17 with the land availability strategy L2 (UGB + 1-mile buffer) and L4 (UGB + 3-mile buffer), respectively. The results show that the optimal ULP and ULE vary, depending on the form of the congestion cost functions, which depend on the parameters K1, K2, b, and d. The ULP and ULE values marked with L and U represent the lower and upper bounds, respectively. The lower bound characterizes the highest density, and the upper bound demonstrates the lowest density.

**Table 16.** Grid Analysis for Optimal ULP and ULE—L2 Case.

| | | | K1 | | | | | |
|---|---|---|---|---|---|---|---|---|
| | | | 0.1 | | 0.3 | | 0.5 | |
| K2 | b | d | ULP | ULE | ULP | ULE | ULP | ULE |
| 0.1 | 1.0 | 1.0 | 0.1000 L | 0.0500 L | 0.1000 L | 0.0500 L | 0.1000 L | 0.0500 L |
| | | 3.0 | 0.1000 L | 0.0805 | 0.1000 L | 0.0805 | 0.1000 L | 0.0805 |
| | | 5.0 | 0.1000 L | 0.2211 | 0.1000 L | 0.2211 | 0.1000 L | 0.2211 |
| | 3.0 | 1.0 | 0.1447 | 0.0500 L | 0.1930 | 0.0500 L | 0.2206 | 0.0500 L |
| | | 3.0 | 0.1443 | 0.0805 | 0.1917 | 0.0803 | 0.2186 | 0.0801 |
| | | 5.0 | 0.1435 | 0.2213 | 0.1876 | 0.2212 | 0.2121 | 0.2209 |
| | 5.0 | 1.0 | 0.3135 | 0.0500 L | 0.5000 U | 0.0500 L | 0.5000 U | 0.0500 L |
| | | 3.0 | 0.3089 | 0.0793 | 0.5000 U | 0.0763 | 0.5000 U | 0.0763 |
| | | 5.0 | 0.2952 | 0.2194 | 0.3594 | 0.2174 | 0.3989 | 0.2160 |
| 0.3 | 1.0 | 1.0 | 0.1000 L | 0.0500 L | 0.1000 L | 0.0500 L | 0.1000 L | 0.0500 L |
| | | 3.0 | 0.1000 L | 0.1108 | 0.1000 L | 0.1108 | 0.1000 L | 0.1108 |
| | | 5.0 | 0.1000 L | 0.2500 U | 0.1000 L | 0.2500 U | 0.1000 L | 0.2500 U |
| | 3.0 | 1.0 | 0.1447 | 0.0500 L | 0.1930 | 0.0500 L | 0.2206 | 0.0500 L |
| | | 3.0 | 0.1439 | 0.1108 | 0.1905 | 0.1105 | 0.2168 | 0.1103 |
| | | 5.0 | 0.1435 | 0.2500 U | 0.1872 | 0.2500 U | 0.2113 | 0.2500 U |
| | 5.0 | 1.0 | 0.3135 | 0.0500 L | 0.5000 U | 0.0500 L | 0.5000 U | 0.0500 L |
| | | 3.0 | 0.3051 | 0.1090 | 0.5000 U | 0.1042 | 0.5000 U | 0.1042 |
| | | 5.0 | 0.2932 | 0.2500 U | 0.3551 | 0.2500 U | 0.3917 | 0.2500 U |
| 0.5 | 1.0 | 1.0 | 0.1000 L | 0.0500 L | 0.1000 L | 0.0500 L | 0.1000 L | 0.0500 L |
| | | 3.0 | 0.1000 L | 0.1285 | 0.1000 L | 0.1285 | 0.1000 L | 0.1285 |
| | | 5.0 | 0.1000 L | 0.2500 U | 0.1000 L | 0.2500 U | 0.1000 L | 0.2500 U |
| | 3.0 | 1.0 | 0.1447 | 0.0500 L | 0.1930 | 0.0500 L | 0.2206 | 0.0500 L |
| | | 3.0 | 0.1438 | 0.1285 | 0.1899 | 0.1283 | 0.2158 | 0.1280 |
| | | 5.0 | 0.1435 | 0.2500 U | 0.1872 | 0.2500 U | 0.2113 | 0.2500 U |
| | 5.0 | 1.0 | 0.3135 | 0.0500 L | 0.5000 U | 0.0500 L | 0.5000 U | 0.0500 L |
| | | 3.0 | 0.3031 | 0.1265 | 0.5000 U | 0.1205 | 0.5000 U | 0.1205 |
| | | 5.0 | 0.2932 | 0.2500 U | 0.3551 | 0.2500 U | 0.3917 | 0.2500 U |

L: Lower bound; U: upper bound.

For example, if K1 = 0.1, K2 = 0.1, b = 1.0, and d = 1.0 in the L2 case, then the optimal ULP and ULE values that minimize the total cost TDC are the lower bounds (ULP: 0.1000; ULE: 0.0500), which means that, for these congestion cost functions, the highest density strategy (compact development) for both residences and workplaces is optimal. However, if the congestion function is characterized by the parameters K1 = 0.1, K2 = 0.1, b = 5.0, and d = 5.0, then the optimal ULP and ULE values (ULP: 0.2952; ULE: 0.2194) demonstrate a lower density. Under this scenario, allowing sprawl to some extent may be the most appropriate strategy. It is also interesting to note that the optimal value of ULE hits its upper bound (ULE: 0.2500) in many cases, especially when d = 5.0 (steep function) and K2 ≥ 0.3 (higher intercept), representing a high congestion cost for workplace locations, which suggests that the suburbanization of workplaces can help to minimize TDC. When b = 1.0 (low steepness) for the congestion cost function for residential locations, ULP hits its lower bound (0.1000), irrespective of the values of the other parameters. Similarly, when d = 1.0 (low steepness) for the congestion cost function for workplace locations, ULE hits its lower bound (0.0500). These lower bounds point to the advisability of compact development for residences and workplaces. The results demonstrate that the form of the congestion function plays an important role in determining the optimal values of ULP and ULE.

**Table 17.** Grid Analysis for Optimal ULP and ULE—L4 Case.

| | | | K1 | | | | | |
|---|---|---|---|---|---|---|---|---|
| | | | 0.1 | | 0.3 | | 0.5 | |
| K2 | b | d | ULP | ULE | ULP | ULE | ULP | ULE |
| 0.1 | 1.0 | 1.0 | 0.1000 L | 0.0500 L | 0.1000 L | 0.0500 L | 0.1000 L | 0.0500 L |
| | | 3.0 | 0.1000 L | 0.0810 | 0.1000 L | 0.0810 | 0.1000 L | 0.0810 |
| | | 5.0 | 0.1000 L | 0.2218 | 0.1000 L | 0.2218 | 0.1000 L | 0.2218 |
| | 3.0 | 1.0 | 0.1476 | 0.0500 L | 0.2064 | 0.0500 L | 0.2412 | 0.0500 L |
| | | 3.0 | 0.1470 | 0.0809 | 0.2048 | 0.0807 | 0.2387 | 0.0806 |
| | | 5.0 | 0.1474 | 0.2223 | 0.2030 | 0.2227 | 0.2346 | 0.2228 |
| | 5.0 | 1.0 | 0.3399 | 0.0500 L | 0.4456 | 0.0500 L | 0.5000 U | 0.0500 L |
| | | 3.0 | 0.3351 | 0.0801 | 0.4273 | 0.0794 | 0.5000 U | 0.0787 |
| | | 5.0 | 0.3248 | 0.2226 | 0.3991 | 0.2220 | 0.4423 | 0.2215 |
| 0.3 | 1.0 | 1.0 | 0.1000 L | 0.0500 L | 0.1000 L | 0.0500 L | 0.1000 L | 0.0500 L |
| | | 3.0 | 0.1000 L | 0.1108 | 0.1000 L | 0.1108 | 0.1000 L | 0.1108 |
| | | 5.0 | 0.1000 L | 0.2500 U | 0.1000 L | 0.2500 U | 0.1000 L | 0.2500 U |
| | 3.0 | 1.0 | 0.1476 | 0.0500 L | 0.2064 | 0.0500 L | 0.2412 | 0.0500 L |
| | | 3.0 | 0.1466 | 0.1107 | 0.2037 | 0.1106 | 0.2368 | 0.1104 |
| | | 5.0 | 0.1480 | 0.2500 U | 0.2036 | 0.2500 U | 0.2351 | 0.2500 U |
| | 5.0 | 1.0 | 0.3399 | 0.0500 L | 0.4456 | 0.0500 L | 0.5000 U | 0.0500 L |
| | | 3.0 | 0.3316 | 0.1097 | 0.4170 | 0.1089 | 0.4783 | 0.1081 |
| | | 5.0 | 0.3243 | 0.2500 U | 0.3973 | 0.2500 U | 0.4391 | 0.2500 U |
| 0.5 | 1.0 | 1.0 | 0.1000 L | 0.0500 L | 0.1000 L | 0.0500 L | 0.1000 L | 0.0500 L |
| | | 3.0 | 0.1000 L | 0.1281 | 0.1000 L | 0.1281 | 0.1000 L | 0.1281 |
| | | 5.0 | 0.1000 L | 0.2500 U | 0.1000 L | 0.2500 U | 0.1000 L | 0.2500 U |
| | 3.0 | 1.0 | 0.1476 | 0.0500 L | 0.2064 | 0.0500 L | 0.2412 | 0.0500 L |
| | | 3.0 | 0.1465 | 0.1281 | 0.2032 | 0.1280 | 0.2360 | 0.1279 |
| | | 5.0 | 0.1480 | 0.2500 U | 0.2036 | 0.2500 U | 0.2351 | 0.2500 U |
| | 5.0 | 1.0 | 0.3399 | 0.0500 L | 0.4456 | 0.0500 L | 0.5000 U | 0.0500 L |
| | | 3.0 | 0.3299 | 0.1272 | 0.4126 | 0.1262 | 0.4676 | 0.1254 |
| | | 5.0 | 0.3243 | 0.2500 U | 0.3973 | 0.2500 U | 0.4391 | 0.2500 U |

L: Lower bound; U: upper bound

## 4. Conclusions and Discussion

Using a Tobit commuting model that was empirically estimated with data for the FAMP region, Virginia, a normative planning model has been developed, incorporating alternative land development and density scenarios. Various growth management policy scenarios have been tested and compared. The results, expressed in terms of population and employment spatial patterns and expected commuting flows, demonstrate that tighter growth control policies increase system-wide commuting costs and flows, and do not necessarily reduce the average trip distances. The density constraints are critical in determining the distribution of populations and employment and must be carefully considered. The results also show that spatial structure variables are indeed important in estimating the Tobit model. The commuting cost minimization model has been expanded to include land development costs and congestion costs. The results demonstrate that the optimal development densities are very sensitive to the form of the congestion cost function and the land availability scenarios (growth boundaries). When the congestion function is not steep, a compact development strategy for residences and workplaces is advisable. However, with increasing steepness and level of the congestion functions, a land development strategy that allows for some sprawl minimizes the total urban development costs. The proposed optimization approach could be used for policy analysis. Since government policies, such as land-use controls and the provision of transportation infrastructure, play a major role in shaping cities, this approach could contribute to a better understanding of the dynamics of urban economies and allows planners to show the implications of policy scenarios to decision makers.

What are the contributions of this research to the literature on land-use pattern studies and policies? In the following section, we discuss three contribution areas, including spatial interaction modeling (SIM), land-use optimization, and the debate on sprawl versus compact development. First, the literature review in Section 1.2 shows that commuting SIMs that use data on commuting flows and land uses have a long history in both statistical estimation and policy simulation. However, we believe that the use of the Tobit model, which accounts for the information embodied in zero flows, the large number of potential residential and employment variables, and the spatial structure variables, all contribute to the innovative nature of this SIM. Second, the optimization framework that accounts for the commuting, land development, and congestion/pollution costs is, we believe, unique in the land-use optimization literature, as discussed in Section 1.3. We are not aware of similar research; therefore, it is not possible to compare the numerical results obtained here with those of similar research, as would be possible with alternative statistical regression models. While the literature presents many land-use simulation models, they are predictive but not normative, and recent land-use optimization models are ecologically oriented, and do not account for commuting interactions and costs. Third, this study sheds light on the complexity of the debate between urban sprawl and compact city development. While the costs of sprawl and congestion have been studied separately and in a discrete fashion (see Section 1.4), they have not been integrated into a comprehensive framework, as was the case in this study. This issue is further discussed below.

Contemporary American planning tends to support compact development strategies in general, because they align with some key planning principles, such as reduced transportation costs, improved public health, provision of affordable housing, reduced urban sprawl, conservation of land, and protection of the natural environment. Therefore, planners often tend to focus on the promotion of condensed development, along with extensive mixed-use development and extensive public transportation systems, as they are often found in European and Asian cities. This study questions the general notion of preferrable compact development strategies when setting up public policy directions and decision-making processes that incorporate all possible contexts and costs. The main arguments against compact development are the deterioration of and stress on infrastructure, lack of open space, negative impact on quality of life, higher housing costs, increased road congestion, increased noise and increased air and water pollution. Compact development can lead to higher land costs because of the limited land available, and it can be more costly to secure land to prepare it for compact development because it often involves building in already developed areas. On the other hand, compact development could lead to lower development costs because it allows for more efficient land use and infrastructure and takes advantage of the existing infrastructure. As for congestion costs, compact development could lead to higher congestion costs when the increased traffic congestion is caused by high population densities. However, compact development could also lead to lower congestion costs when it promotes the use of public and active transportation, which could eventually help to reduce traffic congestion. We believe that this study provides an innovative approach to optimal urban growth boundaries or urban capacity strategies, considering not only commuting costs, but also various other costs, such as land development costs and congestion costs, by applying empirical, analytical, or mathematical modeling approaches. We argue that a full spectrum of cost mechanisms should be examined in developing growth control policies and compact development strategies.

In the following section, we discuss the possible research extensions. First, population and employment could be disaggregated in terms of different industries, income level, ethnic background, gender, etc. Disaggregated model specifications might help us to better understand the spatial structure effects. Second, the model could be extended to include different types of trips, in addition to commuting trips. Shopping trips are prominently included in the Lowry model, but such data are more difficult to obtain. Third, the development cost functions could be extended to include utility, roadways, and other costs that are not reflected in property values. Fourth, congestion cost functions could be empirically estimated, if appropriate data become available. Fifth, the model could incorporate other transportation modes (e.g., transit), which might alter the conclusions reached in this research. Sixth, population and employment densities could vary across spatial units (TAZs), and could be made endogenous to the model, that is, becoming decision variables. This would increase the non-linearity of the optimization model and the complexity of its resolution. Seventh, the SIM could be used to determine the factors underlying the changes in commuting patterns, for instance before, during, and after a pandemic. Finally, the optimization methodology could be expanded to more comprehensively account for environmental factors. In the present approach, environmentally sensitive areas are simply excluded as candidates for development. Ecological indices, based on remote sensing and other data, could be developed, as proposed by Li et al. [53], and could act as constraints in the optimization process. In addition, explicit air quality constraints, as proposed in [54], could be set up to account for the pollution emissions from commuting traffic and economic activities. Such extensions, possibly formulated in a multi-objective optimization framework, could help us to analyze the trade-offs between economic and environmental factors.

**Author Contributions:** Both authors have contributed equally to the research, writing, and revisions of this manuscript. All authors have read and agreed to the published version of the manuscript.

**Funding:** This research received no external funding.

**Institutional Review Board Statement:** Not applicable.

**Informed Consent Statement:** Not applicable.

**Data Availability Statement:** The data that support the findings of this study are available upon request from the corresponding author.

**Conflicts of Interest:** The authors declare no conflict of interest.

## References

1. Lowry, I.S. *A Model of Metropolis*; Rand Corporation: Santa Monica, CA, USA, 1964.
2. Anas, A. Discrete choice theory, information theory and the multinomial logit and gravity models. *Transp. Res. Part B Methodol.* **1983**, *17*, 13–23. [CrossRef]
3. Sen, A.; Smith, T. *Gravity Models of Spatial Interaction Behaviour*; Springer: Berlin/Heidelberg, Germany, 1995.
4. Nijkamp, P.; Ratajczak, W. Gravitational Analysis in Regional Science and Spatial Economics: A Vector Gradient Approach to Trade. *Int. Reg. Sci. Rev.* **2020**, *44*, 400–431. [CrossRef]
5. Fotheringham, A.S. A new set of spatial-interaction models: The theory of competing destinations. *Environ. Plan. A* **1983**, *15*, 15–36. [CrossRef]
6. Stouffer, S.A. Intervening opportunities and competing migrants. *J. Reg. Sci.* **1960**, *2*, 1–26. [CrossRef]
7. Gitlesen, J.P.; Thorsen, I. A Competing Destinations Approach to Modeling Commuting Flows: A Theoretical Interpretation and An Empirical Application of the Model. *Environ. Plan. A Econ. Space* **2000**, *32*, 2057–2074. [CrossRef]
8. Sirmans, C.F. Determinants of journey to work flows: Some empirical evidence. *Ann. Reg. Sci.* **1977**, *11*, 98–108. [CrossRef]
9. Sandow, E. Commuting behaviour in sparsely populated areas: Evidence from northern Sweden. *J. Transp. Geogr.* **2008**, *16*, 14–27. [CrossRef]
10. Sermons, M.; Koppelman, F.S. Representing the differences between female and male commute behavior in residential location choice models. *J. Transp. Geogr.* **2001**, *9*, 101–110. [CrossRef]

11. Prashker, J.; Shiftan, Y.; Hershkovitch-Sarusi, P. Residential choice location, gender and the commute trip to work in Tel Aviv. *J. Transp. Geogr.* **2008**, *16*, 332–341. [CrossRef]
12. O'Kelly, M.E.; Niedzielski, M.A.; Gleeson, J. Spatial interaction models from Irish commuting data: Variations in trip length by occupation and gender. *J. Geogr. Syst.* **2012**, *14*, 357–387. [CrossRef]
13. Lin, D.; Allan, A.; Cui, J. The impacts of urban spatial structure and socio-economic factors on patterns of commuting: A review. *Int. J. Urban Sci.* **2015**, *19*, 238–255. [CrossRef]
14. So, K.S.; Orazem, P.F.; Otto, D.M. The effects of housing prices, wages, and commuting time on joint residential and job location choices. *Am. J. Agric. Econ.* **2001**, *83*, 1036–1048. [CrossRef]
15. Wu, K.-L. Employment and Housing Development and Their Impacts on Metropolitan Commuting: An Empirical Study of the Development of the Silicon Valley Region of the San Francisco Bay Area. Ph.D. Thesis, City and Regional Planning. University of California at Berkeley, Berkeley, CA, USA, 1997.
16. Glenn, P.T.I.; Ubøe, J. Wage payoffs and distance deterrence in the journey to work. *Transp. Res. Part B* **2004**, *38*, 853–867. [CrossRef]
17. Ahrens, A.; Lyons, S. Do rising rents lead to longer commutes? A gravity model of commuting flows in Ireland. *Urban Stud.* **2020**, *58*, 264–279. [CrossRef]
18. Sohn, J. Are commuting patterns a good indicator of urban spatial structure? *J. Transp. Geogr.* **2005**, *13*, 306–317. [CrossRef]
19. Wilson, A.G.; Coelho, J.D.; Macgill, S.M.; Williams, H.C.W.L. *Optimization in Locational and Transport Analysis*; John Wiley: New York, NY, USA, 1981.
20. Kim, T.J. Alternative transportation modes in an urban land use model: A general equilibrium approach. *J. Urban Econ.* **1979**, *6*, 197–215. [CrossRef]
21. Boyce, D.E.; Lundqvist, L. Network equilibrium models of urban location and travel choices: Alternative formulations for the stockholm region. *Pap. Reg. Sci.* **2005**, *61*, 93–104. [CrossRef]
22. Kim, T.J. *Integrated Urban System modeling: Theory and Practice*; Martinus Nijhoff: Norwell, MA, USA, 1989.
23. Prastacos, P. An integrated land-use-transportation model for the San Francisco Region: 1. Design and mathematical structure. *Environ. Plan. A* **1986**, *18*, 307–322. [CrossRef]
24. Prastacos, P. An integrated land-use-transportation model for the San Francisco Region: 2. Empirical estimation and results. *Environ. Plan. A* **1986**, *18*, 511–528. [CrossRef]
25. Caindec, E.K.; Prastacos, P. *A Description of POLIS. The Projective Optimization Land Use Information System*; Working Paper 95-1; Association of Bay Area Governments: Oakland, CA, USA, 1995.
26. Barber, G.M. Urban population distribution planning. *Ann. Assoc. Am. Geogr.* **1977**, *67*, 239–245. [CrossRef]
27. Garin, R. A matrix formulation of the lowry model for intrametropolitan activity allocation. *J. Am. Inst. Plan.* **1966**, *32*, 361–364. [CrossRef]
28. Barber, G.M. Locating employment growth in urban areas to minimize travel time. *Prof. Geogr.* **1978**, *30*, 149–155. [CrossRef]
29. Ma, S.; Zhang, Y.; Sun, C. Optimization and Application of Integrated Land Use and Transportation Model in Small- and Medium-Sized Cities in China. *Sustainability* **2019**, *11*, 2555. [CrossRef]
30. Samani, A.R.; Mishra, S.; Lee, D.J.-H.; Golias, M.M.; Everett, J. A new approach to develop large-scale land-use models using publicly available data. *Environ. Plan. B Urban Anal. City Sci.* **2021**, *49*, 169–187. [CrossRef]
31. Putnam, S.H. *Integrated Urban Models: Policy Analysis of Transportation and Land Use*; Pion Limited: London, UK, 1983.
32. Wu, J.; Yi, T.; Wang, H.; Wang, H.; Fu, J.; Zhao, Y. Evaluation of Medical Carrying Capacity for Megacities from a Traffic Analysis Zone View: A Case Study in Shenzhen, China. *Land* **2022**, *11*, 888. [CrossRef]
33. Zhang, D.; Zhou, C.; Sun, D.; Qian, Y. The influence of the spatial pattern of urban road networks on the quality of business environments: The case of Dalian City. *Environ. Dev. Sustain.* **2021**, *24*, 9429–9446. [CrossRef]
34. Yuan, M.; Liu, Y.; He, J.; Liu, D. Regional land-use allocation using a coupled MAS and GA model: From local simulation to global optimization, a case study in Caidian District, Wuhan, China. *Cartogr. Geogr. Inf. Sci.* **2014**, *41*, 363–378. [CrossRef]
35. Wang, Z.; Gao, Y.; Wang, X.; Lin, Q.; Li, L. A new approach to land use optimization and simulation considering urban development sustainability: A case study of Bortala, China. *Sustain. Cities Soc.* **2022**, *87*, 104135. [CrossRef]
36. Clark, W.V.W.; Wang, W. Job access and commuting patterns: Balancing work and residence in Los Angeles. *Urban Geogr.* **2005**, *26*, 610–626. [CrossRef]
37. Ewing, R.; Pendall, R.; Chen, D. Measuring Sprawl and Its Transportation Impacts. *Transp. Res. Rec. J. Transp. Res. Board* **2003**, *1831*, 175–183. [CrossRef]
38. Weber, J.; Sultana, S. Employment Sprawl, Race and the Journey to Work in Birmingham, Alabama. *Southeast. Geogr.* **2008**, *48*, 53–74. [CrossRef]
39. Sultana, S. Racial variation in males commuting time: What does the evidence suggest? *Prof. Geogr.* **2005**, *57*, 66–82. [CrossRef]
40. Dunphy, T.; Fisher, K. Transportation, Congestion, and Density: New Insights. *Transp. Res. Rec.* **1996**, *1552*, 89–96. [CrossRef]
41. Levinson, D.M.; Kumar, A. Density and the Journey to Work. *Growth Change* **1997**, *28*, 147–172. [CrossRef] [PubMed]
42. O'Toole, R. The Myth of the Compact City: Why Compact Development Is Not the Way to Reduce Carbon Dioxide Emissions. *Cato Inst. Policy Anal.* **2009**, *653*. Available online: https://ssrn.com/abstract=1543980 (accessed on 18 January 2023). [CrossRef]
43. Cambridge Systematics, Inc. *Moving Cooler: An Analysis of Transportation Strategies for Reducing Greenhouse Gas Emissions*; Urban Land Institute: Washington, DC, USA, 2009.

44. Stevens, M.R. Does compact development make people drive less? *J. Am. Plan. Assoc.* **2017**, *83*, 7–18. [CrossRef]
45. Emrath, P.; Liu, F. Vehicle carbon dioxide emissions and the compactness of residential development. *Cityscape A J. Policy Dev. Res.* **2008**, *10*, 185–202.
46. Stone, B., Jr. Urban sprawl and air quality in large US cities. *J. Environ. Manag.* **2008**, *86*, 688–698. [CrossRef]
47. Schindler, M.; Caruso, G. Urban compactness and the trade-off between air pollution emission and exposure: Lessons from a spatially explicit theoretical model. *Comput. Environ. Urban Syst.* **2014**, *45*, 13–23. [CrossRef]
48. Zhang, D.; Zhou, C.; He, B.-J. Spatial and temporal heterogeneity of urban land area and PM2.5 concentration in China. *Urban Clim.* **2022**, *45*, 101268. [CrossRef]
49. Tobin, J. Estimation of relationships for limited dependent variables. *Econom. J. Econom. Soc.* **1958**, *26*, 24–36. [CrossRef]
50. Guldmann, J.-M. Competing destinations and intervening opportunities interaction models of inter-city telecommunication flows. *Pap. Reg. Sci.* **1999**, *78*, 179–194. [CrossRef]
51. Veall, M.R.; Zimmermann, K.F. Pseudo-$R^2$ measures for some common limited dependent variable models. *J. Econ. Surv.* **1996**, *10*, 241–259. [CrossRef]
52. He, F.; Yang, J.; Zhang, Y.; Sun, D.; Wang, L.; Xiao, X.; Xia, J. Offshore Island Connection Line: A new perspective of coastal urban development boundary simulation and multi-scenario prediction. *GIScience Remote Sens.* **2022**, *59*, 801–821. [CrossRef]
53. Li, J.; Gong, J.; Guldmann, J.-M.; Yang, J. Assessment of Urban Ecological Quality and Spatial Heterogeneity Based on Remote Sensing: A Case Study of the Rapid Urbanization of Wuhan City. *Remote Sens.* **2021**, *13*, 4440. [CrossRef]
54. Gulmann, J.-M. Urban Land Use Allocation and Environmental Pollution Control: An Intertemporal Optimization Approach. *Socio-Econ. Plan. Sci.* **1979**, *13*, 71–86. [CrossRef]

land

Article

# Housing Prices and the Characteristics of Nearby Green Space: Does Landscape Pattern Index Matter? Evidence from Metropolitan Area

Yiyi Chen [1,2,3], Colin A. Jones [4,*], Neil A. Dunse [4], Enquan Li [5] and Ye Liu [1,2,3,*]

[1] School of Geography and Planning, Sun Yat-Sen University, Guangzhou 510006, China
[2] Guangdong Provincial Engineering Research Center for Public Security and Disaster, Guangzhou 510275, China
[3] Guangdong Key Laboratory for Urbanization and Geo-Simulation, Sun Yat-Sen University, Guangzhou 510275, China
[4] The Urban Institute, Heriot-Watt University, Edinburgh EH14 4AS, UK
[5] School of Mathematics, University of Birmingham, Birmingham B15 2TT, UK
* Correspondence: c.a.jones@hw.ac.uk (C.A.J.); liuye25@mail.sysu.edu.cn (Y.L.)

**Abstract:** This study aimed to examine the association between housing prices and green space characteristics with a special focus on exploring the effects of the shape pattern index. The research was based on a hedonic price model across two main distance buffers from residential properties to urban green spaces. Green spaces were characterized by size and shape measured by a landscape shape index (LSI). This study was based on 16,222 housing transaction data obtained from the website of real estate agencies during December 2019 in the Metropolitan Area of Beijing. Linear regression and semi-log regression analysis were used to examine the associations between independent housing and neighborhood characteristic variables and housing prices. The results suggested that a one-unit increase in the natural logarithm of the landscape shape index (LSI) can increase housing prices by 4% (5543 CNY ≈ 826 USD). Such marginal effects were more pronounced for residences located close to urban green spaces and tended to decay as the distance from residences to green spaces increased. Additional analysis captured the marginal effects of the natural logarithm of the landscape shape index (LSI > 1.3) on achieving the maximum monetary evaluation of the property. The findings of this study suggest that the effects of specific green space characteristics on housing prices should be taken into account in landscape and urban design.

**Keywords:** green space; shape pattern index; landscape shape index (LSI); hedonic price model; housing price; distance decay effect

**Citation:** Chen, Y.; Jones, C.A.; Dunse, N.A.; Li, E.; Liu, Y. Housing Prices and the Characteristics of Nearby Green Space: Does Landscape Pattern Index Matter? Evidence from Metropolitan Area. *Land* **2023**, *12*, 496. https://doi.org/10.3390/land12020496

Academic Editors: Bindong Sun, Tinglin Zhang, Wan Li, Chun Yin and Honghuan Gu

Received: 11 January 2023
Revised: 12 February 2023
Accepted: 14 February 2023
Published: 16 February 2023



## 1. Introduction

The value of urban green space is well established not only in moderating urban heat island effects [1,2], improving air quality [3,4], promoting people's mental health and self-rated health [5–7], wellbeing [8–10], and also in contributing to housing prices [11,12]. The influence of parks on house prices has long been recognized. In the 1850s, the New York City Council argued that the creation of Central Park would increase nearby property prices [13]. More recent empirical studies have pointed out that residential proximity to green spaces as well as parks has positive impacts on housing prices across the world. In the UK for example, residents have the willingness to pay an extra premium to live close to urban parks [14]. Similar results have been found in Finland, where residents were found to be willing to pay 4.9 percent more to reside adjacent to natural greenness [15]. Studies in China have reported equivalent results to western countries: in Guangzhou, Jim and Chen [11] found that residential proximity to water bodies and the visibility of green spaces had positive influences on housing prices.

Later studies have focused on how the characteristics of a green space can influence house prices. Clearly, the nature of the green space is a factor in the desirability of a neighborhood. There are all sorts of green spaces, including parks that are open to the public, private golf courses, gardens, and recreational facilities. Not all of these have an equivalent positive influence on house prices. Large, flat, open spaces which are used primarily for sports activities are much less preferred than natural areas that encompass water features [16]. Some green spaces will be of such poor quality, for example, in terms of standard of upkeep, that they may create negative impacts on local house prices [13]. The size and shape of the green space are also likely to impact nearby house prices. While the role of size is arguably straightforward, the influence of shape is less obviously defined. An irregular shape could be seen as contributing to a more distinctive neighborhood, and hence to higher house prices.

Evidence on the role of the shape and size of green spaces on the housing market is limited and mixed. One underlying mechanism is that landscape shape and size are correlated with land use types and patterns that have significant changes, along with the urbanization processes of the metropolitan area in China [17,18]. The continuous changes in land use patterns play a substantial role in influencing the housing price [19]. Several studies have found that the size of a green space had a positive influence on housing prices [20–22]. However, one study found no significant association between park size and housing price, and indicated that smaller parks distributed equally across the urban area might contribute better to housing prices compared to larger parks [12]. This issue of the macro-pattern of green spaces within cities has also been considered by other studies. Xiao, et al. [23] evaluated the effects of different types of green spaces on housing prices. They found that a one-unit increase in a residential development's ratio of green to the total area would lead to an increase of 8.7 percent in housing prices in Shanghai. Similarly, Jiao and Liu [24] quantified that city-level parks had significant influences on amenity values, while district-level parks did not.

In addition to the general effect of residential proximity on the housing premium, some studies have applied more complex measures of accessibility than simple Euclidean distance. Park, et al. [25] undertook a study of the impact of a park in Seoul, taking into account the shortest walking distance while considering crossroads and park entrances. Property values more generally were found to incrementally increase with residential proximity to urban green spaces at specific distance intervals. Crompton [13] reviewed studies of the effects of residential proximity to parks and open spaces on housing value and found that they contribute to housing prices differentially through distance buffers. Specifically, these studies found that the effects on house price of adjacency to a park and open space tend to decay at different rates within and beyond a 500-foot distance buffer. In another recent study exploring the effects of urban green space on housing values in Germany [26], researchers found that the influence of urban parks on housing prices increased with closeness through buffers. For instance, a one unit increase in the residential distance to parks would lead to a 0.1 percent increase in housing price within a 500 m buffer zone, compared to a 0.19 percent increase in housing price within a 1000 m buffer zone.

Many studies have therefore indicated that accessibility to green space, as discussed above, has played an essential role in affecting housing values. Yet, limited studies have focused on exploring the direct relationship between the size and especially the shape configuration characteristics of urban spaces and housing prices. This paper addressed this vacuum by assessing the role of the size and shape of green spaces on housing prices in urban Beijing. In particular, its premise was that irregular large green spaces have a positive impact on house prices. To assess this influence, the paper adopted the use of a shape index developed by ecological studies. Such studies have applied a shape index to green spaces to examine the relationship between fragmentation characteristics and species richness [27,28].

To address this task, the paper applied a hedonic price model. The hedonic price approach has been the conventional method for exploring the association between housing characteristics and value, which can be directly estimated as the marginal willingness to pay for an additional attribute of each property [29,30]. Hedonic models enable the impact of housing structural characteristics, locational/accessibility relationships, and neighborhood and environmental features on housing prices to be quantified [11,23]. The associations between different characteristics and housing prices can be complex and diverse. For housing structural characteristics, traditional studies have found a significant association between housing age, size, storey, number of bedrooms and bathrooms, window orientation, elevators, and housing prices [11,31–33]. Neighborhood characteristics can encompass the accessibility to the nearest social amenities (e.g., schools), train stations, central business districts, and natural landscapes (e.g., parks, forests, urban green space).

Building on the above, we therefore applied a hedonic housing price model that encompassed housing attributes and neighborhood characteristics including green space features. This study provides new insights into examining the impact of specific landscape patterns on the housing market, which reflects the potential correlations between changing land use patterns attributed to urbanization processes and housing prices. This study assessed the effects of specific green space configurations (size and landscape shape index) on housing prices in selected districts of the city of Beijing. The specific aims of this study were to: (1) explore the overall relationship between housing attributes, neighborhood characteristics and green space characteristics and housing price; (2) examine residents' willingness to pay for green space characteristics, notably in terms of its size, proximity level, and its green space shape (using a landscape shape index); abd (3) assess whether residents' willingness to pay for green space characteristics vary incrementally with distance buffers.

The remainder of this paper is organized as follows: Section 2 outlines the materials and methodology. Results are outlined in Section 3. Sections 4 and 5 report the discussions and conclusion.

## 2. Materials and Methods

### 2.1. Study Area

The scope of this study was primarily focused on selected urban districts of Beijing, namely Xicheng, Dongcheng, Haidian, Chaoyang, Fengtai and Shijingshan. The distribution of green spaces and the boundary of selected districts are illustrated in Figure 1. It is worth noting that the municipal government of Beijing has massively invested in urban greenness infrastructure in order to provide a better quality of life and potential economic benefits for residents. In total, 31 parks and 13 urban forests were added in 2019, contributing to an 83% coverage rate of parks within a radius of 500 m from residences according to the Beijing Gardening and Greening Bureau (Xinhua, 2019). The 13th Five-Year Plan (Bureau, 2020) expects the coverage rate of green space and parks within 500-m of residences to reach 48.5% and 85%, respectively. A total of 700 hectares of new urban green space, 41 new leisure parks, 13 urban forests and 50 pocket parks are planned to be built in 2020. These policies that continue to expand urban greenness provide a potential underlying pathway for raising the nearby housing premium.

### 2.2. Housing Data Acquisition

The housing data used in this study were captured from the website of a real estate agency (https://bj.lianjia.com (accessed on 28 December 2019)). Real estate agencies provide access to accurate housing prices at the household level [34]. We captured a list of housing sales prices and structural variables through December 2019 from Lianjia.com by conducting the web-crawling Python program. The original data on residential properties were first accessed by the request methods in the Beautifulsoup library which refers to a Python library for extracting data out of HTML and XML files (https://beautiful-soup-4.readthedocs.io/en/latest (accessed on 2 January 2020)). The data were extracted by selecting a useful keyword stored within the DataFrame format of the Pandas library. Housing price was captured in CNY per square meter per dwelling. Lianjia is one of the largest real estate agents in urban China, with a 50 percent market share in Beijing, providing relatively precise housing information [35]. Housing variables related to structural characteristics were captured, including housing age, housing asking price, the age of the properties, number of bedrooms, number of bathrooms, size of the property, whether the properties have an elevator, the number of storeys in the properties, and window orientation. These are equivalent to previous studies [36].

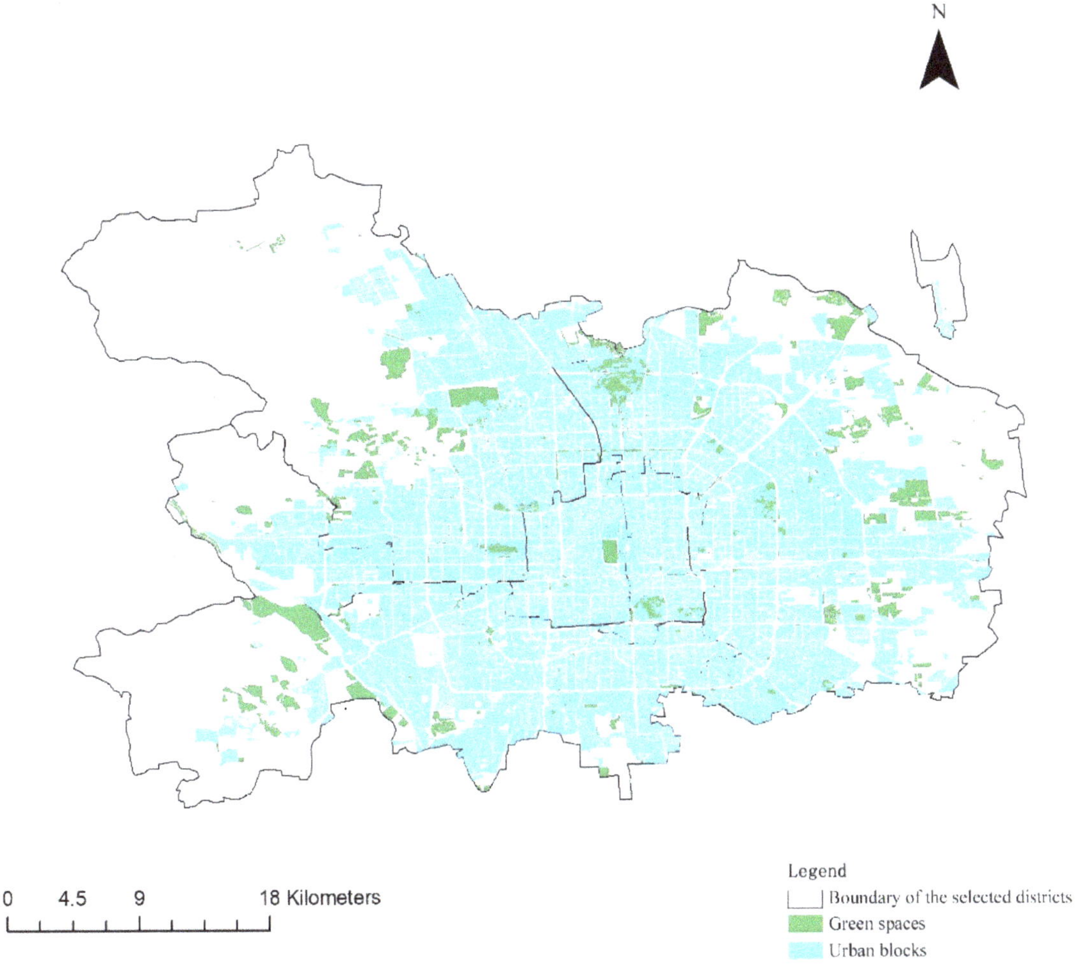


**Figure 1.** Displays the distribution of boundary of selected districts, green spaces, and urban blocks in Beijing.

Additionally, we captured the coordinates of the latitude and longitude of each property in Baidu Map API (Baidu Developer: https://lbsyun.baidu.com (accessed on 2 January 2020)) and transformed them into WGS-84 coordinates in GIS. To ensure the consistency of housing structures in the study, we excluded properties that were not normal commodity housing, such as villas, loft apartments and Soho apartments. After these exclusions, 16,222 asking price transaction records were included in this study.

*2.3. Data on Green Space and Neighbourhood Characteristics*

The green space data were obtained from Beijing City Lab (http://www.beijingcitylab.com (accessed on 2 January 2020)), which shared information on 16,721 urban green lands in 287 Chinese cities in 2017 [37]. We narrowed down the scope of the data and selected

1542 green space polygons within the Beijing administrative boundary (see Figure 1). The data included the size of the green spaces while a 'landscape shape index' (hereafter LSI) was calculated for each polygon based on the approach [38,39]. The concept of shape index was first proposed as a diversity index by [39] to quantify habitat edges for wildlife species. LSI has been frequently undertaken to identify the divergence of the shape of a landscape patch from the ideal circle [39,40].

The use of this index is the key contribution of this paper towards understanding the spatial pattern of house prices. In this study, we applied the LSI as a landscape metric to examine the effect of shape configuration characteristics of green space patches on any housing premium. The LSI quantifies the shape of the green space by taking the total length of the green space divided by the total area and adjusting by a constant for a square standard. The equation of LSI can be written as follows:

$$\text{LSI} = \frac{P}{\sqrt[2]{\pi \times A}} \quad (1)$$

where $P$ refers to the perimeter of the patches. $A$ refers to the total landscape area ($m^2$). If the landscape shape refers to a square, then the LSI equals 1. The larger the LSI, the greater the complexity of the landscape shape. While LSI values are not associated with definitive shapes, higher values suggest more attractive intricate landscapes. Figure 2 displays the shape series with increasing LSI values. All shapes are derived from the green spaces' polygon database.

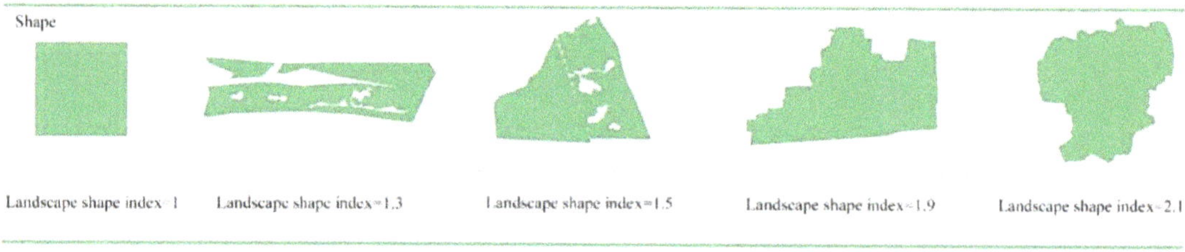


**Figure 2.** Landscape shapes with increasing LSI value.

To examine whether the marginal effects of LSI on housing prices would decay as the distance from residences to green spaces increased, we defined two key buffer variables. Accessibility to green space from residences was considered in terms of intervals, within 300 m and from 300 m up to 1000 m, as studies have found that these are walkable thresholds for residents' access to natural greenness [41]. We calculated the distance buffers based on residents' home addresses. Two main buffers were therefore generated for access to the nearest green spaces. The definition of the intervals is illustrated in Figure 3.

Regarding the neighborhood characteristics, we used the spatial 'Near' tool in ArcGIS to calculate the direct distance from each property to the nearest green space. We also measured the direct distance to the nearest central business district (CBD) and the nearest subway station that may potentially contribute to the housing premium.

### 2.4. *Construction of the Hedonic Pricing Model*

Two functional forms of the hedonic pricing model were employed to assess the associations between housing price per square meter and the accessibility of the green space and its size and shape (LSI), namely, the linear regression [42] and semi-log regression [43,44]. Accordingly, it is not only feasible to interpret the implicit property value [45] but also to identify the percentage of change in the property value. Semi-logarithmic forms have been frequently formulated in hedonic models [46]. In this study, when the association was assumed linear, the equation can be written as Equation (2):

$$\text{P} = \alpha_0 + H\beta + G\mu + N\chi + \varepsilon \quad (2)$$

where P denotes the asking price at the residential unit level. $H$ denotes the matrix of housing structure characteristics. $G$ denotes green space characteristics including our main interest landscape shape index that it is measured in the Natural logarithm form. $N$ denotes the neighborhood characteristics. $\alpha$, $\beta$, $\mu$, $\chi$ are associated parameter vectors. $\varepsilon$ is a vector of random error terms. When the association was assumed semi-log, the model was specified by a natural log transformation of the housing price and the equation can be presented as Equation (3):

$$\text{LnP} = \alpha_0 + H\beta + G\mu + N\chi + \varepsilon \tag{3}$$

In accordance with the study [47], we separated our model into two main categories. One set of the model included direct distance and the other one included the distance buffer. Both the distance metric and distance buffer were treated as proxies for the accessibility of green space [48]. The rationale behind this was that if the residence is located at a close distance to a specific green space, it might also be located within a closed-distance buffer of a green space [49].

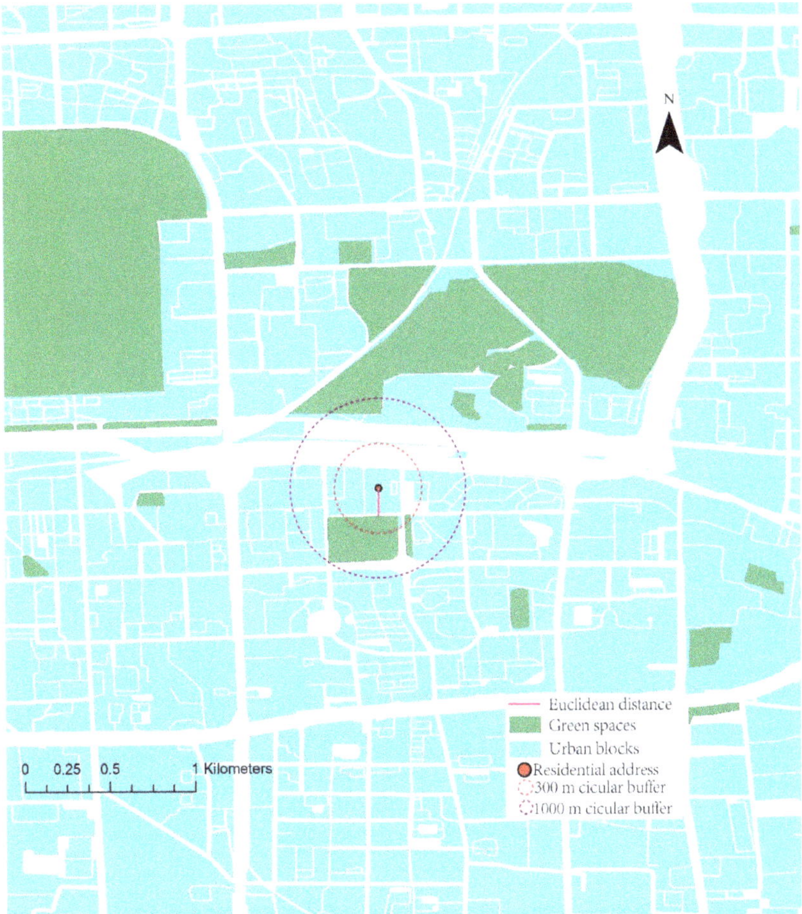


**Figure 3.** The definition of distance buffers with green spaces around the residences.

# 3. Results

## *3.1. Descriptive Analysis*

Table 1 presents the descriptive characteristics of the sample (n = 16,222). The average mean of the housing asking price is 80,249.6 CNY (≈11,957.2 USD) per square meter. Generally, the housing price decays with the distance from the central business district. The housing prices of Dongcheng and Xicheng districts, which are closest to the city center, have the highest average prices among the districts, with 113,592.9 CNY (≈16,925.3 USD) and 97,468.2 CNY (≈14,522.8 USD) per square meter, respectively. These are followed by Haidian and Chaoyang districts, with 84,605.3 CNY (≈12,606.2 USD) and 66,710.4 CNY (≈9939.8 USD) per square meter, respectively.

**Table 1.** Statistical description.

| Variables | Description | n | Mean | SD |
|---|---|---|---|---|
| Dependent variable | | | | |
| Price | Selling price per square meter in CNY | 16,222 | 80,249.60 | 28,219.83 |
| Ln price | Logarithmic form of the selling price in 10,000 CNY (Chinese currency, US $1 = 6.9 CNY) | 16,222 | 11.23 | 0.35 |
| Price in districts | | | | |
| Price in Shijingshan district | Selling price per square meters in CNY | 1862 | 52,854.39 | 11,997.59 |
| Price in Fengtai district | Selling price per square meters in CNY | 2985 | 57,166.82 | 11,324.8 |
| Price in Chaoyang district | Selling price per square meters in CNY | 2912 | 66,710.38 | 16,588.28 |
| Price in Haidian district | Selling price per square meters in CNY | 2672 | 84,605.26 | 19,498.39 |
| Price in Xicheng district | Selling price per square meters in CNY | 2815 | 97,468.18 | 18,376.97 |
| Price in Dongcheng district | Selling price per square meters in CNY | 2976 | 113,592.9 | 24,492.97 |
| Housing characteristics | | | | |
| Housing age | 2019 minus the construction date of the properties | 16,222 | 26.52 | 38.04 |
| Bedroom | Number of bedrooms | 16,222 | 2.24 | 0.82 |
| Bathroom | Number of bathrooms | 16,222 | 1.28 | 0.55 |
| Housing size | The average size of the property ($m^2$) | 16,222 | 92.25 | 47.26 |
| Elevator | Dummy variable, 1 if the property has an elevator | 16,222 | 0.63 | 0.48 |
| Storey | Category variable:<br>0 equals to basement<br>1 equals to the bottom storey (below the 3rd floor)<br>2 equals to the low storey (between the 4th and 6th floor)<br>3 equals to the middle storey (between the 7th and 10th floor)<br>4 equals to the high storey (between the 10th and 17th floor)<br>5 equals to the top storey (higher than the 18th floor) | 16,222 | 3.05 | 1.19 |
| Window orientation | Dummy variable, 1 if the residence has north-south facing windows | 16,222 | 0.38 | 0.49 |
| District | Category variable:<br>1 represents the Shijingshan district<br>2 represents the Fengtai district<br>3 represents the Chaoyang district<br>4 represents the Haidian district<br>5 represents the Dongcheng district<br>6 represents the Xicheng district | 16,222 | 3.65 | 1.65 |
| Neighbourhood characteristics | | | | |
| Distance to the nearest green space | Road distance to nearest green space (m) | 16,222 | 564.58 | 425.19 |
| Distance to subway | Road distance to the nearest subway station (m) | 16,222 | 5302.78 | 3015.51 |
| Distance to CBD | Road distance to the central business district (m) | 16,222 | 12,217.69 | 7829.55 |
| Green space characteristics | | | | |
| Natural logarithm of landscape shape index | The total length of the green space divided by the total area, adjusted by a constant for a square standard | 16,222 | 0.7 | 0.16 |

**Table 1.** *Cont.*

| Variables | Description | n | Mean | SD |
|---|---|---|---|---|
| Green spaces' size (ha) | The size of the green space (ha) | 16,222 | 17.51 | 55.21 |
| Residence buffer (300 m) | Dummy variable, 1 if housing is within 300 m buffers from the nearest green space. | 16,222 | 0.26 | 0.44 |
| Residence buffer (300–1000 m) | Dummy variable, 1 if housing is within 300–1000 m buffers from the nearest green space. | 16,222 | 0.61 | 0.49 |

Regarding the housing characteristics, the average housing age is 27 years, and the average housing size is 92 square meters. Most of the housing comprises two bedrooms and one bathroom. Approximately 63% of the houses have elevators and most of the residences are in the middle storey of buildings, with 38% of residences having north-south facing windows.

In addition, the average size of green spaces and the average distance to the nearest green space are 17.51 ha and 564.58 m, respectively, which is in line with the policy requirement of the Classification Standard of Urban Green Space established by the Minister of Construction. This policy has required that the service ring buffer of green space around residences should be at least 1000 m if the size of green space is between 2 ha and 20 ha. The average natural logarithm of LSI is 0.7, which refers to a raw LSI of 2.0. While 26% of housing is within 300 m of the green space, 61% of homes in the sample lie within a 300–1000 m distance from the green space.

To detect whether there is potential multicollinearity between dependent variables, we applied the variation inflation factor (VIF) diagnostics in Stata version 14 for Windows. The results suggested that all the VIFs were lower than 4.3, confirming that there were no serious collinearity issues in the ordinary least squares regression model [50].

### 3.2. Baseline Results

In accordance with our aims, Table 2 presents the baseline results by exploring the effects of green space characteristics on housing prices while controlling for housing attributes and neighborhood characteristics in four columns. Results of the estimation for the linear regression model are presented in Column 1 and 2, whereas results of the estimation for the semi-log model are presented in Column 3 and 4. Two measures of direct distance and distance buffers are, respectively, applied in Columns 1 and 3 and Columns 2 and 4.

Before looking at the relationship between green space and house price it is useful to examine the wider housing and neighborhood characteristic coefficients. They indicate that the numbers of bedrooms and bathrooms, the residence's elevator and window orientation are significantly, and as expected, positively associated with housing price. Specifically, in terms of the results from column 2, an additional elevator provided by the housing would lead to an increase in housing price by 5.5% on an average 3249 CNY (≈484 USD). Every additional bedroom and bathroom in a housing unit will lead to a 1% (550 CNY ≈ 82 USD) and 9% (6691 CNY ≈ 997 USD) increase in the housing price. Interestingly, we find that the size and storey of the housing unit have negative effects on housing price. One possible explanation might be that we used the unit price of each housing unit, which is consistent with one study conducted in Korea where cities in Korea have similar highly mixed land uses and densities [51]. Additionally, we find that housing in different districts plays a significant role in influencing housing prices.

Locations in Haidian, Xicheng and Dongcheng districts were positively associated with housing price compared to residents living in Haidian and Fengtai districts. This is logical, since those districts with higher asking prices are located at a relatively close distance to CBD, which potentially promotes the properties' values. Second, we found that residents are willing to pay for better access to green space and CBD, although the implicit price stays at a low level, accounting for 5.0 CNY (≈0.7 USD) and 1.4 CNY (≈0.2 USD), respectively. In addition, we found a negative effect of proximity to the nearest subway station on housing prices.

**Table 2.** Hedonic price model incorporating green space characteristics.

| | (1) | (2) | (3) | (4) |
|---|---|---|---|---|
| | Price | Price | Lnprice | Lnprice |
| Age | 5.521 | 5.105 | 0.000 | 0.000 |
| | (3.556) | (3.560) | (0.000) | (0.000) |
| Bedroom | 497.053 ** | 550.176 ** | 0.006 ** | 0.007 ** |
| | (241.492) | (241.627) | (0.003) | (0.003) |
| Bathroom | 6649.785 *** | 6691.339 *** | 0.086 *** | 0.087 *** |
| | (419.859) | (420.137) | (0.005) | (0.005) |
| Housing size | −64.680 *** | −65.344 *** | −0.001 *** | −0.001 *** |
| | (6.016) | (6.025) | (0.000) | (0.000) |
| Elevator | 3285.481 *** | 3249.162 *** | 0.056 *** | 0.055 *** |
| | (328.368) | (328.635) | (0.004) | (0.004) |
| Storey | −369.724 *** | −373.778 *** | −0.005 *** | −0.005 *** |
| | (111.162) | (111.239) | (0.001) | (0.001) |
| Window orientation | 3792.511 *** | 3783.455 *** | 0.047 *** | 0.046 *** |
| | (304.740) | (304.974) | (0.004) | (0.004) |
| Shijingshan district | Reference | Reference | Reference | Reference |
| Fengtai district | −8755.901 *** | −9198.544 *** | −0.109 *** | −0.116 *** |
| | (611.329) | (611.727) | (0.007) | (0.007) |
| Chaoyang district | −12,938.66 *** | −13,733.41 *** | −0.156 *** | −0.167 *** |
| | (825.808) | (824.292) | (0.010) | (0.010) |
| Hadian district | 22,216.37 *** | 21,572.51 *** | 0.327 *** | 0.318 *** |
| | (565.583) | (565.449) | (0.007) | (0.007) |
| Dongcheng district | 13,629.11 *** | 12,507.49 *** | 0.171 *** | 0.156 *** |
| | (914.015) | (913.559) | (0.011) | (0.011) |
| Xicheng district | 41,865.95 *** | 41,302.11 *** | 0.487 *** | 0.479 *** |
| | (710.261) | (708.980) | (0.008) | (0.008) |
| Distance to green spaces | −5.004 *** | | −0.000 *** | |
| | (0.315) | | (0.000) | |
| Distance to subway | 1.111 *** | 1.115 *** | 0.000 *** | 0.000 *** |
| | (0.057) | (0.057) | (0.000) | (0.000) |
| Distance to CBD | −1.351 *** | −1.396 *** | −0.000 *** | −0.000 *** |
| | (0.034) | (0.034) | (0.000) | (0.000) |
| Natural logarithm of landscape shape index | 5439.223 *** | 5543.887 *** | 0.040 *** | 0.041 *** |
| | (829.323) | (830.391) | (0.010) | (0.010) |
| Green spaces' size | 18.411 *** | 17.363 *** | 0.000 *** | 0.000 *** |
| | (2.443) | (2.443) | (0.000) | (0.000) |
| Distance buffer (0–300 m) | | 7052.248 *** | | 0.088 *** |
| | | (462.638) | | (0.005) |
| Distance buffer (300–1000 m) | | 5068.837 *** | | 0.062 *** |
| | | (416.700) | | (0.005) |
| _cons | 73,614.84 *** | 66,869.96 *** | 11.193 *** | 11.109 *** |
| | (1172.639) | (1247.787) | (0.014) | (0.015) |
| N | 16,222 | 16,222 | 16,222 | 16,222 |
| AIC | 361,486.8 | 361,508 | −6927.924 | −6870.389 |
| BIC | 361,625.3 | 361,654.2 | −6789.429 | −6724.201 |

Notes: Standard errors in parentheses ** $p < 0.05$, *** $p < 0.01$. AIC: Akaike information criterion; BIC: Bayesian information criterion.

Third, as expected, all green space characteristics were statistically significant at the 1% level and the direction of the impact was consistent with our expectations. Among these variables, every one-unit increase in the natural logarithm of LSI can increase house value by 4% 5544 CNY (≈826 USD) on average, which indicates that an improvement in LSI can significantly increase the nearby housing price. When the size of the green space increased by 1 ha, the relative housing price increased by 17 CNY (≈2.5 USD). Two distance buffers from the residence to the green space were positively associated with housing prices. In particular, a one percent increase of green space in 300 m and a 300–1000 m buffer would lead to an increase in housing prices of 9% (3217 CNY ≈ 479 USD) and 6% (2589 CNY ≈ 386 USD)m respectively. This finding is largely aligned with one case study conducted in Beijing suggesting that a 0.5% to 14.1% increase would be observed with respect to housing located 850–1604 m away from green spaces [52].

### 3.3. Comparison between Two Key Distance Buffers from Residential Property to Green Spaces

In the baseline model, housing price decays with the increasing distance from green spaces. In this subsection, we therefore further explored whether the effects of green space characteristics on housing prices will vary depending on different distance buffers from green spaces. The results from Table 3 are presented in four columns. Columns 1 and 2 present the results by conducting the linear regression model. Columns 3 and 4 report the results by using the semi-log regression model. Notably, Column 1 and Column 3 examine the associations between housing characteristics, green space characteristics, neighborhood characteristics and housing prices within 300 m distance buffers from residential property to green spaces whereas Column 2 and Column 4 explore such associations within 300–1000 m distance buffers from residential property to green spaces.

**Table 3.** Hedonic price model with different distance intervals from residence to green spaces.

| | (1) | (2) | (3) | (4) |
|---|---|---|---|---|
| | <300 M | 300–1000 M | <300 M | 300–1000 M |
| Age | 91.027 *** | 2.306 | 0.001 *** | 0.000 |
| | (24.954) | (3.609) | (0.000) | (0.000) |
| Bedroom | −1433.49 *** | 1568.546 *** | −0.015 *** | 0.019 *** |
| | (438.593) | (319.400) | (0.005) | (0.004) |
| Bathroom | 5982.370 *** | 7186.522 *** | 0.074 *** | 0.094 *** |
| | (778.738) | (542.004) | (0.009) | (0.006) |
| Housing size | −23.694 ** | −94.286 *** | −0.000 *** | −0.001 *** |
| | (10.273) | (8.157) | (0.000) | (0.000) |
| Elevator | 4956.562 *** | 3153.681 *** | 0.067 *** | 0.056 *** |
| | (644.594) | (425.682) | (0.008) | (0.005) |
| Storey | −418.954 ** | −388.525 *** | −0.005 ** | −0.006 *** |
| | (209.835) | (141.001) | (0.002) | (0.002) |
| Window direction | 4152.365 *** | 3667.148 *** | 0.046 *** | 0.047 *** |
| | (564.949) | (390.587) | (0.007) | (0.005) |
| Shijingshan district | Reference | Reference | Reference | Reference |
| Fengtai district | −2059.774 * | −9764.051 *** | −0.016 | −0.132 *** |
| | (1167.804) | (803.446) | (0.014) | (0.009) |
| Chaoyang district | −963.475 | −15,001.25 *** | 0.020 | −0.200 *** |
| | (1544.736) | (1054.562) | (0.018) | (0.012) |
| Haidian district | 29,679.8 *** | 21,084.81 *** | 0.425 *** | 0.303 *** |
| | (1080.593) | (729.712) | (0.013) | (0.009) |
| Dongcheng district | 25,177.31 *** | 11,403.26 *** | 0.335 *** | 0.124 *** |
| | (1698.578) | (1195.870) | (0.020) | (0.014) |
| Xicheng district | 53,808.15 *** | 42,290.27 *** | 0.625 *** | 0.474 *** |
| | (1418.500) | (923.819) | (0.017) | (0.011) |
| Distance to green spaces | −8.365 *** | −3.762 *** | −0.000 *** | −0.000 *** |
| | (3.017) | (0.893) | (0.000) | (0.000) |
| Distance to subway | 0.642 *** | 1.359 *** | 0.000 *** | 0.000 *** |
| | (0.111) | (0.070) | (0.000) | (0.000) |
| Distance to CBD | −0.952 *** | −1.384 *** | −0.000 *** | −0.000 *** |
| | (0.068) | (0.046) | (0.000) | (0.000) |
| Natural logarithm of landscape shape index | 7130.456 *** | 5418.881 *** | 0.061 *** | 0.044 *** |
| | (1485.018) | (1063.307) | (0.017) | (0.012) |
| Green spaces' size (ha) | −8.799 ** | 28.652 *** | −0.000 *** | 0.000 *** |
| | (4.411) | (3.184) | (0.000) | (0.000) |
| _cons | 60,703.68 *** | 72,922.49 *** | 11.031 *** | 11.200 *** |
| | (2412.823) | (1637.938) | (0.028) | (0.019) |
| N | 4288.000 | 9928.000 | 4288.000 | 9928.000 |
| AIC | 95,320.89 | 221,208.8 | −2062.173 | −4322.276 |
| BIC | 95,435.43 | 221,338.5 | −1947.629 | −4192.62 |

Notes: Standard errors in parentheses * $p < 0.10$, ** $p < 0.05$, *** $p < 0.01$. AIC: Akaike information criterion; BIC: Bayesian information criterion.

Regarding the housing characteristics, we found that housing price decreases by 1.1% (1803 CNY ≈ 267 USD) and 0.1% (485 CNY ≈ 72 USD), respectively, with regard to the elevator and window orientation variables if residents are living within distance buffers changing from 300 m to 300–1000 m. Similarly, we found that housing price decreases by 71 CNY (≈11 USD) in terms of housing size variables within the same distance buffer, though the estimated effects are relatively small. Conversely, housing price increases by 3.4% (2969 CNY ≈ 442 USD) and 2% (1204 CNY ≈ 179 USD), respectively, with respect to the number of bedrooms and bathrooms when residents are living within a distance buffer change from 300 m to 300–1000 m.

The results indicate a mixed pattern in terms of neighborhood characteristics. Consistent with the results from the baseline model, there was a positive effect of residential proximity to green spaces on housing prices within a 300 m distance buffer from green spaces. Such effects became less pronounced in terms of residents living within a 300–1000 m distance buffer from green spaces. Similarly, a positive association between residential proximity to CBD and housing prices was found with regards to residents living within a 300 m distance buffer from a residential property to green spaces. A slight decrease in housing prices was found in terms of residents living within a 300–1000 m distance buffer from a residential property to green spaces. Conversely, residential proximity to the subway station negatively influenced the housing price in terms of residents living within a 300 m distance buffer from a residential property to green spaces. Regarding the green space characteristics, housing price decreased by 1.7% (1711 CNY ≈ 255 USD) in relation to the natural logarithm of LSI if residents were living within a distance buffer ranging from 300 m to 300–1000 m. Additionally, housing price increased by 37 CNY (≈6 USD) with respect to the size of the green spaces when residents were living within the distance buffer ranging from 300 m to 1000 m.

### 3.4. Further Analysis

It is noteworthy that the results from prior analyses do not take into consideration neighborhood heterogeneity and are only based on the mean impact of the whole sample. Consequently, this section studied the heterogeneity between the natural logarithm of LSI and three variables: distance buffers (0–300 m), distance buffer (300–1000 m), and green spaces' size dummy.

Findings from the first three rows of the first column of Table 4 show that the cross-term coefficient between the natural logarithm of LSI and distance buffer variables is considered positive when the natural logarithm of the LSI value is greater than 1.3, indicating that, in comparison to residences located outside 300 m, the natural logarithm of LSI is more beneficial in promoting residents' housing prices. Such effects become more pronounced when the value of the natural logarithm of LSI increases. Nonetheless, the cross-term coefficient between the natural logarithm of LSI and distance buffer variables is still positive in terms of the value if the natural logarithm of LSI is greater than 1.3, and the marginal effects of the interaction term on housing prices decreased when the buffer range expanded from 300 m to 300–1000 m.

**Table 4.** Interaction term between distance buffers' dummy, green spaces' size dummy and LSI dummy.

| | (1) | (2) |
|---|---|---|
| Variables | | |
| Natural logarithm of LSI (LSI ≤ 1.3) × Distance buffer (0–300 m) | −2151.502 | |
| | (2039.206) | |
| Natural logarithm of LSI (1.3 < LSI ≤ 2.0) × Distance buffer (0–300 m) | 1921.477 ** | |
| | (808.856) | |
| Natural logarithm of LSI (2.0 < LSI ≤ 2.7) × Distance buffer (0–300 m) | 4316.784 *** | |
| | (803.343) | |

**Table 4.** *Cont.*

| | (1) | (2) |
|---|---|---|
| Natural logarithm of LSI (LSI ≤ 1.3) × Distance buffer (300–1000 m) | −3381.739 * | |
| | (1996.460) | |
| Natural logarithm of LSI (1.3 < LSI ≤ 2.0) × Distance buffer (300–1000 m) | 1417.924 ** | |
| | (602.240) | |
| Natural logarithm of LSI (2.0 < LSI ≤ 2.7) × Distance buffer (300–1000 m) | 3526.776 *** | |
| | (593.236) | |
| Natural logarithm of LSI (LSI ≤ 1.3) × Green spaces' size dummy | | −1872.362 |
| | | (1940.355) |
| Natural logarithm of LSI (1.3 < LSI ≤ 2.0) × Green spaces' size dummy | | 1256.991 ** |
| | | (501.544) |
| Natural logarithm of LSI (2.0 < LSI ≤ 2.7) × Green spaces' size dummy | | 1335.093 ** |
| | | (577.166) |
| _cons | 74,392.93 *** | 77,146.92 *** |
| | (1241.136) | (1241.136) |
| Control | Yes | Yes |
| N | 16,222 | 16,222 |
| Aic | 361,447 | 361,523 |
| Bic | 361,623.9 | 361,676.8 |

Notes: Column 1 refers to explore the interaction term between the distance buffers' dummy and LSI dummy. Column 2 refers to explore the interaction term between green spaces' size dummy and LSI dummy. Controls refer to adjustments for housing structure characteristics, green space characteristics and neighbourhood characteristics. Standard errors in parentheses * $p < 0.10$, ** $p < 0.05$, *** $p < 0.01$.

The second column focuses on the interaction influences between the natural logarithm of LSI and green spaces' size buffer. The results indicate that the natural logarithm of LSI is more beneficial to promote residents' housing prices if residential proximity to a larger size of green space compared to those living adjacent to a smaller size of green space. Similarly, the impact becomes more obvious with the increase of the natural logarithm of LSI.

### 3.5. Robustness Check

The results of the robustness check are reported in Table 5. In accordance with the general distance decay theory [53], we conducted a different scale of distance buffers to examine whether the influence of natural logarithm of LSI on housing prices is robust. We found that the natural logarithm of LSI is positively associated with the nearby housing price across different distance buffers, which is consistent with the findings from our baseline model. The findings are trustworthy as the positive association between the natural logarithm of LSI and housing price is still statistically robust after changing the distance buffers. In addition, four distance buffers from the residence to the green space are statistically significant at the 1% level, and the direction of the impact is consistent with our baseline model.

**Table 5.** Results of robustness check.

| | (1) | (2) | (3) | (4) |
|---|---|---|---|---|
| | Price | Lnprice | Price | Lnprice |
| Natural logarithm of Landscape shape index | 5620.815 *** | 0.043 *** | 5271.238 *** | 0.038 *** |
| | (832.312) | (0.010) | (830.445) | (0.010) |
| Distance buffer (0–500 m) | 7008.434 *** | 0.106 *** | | |
| | (974.587) | (0.011) | | |

**Table 5.** *Cont.*

| | (1) | (2) | (3) | (4) |
|---|---|---|---|---|
| Distance buffer (500–1000 m) | 2948.139 *** | 0.053 *** | | |
| | (1003.952) | (0.012) | | |
| Distance buffer (0–800 m) | | | 6201.011 *** | 0.076 *** |
| | | | (425.316) | (0.005) |
| Distance buffer (800–1500 m) | | | 4882.665 *** | 0.061 *** |
| | | | 6201.011 *** | 0.076 *** |
| Control | Yes | Yes | Yes | Yes |
| N | 16,222 | 16,222 | 16,222 | 16,222 |
| AIC | 361,554.9 | −6854.140 | 361,528.1 | −6840.925 |
| BIC | 361,701.1 | −6707.952 | 361,674.3 | −6694.737 |

Notes: Controls refer to adjustments for housing structure characteristics and neighborhood characteristics. Standard errors in parentheses *** $p < 0.01$. AIC: Akaike information criterion; BIC: Bayesian information criterion.

## 4. Discussion

This paper explored the association between housing prices and green space characteristics by standardizing for housing/neighborhood characteristics. It also compared the difference in housing prices between distance buffers from residential property to green spaces. The evidence highlights the fact that the impact of the shape of green space (LSI) on housing prices varied across distance buffers and the size of the green spaces. The maximum impact on property value occurred when residential properties are located at a close distance to nearby large green spaces with a higher value for LSI.

### *4.1. The Effect of Green Space Characteristics on Housing Price*

As expected, we found that the size of a green space has a positive impact on housing prices, which aligns with results found in previously in the literature [54]. Moreover, we found a positive effect for residential property prices within a 300 m distance interval from green spaces. The results further indicate that the housing price then decays with the distance from green spaces, which is in line with other findings [26]. The landscape shape index (LSI), which is our key variable, is positively associated with housing prices, at the statistically significant 1% level. Additionally, our finding provides substantial evidence supporting the view that LSI as an independent shape configuration of a green space patch influences residential value. It is clear that a higher LSI would contribute to higher housing prices nearby, and a curved greenness value contributes more than a square one. One possible explanation might be that irregularly shaped boundaries of patches play an important role in creating sustainable and ecologically sound landscape conditions which contribute to housing premiums [55], while beautiful ecological landscapes bring developers great benefits and willingness for ecological protection, thus promoting the rise of housing prices [56]. On the other hand, the services individuals perceived from the green spaces may be split into those that can be identified as direct-use values and those that can be identified as indirect use values. Direct-use values denote benefits individuals can directly obtain from the green space. For example, amenities provided by that green space can attract more people to access green spaces, thereby prompting people to purchase houses nearby [57]. Indirect-use values refer to values provided by green space that sustain ecological landscape conditions through services such as the maintenance of water quality and indigenous biodiversity. Such a potential non-use value might contribute to housing premiums in the long term, which should not be overlooked. Another way to look at this relationship is that, given that plant species richness is associated with a high LSI score [28], so home buyers are willing to pay more for residential units with this experience.

### 4.2. Further Analysis and Robustness Check

Findings from this analysis demonstrate that different interaction terms between buffers and size and LSI quantify the evidence that the larger the size of the green spaces, the closer properties are to green spaces, the higher a value the property can achieve, whereas our study suggests that the value of the natural logarithm of LSI greater than 1.3 might provide benefits to promote the housing premium. This finding sheds new light on the underlying mechanism that configuration shape index should be taken into account in making policy decisions related to green space preservation and allocation. Planners and designers should comprehensively consider this index when conducting site selections of urban greenness, rather than simply calculating the size of green spaces, or measuring the direct distance to green spaces. In addition, more commercial housing should be encouraged to build in the suburban area instead of the city center, since limited urban green spaces are located in the urban area whereas a large-scale vegetation-dominated landscape is located in the rural area [58]. Results from the robustness check indicate that the positive association between the natural logarithm of LSI and housing price is statistically robust after changing the distance buffers. Specifically, we found that the LSI-housing price association decays as the distance buffer increases. Such a distance-decay effect is largely aligned with one study suggesting that different types of open spaces have implicit marginal price functions that decrease as the distance from the home address to the open space rises [59].

### 4.3. Contributions and Limitations

Our findings provide empirical evidence of the capitalization of green space shape configuration characteristics in housing prices. This study broadens the horizon of the urban designer and planner and recommends that more attention be paid to the shape and size of green spaces. It also promotes developers' awareness of achieving the potential property-implicit value by providing technical support for the construction of landscape diversity. Based on the big data sample in urban Beijing, this study provides evidence that landscape shape index (LSI) can be used as an independent reference index for landscape and urban planning and can also be applied as a shaped patch in the evaluation of a housing premium. This study will helps urban designers and planners realize that they should focus more on reconsidering the shape configuration of an environmental feature and not just its size, for example, in cities where green spaces are dominated by small patches [60]. Additionally, it is worth noting that the LSI can serve as the reference for megacities with respect to the design standard and distribution of green space resources.

In addition to these contributions, the limitations of this study are threefold. First, it has been suggested that environmental and visual contacts with green space play an essential role in contributing to housing prices [61]. For example, environmental quality [62], air quality [63,64] as well as the noise level of the neighborhood [65] can positively influence the housing price. However, we have not controlled for housing price effects of environmental quality and visual amenities in the potential price function due to data limitations. Second, empirical studies have suggested different types of natural greenness may have different impacts on housing prices [20,66]. However, we do not have the layout of the types of green spaces in our dataset due to data restrictions. These issues warrant further study. Lastly, the spillover effects of the landscape pattern index on housing prices might be overestimated, as we used the offer price instead of the transaction price. Evidence regarding the relationship between the landscape pattern index and transaction prices should be quantified if the transaction data are available. Future studies should shed new light on matching spatial POI data with environmental quality to provide potential information on the economic value of urban green spaces for stockholders.

## 5. Conclusions

To the best of our knowledge, this is the first study to evaluate the influence of green space shape configuration characteristics on housing prices within the metropolitan area. The overall findings from this study indicate that housing characteristics, neighborhood characteristics and green space characteristics are positively associated with housing prices. These effects are more pronounced for residents living close to urban green spaces and tend to decay as the distance from residences to green spaces increases. The landscape shape index plays a significant role in influencing the spatial pattern of housing prices where a one-unit increase in the natural logarithm of the LSI can increase housing prices by 4% (5543 CNY $\approx$ 826 USD). The LSI tends to be more sensitive to supporting a housing premium within relatively close distance margins of urban green spaces. These results highlight the importance of considering the direct effect of green space shape configuration on housing prices. It provides potential evidence for stakeholders to consider specific green space characteristics in urban planning and design.

**Author Contributions:** Conceptualization, Y.C. and C.A.J.; Data curation, Y.C. and E.L.; Formal analysis, Y.C.; Funding acquisition, Y.C. and Y.L.; Methodology, Y.C. and C.A.J.; Software, Y.C. and E.L.; Supervision, C.A.J. and Y.L.; Validation, Y.C., C.A.J., N.A.D. and Y.L.; Visualization, Y.C.; Writing—original draft, Y.C.; Writing—review & editing, Y.C., C.A.J., N.A.D. and Y.L. All authors have read and agreed to the published version of the manuscript.

**Funding:** This work was supported by the National Natural Science Foundation of China (No. 42201203; No. 42171196).

**Data Availability Statement:** Data sharing is not available.

**Acknowledgments:** The authors are thankful to the editor and three anonymous reviewers for their insightful comments and suggestions on the previous version of the manuscript. The authors wish to express their appreciation to the National Natural Science Foundation of China.

**Conflicts of Interest:** The authors declare no conflict of interest.

## References

1. Park, K.-H.; Jung, S.-K. Analysis on urban heat island effects for the metropolitan green space planning. *J. Korean Assoc. Geogr. Inf. Stud.* **1999**, *2*, 35–45.
2. Zhang, Y.; Murray, A.T.; Turner, B.L. Optimizing green space locations to reduce daytime and nighttime urban heat island effects in Phoenix, Arizona. *Landsc. Urban Plan.* **2017**, *165*, 162–171. [CrossRef]
3. De Ridder, K.; Adamec, V.; Bañuelos, A.; Bruse, M.; Bürger, M.; Damsgaard, O.; Dufek, J.; Hirsch, J.; Lefebre, F.; Pérez-Lacorzana, J. An integrated methodology to assess the benefits of urban green space. *Sci. Total Environ.* **2004**, *334*, 489–497. [CrossRef] [PubMed]
4. Smith, V.K.; Huang, J.-C. Can markets value air quality? A meta-analysis of hedonic property value models. *J. Political Econ.* **1995**, *103*, 209–227. [CrossRef]
5. Beyer, K.M.; Kaltenbach, A.; Szabo, A.; Bogar, S.; Nieto, F.J.; Malecki, K.M. Exposure to neighborhood green space and mental health: Evidence from the survey of the health of Wisconsin. *Int. J. Environ. Res. Public Health* **2014**, *11*, 3453–3472. [CrossRef] [PubMed]
6. Chen, Y.; Stephens, M.; Jones, C.A. Does residents' satisfaction with the neighbourhood environment relate to residents' self-rated health? Evidence from Beijing. *Int. J. Environ. Res. Public Health* **2019**, *16*, 5051. [CrossRef] [PubMed]
7. Nutsford, D.; Pearson, A.; Kingham, S. An ecological study investigating the association between access to urban green space and mental health. *Public Health* **2013**, *127*, 1005–1011. [CrossRef]
8. Groenewegen, P.P.; Van den Berg, A.E.; De Vries, S.; Verheij, R.A. Vitamin G: Effects of green space on health, well-being, and social safety. *BMC Public Health* **2006**, *6*, 149. [CrossRef]
9. Groenewegen, P.P.; Van Den Berg, A.E.; Maas, J.; Verheij, R.A.; De Vries, S. Is a green residential environment better for health? If so, why? *Ann. Assoc. Am. Geogr.* **2012**, *102*, 996–1003. [CrossRef]
10. Maas, J.; Verheij, R.A.; Groenewegen, P.P.; De Vries, S.; Spreeuwenberg, P. Green space, urbanity, and health: How strong is the relation? *J. Epidemiol. Community Health* **2006**, *60*, 587–592. [CrossRef]
11. Jim, C.Y.; Chen, W.Y. Impacts of urban environmental elements on residential housing prices in Guangzhou (China). *Landsc. Urban Plan.* **2006**, *78*, 422–434. [CrossRef]
12. Morancho, A.B. A hedonic valuation of urban green areas. *Landsc. Urban Plan.* **2003**, *66*, 35–41. [CrossRef]
13. Crompton, J.L. The impact of parks on property values: A review of the empirical evidence. *J. Leis. Res.* **2001**, *33*, 1–13. [CrossRef]
14. Dehring, C.; Dunse, N. Housing density and the effect of proximity to public open space in Aberdeen, Scotland. *Real Estate Econ.* **2006**, *34*, 553–566. [CrossRef]

15. Tyrväinen, L.; Miettinen, A. Property prices and urban forest amenities. *J. Environ. Econ. Manag.* **2000**, *39*, 205–223. [CrossRef]
16. Kaplan, R.; Kaplan, S. *The Experience of Nature: A Psychological Perspective*; Cambridge University Press: Cambridge, UK, 1989.
17. Qin, F.; Fukamachi, K.; Shibata, S. Land-Use/Landscape Pattern Changes and Related Environmental Driving Forces in a Dong Ethnic Minority Village in Southwestern China. *Land* **2022**, *11*, 349. [CrossRef]
18. Huang, J.; Tu, Z.; Lin, J. Land-use dynamics and landscape pattern change in a coastal gulf region, southeast China. *Int. J. Sustain. Dev. World Ecol.* **2009**, *16*, 61–66. [CrossRef]
19. Yii, K.-J.; Tan, C.-T.; Ho, W.-K.; Kwan, X.-H.; Nerissa, F.-T.S.; Tan, Y.-Y.; Wong, K.-H. Land availability and housing price in China: Empirical evidence from nonlinear autoregressive distributed lag (NARDL). *Land Use Policy* **2022**, *113*, 105888. [CrossRef]
20. Bolitzer, B.; Netusil, N.R. The impact of open spaces on property values in Portland, Oregon. *J. Environ. Manag.* **2000**, *59*, 185–193. [CrossRef]
21. Poudyal, N.C.; Hodges, D.G.; Merrett, C.D. A hedonic analysis of the demand for and benefits of urban recreation parks. *Land Use Policy* **2009**, *26*, 975–983. [CrossRef]
22. Tyrväinen, L. The amenity value of the urban forest: An application of the hedonic pricing method. *Landsc. Urban Plan.* **1997**, *37*, 211–222. [CrossRef]
23. Xiao, Y.; Li, Z.; Webster, C. Estimating the mediating effect of privately-supplied green space on the relationship between urban public green space and property value: Evidence from Shanghai, China. *Land Use Policy* **2016**, *54*, 439–447. [CrossRef]
24. Jiao, L.; Liu, Y. Geographic field model based hedonic valuation of urban open spaces in Wuhan, China. *Landsc. Urban Plan.* **2010**, *98*, 47–55. [CrossRef]
25. Park, J.H.; Lee, D.K.; Park, C.; Kim, H.G.; Jung, T.Y.; Kim, S. Park accessibility impacts housing prices in Seoul. *Sustainability* **2017**, *9*, 185. [CrossRef]
26. Kolbe, J.; Wüstemann, H. *Estimating the Value of Urban Green Space: A Hedonic Pricing Analysis of the Housing Market in Cologne, Germany*; EconStor: Berlin, Germany, 2014.
27. Li, B.; Zhang, W.; Shu, X.; Pei, E.; Yuan, X.; Wang, T.; Wang, Z. Influence of breeding habitat characteristics and landscape heterogeneity on anuran species richness and abundance in urban parks of Shanghai, China. *Urban For. Urban Green.* **2018**, *32*, 56–63. [CrossRef]
28. Moser, D.; Zechmeister, H.G.; Plutzar, C.; Sauberer, N.; Wrbka, T.; Grabherr, G. Landscape patch shape complexity as an effective measure for plant species richness in rural landscapes. *Landsc. Ecol.* **2002**, *17*, 657–669. [CrossRef]
29. Kwak, S.-J.; Yoo, S.-H.; Han, S.-Y. Estimating the public's value for urban forest in the Seoul Metropolitan Area of Korea: A contingent valuation study. *Urban Stud.* **2003**, *40*, 2207–2221. [CrossRef]
30. Tyrväinen, L.; Väänänen, H. The economic value of urban forest amenities: An application of the contingent valuation method. *Landsc. Urban Plan.* **1998**, *43*, 105–118. [CrossRef]
31. He, C.; Wang, Z.; Guo, H.; Sheng, H.; Zhou, R.; Yang, Y. Driving forces analysis for residential housing price in Beijing. *Procedia Environ. Sci.* **2010**, *2*, 925–936. [CrossRef]
32. Mora-Garcia, R.-T.; Cespedes-Lopez, M.-F.; Perez-Sanchez, V.R.; Marti, P.; Perez-Sanchez, J.-C. Determinants of the price of housing in the province of alicante (spain): Analysis using quantile regression. *Sustainability* **2019**, *11*, 437. [CrossRef]
33. Zhang, Y.; Dong, R. Impacts of street-visible greenery on housing prices: Evidence from a hedonic price model and a massive street view image dataset in Beijing. *ISPRS Int. J. Geo-Inf.* **2018**, *7*, 104. [CrossRef]
34. Tan, J.; Cheong, S.A. The Regime Shift Associated with the 2004–2008 US Housing Market Bubble. *PLoS ONE* **2016**, *11*, e0162140. [CrossRef] [PubMed]
35. Li, H.; Wei, Y.D.; Wu, Y.; Tian, G. Analyzing housing prices in Shanghai with open data: Amenity, accessibility and urban structure. *Cities* **2019**, *91*, 165–179. [CrossRef]
36. McLeod, P.B. The demand for local amenity: An hedonic price analysis. *Environ. Plan. A* **1984**, *16*, 389–400. [CrossRef]
37. Lab, B.C. *Data 40 Urban Green Lands in Main Chinese Cities 2017*; Beijing City Lab: Beijing, China, 2019.
38. Li, F.; Li, F.; Li, S.; Long, Y. Deciphering the recreational use of urban parks: Experiments using multi-source big data for all Chinese cities. *Sci. Total Environ.* **2020**, *701*, 134896. [CrossRef]
39. Patton, D.R. A diversity index for quantifying habitat "edge". *Wildl. Soc. Bull.* **1975**, *3*, 171–173.
40. Yu, H.; Kong, B.; He, Z.-W.; Wang, G.; Wang, Q. The potential of integrating landscape, geochemical and economical indices to analyze watershed ecological environment. *J. Hydrol.* **2020**, *583*, 124298. [CrossRef]
41. Barker, G. *A Framework for the Future: Green Networks with Multiple Uses in and around Towns and Cities*; English Nature Peterborough: Peterborough, UK, 1997.
42. Kong, F.; Nakagoshi, N. Spatial-temporal gradient analysis of urban green spaces in Jinan, China. *Landsc. Urban Plan.* **2006**, *78*, 147–164. [CrossRef]
43. Geoghegan, J. The value of open spaces in residential land use. *Land Use Policy* **2002**, *19*, 91–98. [CrossRef]
44. Jim, C.Y.; Chen, W.Y. Consumption preferences and environmental externalities: A hedonic analysis of the housing market in Guangzhou. *Geoforum* **2007**, *38*, 414–431. [CrossRef]
45. Acharya, G.; Bennett, L.L. Valuing open space and land-use patterns in urban watersheds. *J. Real Estate Financ. Econ.* **2001**, *22*, 221–237. [CrossRef]
46. Sirmans, S.; Macpherson, D.; Zietz, E. The composition of hedonic pricing models. *J. Real Estate Lit.* **2005**, *13*, 1–44. [CrossRef]

47. Schipperijn, J.; Ekholm, O.; Stigsdotter, U.K.; Toftager, M.; Bentsen, P.; Kamper-Jørgensen, F.; Randrup, T.B. Factors influencing the use of green space: Results from a Danish national representative survey. *Landsc. Urban Plan.* **2010**, *95*, 130–137. [CrossRef]
48. Ambrey, C.; Fleming, C. Public greenspace and life satisfaction in urban Australia. *Urban Stud.* **2014**, *51*, 1290–1321. [CrossRef]
49. Krekel, C.; Kolbe, J.; Wüstemann, H. The greener, the happier? The effect of urban land use on residential well-being. *Ecol. Econ.* **2016**, *121*, 117–127. [CrossRef]
50. O'brien, R.M. A caution regarding rules of thumb for variance inflation factors. *Qual. Quant.* **2007**, *41*, 673–690. [CrossRef]
51. Kim, H.-S.; Lee, G.-E.; Lee, J.-S.; Choi, Y. Understanding the local impact of urban park plans and park typology on housing price: A case study of the Busan metropolitan region, Korea. *Landsc. Urban Plan.* **2019**, *184*, 1–11. [CrossRef]
52. Biao, Z.; Gaodi, X.; Bin, X.; Canqiang, Z. The effects of public green spaces on residential property value in Beijing. *J. Resour. Ecol.* **2012**, *3*, 243–252. [CrossRef]
53. Se Can, A.; Megbolugbe, I. Spatial dependence and house price index construction. *J. Real Estate Financ. Econ.* **1997**, *14*, 203–222. [CrossRef]
54. Xiao, Y.; Lu, Y.; Guo, Y.; Yuan, Y. Estimating the willingness to pay for green space services in Shanghai: Implications for social equity in urban China. *Urban For. Urban Green.* **2017**, *26*, 95–103. [CrossRef]
55. Kim, J.-H.; Li, W.; Newman, G.; Kil, S.-H.; Park, S.Y. The influence of urban landscape spatial patterns on single-family housing prices. *Environ. Plan. B Urban Anal. City Sci.* **2018**, *45*, 26–43. [CrossRef] [PubMed]
56. Liu, G.; Wang, X.; Gu, J.; Liu, Y.; Zhou, T. Temporal and spatial effects of a 'Shan Shui'landscape on housing price: A case study of Chongqing, China. *Habitat Int.* **2019**, *94*, 102068. [CrossRef]
57. Lin, I.-H.; Wu, C.; De Sousa, C. Examining the economic impact of park facilities on neighboring residential property values. *Appl. Geogr.* **2013**, *45*, 322–331. [CrossRef]
58. Richardson, E.A.; Mitchell, R.; Hartig, T.; De Vries, S.; Astell-Burt, T.; Frumkin, H. Green cities and health: A question of scale? *Epidemiol Community Health* **2012**, *66*, 160–165. [CrossRef]
59. Cho, S.-H.; Lambert, D.M.; Kim, S.G.; Roberts, R.K.; Park, W.M. Relationship between value of open space and distance from housing locations within a community. *J. Geogr. Syst.* **2011**, *13*, 393–414. [CrossRef]
60. Li, X.; Zhou, W.; Ouyang, Z.; Xu, W.; Zheng, H. Spatial pattern of greenspace affects land surface temperature: Evidence from the heavily urbanized Beijing metropolitan area, China. *Landsc. Ecol.* **2012**, *27*, 887–898. [CrossRef]
61. Jim, C.Y.; Chen, W.Y. External effects of neighbourhood parks and landscape elements on high-rise residential value. *Land Use Policy* **2010**, *27*, 662–670. [CrossRef]
62. Brasington, D.M.; Hite, D. Demand for environmental quality: A spatial hedonic analysis. *Reg. Sci. Urban Econ.* **2005**, *35*, 57–82. [CrossRef]
63. Dai, J.; Lv, P.; Ma, Z.; Bi, J.; Wen, T. Environmental risk and housing price: An empirical study of Nanjing, China. *J. Clean. Prod.* **2020**, *252*, 119828. [CrossRef]
64. Ridker, R.G.; Henning, J.A. The determinants of residential property values with special reference to air pollution. *Rev. Econ. Stat.* **1967**, 246–257. [CrossRef]
65. Duarte, C.M.; Tamez, C.G. Does noise have a stationary impact on residential values? *J. Eur. Real Estate Res.* **2009**, *2*, 259–279. [CrossRef]
66. Lutzenhiser, M.; Netusil, N.R. The effect of open spaces on a home's sale price. *Contemp. Econ. Policy* **2001**, *19*, 291–298. [CrossRef]

MDPI
St. Alban-Anlage 66
4052 Basel
Switzerland
www.mdpi.com

*Land* Editorial Office
E-mail: land@mdpi.com
www.mdpi.com/journal/land

www.ingramcontent.com/pod-product-compliance
Lightning Source LLC
Chambersburg PA
CBHW040826250926
51850CB00007BA/1267